The Logical Processing
of Digital Signals

Computer Systems Engineering Series
Douglas Lewin, *Editor*

Computer Interfacing and On-Line Operation
J. C. Cluley

Interactive Computer Graphics
B. S. Walker, J. R. Gurd, *and* E. A. Drawneek

Automata Theory: Fundamentals and Applications
Igor Aleksander *and* F. Keith Hanna

Computer-Aided Design of Digital Systems
Douglas Lewin

Real-Time Computer Systems
A. L. Freedman *and* R. A. Lees

Fuzzy Switching and Automata: Theory and Applications
Abraham Kandel *and* Samuel C. Lee

Logical Processing of Digital Signals
S. L. Hurst

Programming for Minicomputers
J. C. Cluley

The Logical Processing of Digital Signals

S. L. Hurst
Senior Lecturer
School of Electrical Engineering
University of Bath, England

Crane, Russak
& Company, Inc.
NEW YORK

Edward Arnold
LONDON

Logical Processing of Digital Signals
Published in the United States by
   Crane, Russak & Company, Inc.
   347 Madison Avenue
   New York, New York 10017
ISBN 0-8448-0907-1
LC 76-41553

Published in Great Britain by
   Edward Arnold (Publishers) Ltd.
   41 Bedford Square
   London WC1B 3DP
ISBN 0 7131 2703 4

Copyright © 1978 Crane, Russak & Company, Inc.

No part of this publication may be reproduced, stored in a retrieval system, or transmitted in any form or by any means, electronic, mechanical, photocopying, recording, or otherwise, without the prior written permission of the publisher.

Printed in the United States of America

Contents

Editor's Foreword vii
Preface ix
Acknowledgments xi
List of Symbols xiii

CHAPTER 1
A Survey of Binary and
Higher-Valued Logic 1

CHAPTER 2
The Classification of
Logic Functions 76

CHAPTER 3
Geometric Constructions for
Binary and Non-Binary Functions 142

CHAPTER 4
The Design of Binary Logic Networks
Using Principally Threshold-Logic Gates 180

CHAPTER 5
The Design of Binary Logic Networks
Using Spectral and Other Techniques 270

CHAPTER 6
The Design of
Ternary Logic Networks 407

CHAPTER 7
Circuit Designs for
Digital Logic Functions 471

Appendices 531
Index 575

Ἐκδιδάσκει πάνθ'ὁ
γηράσκων χρόνος

AESCHYLUS
(522–456 B.C.)

Editor's Foreword

The continuing expansion of knowledge in the computer sciences means that it is now more important than ever for the professional computer engineer to keep abreast of the latest developments in his field. Moreover, due to the rapid assimilation of computer techniques into all areas of science and engineering, non-specialists are also finding it essential to acquire expertise in these disciplines. Thus, there exists a need for readable, up-to-date texts on relevant specialist topics of computer engineering which can form authoritative source books for both the practicing engineer and the academic.

This series is an attempt to fulfill such a requirement and is directed primarily at the professional engineer and graduate student in computer technology; in many cases the books will also meet the needs of specialist options offered in undergraduate courses.

The texts will embrace all aspects of computer systems design with an overall emphasis on the engineering of integrated hardware–software systems. In general the series will present established theory and techniques which have found direct application in systems design. However, promising new theoretical methods will also be covered.

All the books will follow a similar basic pattern of a review of the fundamental aspects of the subject, followed by a survey of the current state of the art, and, where applicable, design examples. An important feature will be the associated bibliography and references, which will select the more important and fundamental publications. The objective is to bring the reader up to a level in the subject where he can read current technical papers and apply the results to his own research and design activities. In the main, authors will be chosen from specialists in their field, drawn from both industrial and academic environments, and experienced in communicating technical ideas.

New books will be regularly added to the series to provide an up-to-date source of specialist texts in Computer Systems Engineering.

DOUGLAS LEWIN
Brunel University

Preface

The explosion in the use of digital circuits and equipments, particularly during the past decade, has been one of the outstanding engineering features of technological society. The digital computer has invaded the lives of virtually all people in some aspect or other, and frequently has to take the blame for human absurdities rather than engineering failures.

The increase in the use of digital equipments has of course accelerated with the advent of microelectronics. To date, however, the term "digital" is often taken without thought or qualification to mean two-valued, i.e. binary, circuits and systems, whereas the true definition of digital relates to any discrete-valued system, whether two-valued, three-valued, or higher-valued. This book, therefore, will not restrict itself entirely to two-valued considerations, but because of the vastly greater amount of research and development which has been concentrated on binary work, together with the inherent natural preference for two-state operation of conventional switching devices, binary considerations will perforce occupy the major part of the text. Future development of integrated circuit topologies may, however, lead into the adoption of three- or higher-valued logic as a practical possibility.

Present-day industrial design of binary systems is peculiar in that complex systems may be and frequently are designed without the aid of any sophisticated design or signal-processing procedure. Frequently a logical statement of the problem may be translated directly into hardware, with intuitive and/or past design experience as an aid possibly to refine the final design. Because of this innate ability to design workable logic networks, the case is often made that more sophistication is merely an academic exercise. However, intuitive design is based upon Boolean AND/OR decisions, and any attempt to introduce more powerful logic relationships, such as the exclusive-OR for example, is generally impossible. This is where more advanced non-Boolean techniques may yield powerful advantages and enable more compact systems finally to be manufactured.

A matching aspect where more powerful design tools would be advantageous is to minimize the silicon area of microelectronic circuits or to enable more logic to be contained on a given area. Minimization of area for a given duty means higher production yields, whilst more logic power per chip is a continuous demand. While we are all used to the almost inconceivable number of active devices which may be laid down on a chip, it is the interconnections of them which to a large degree determine the final chip size. A more powerful processing of the total logic has great potential in areas of other than simple repetitive topology.

These and other associated practical aspects therefore lie behind the topics which will be discussed in this book. The choice of topics is perforce somewhat arbitrary, as there have been a large number of contributions to basic switching concepts propounded in the past 10 to 20 years, some of which will be superseded by revised ideas and concepts. If there is a general theme for this book it is that the Boolean world of AND/OR, NAND/NOR gates and associated Boolean equations and constructions has served us very faithfully up to now, but with the growth of system complexity and demand for greater reliability it is now appropriate to look at newer design techniques and logic gates for the next generations of digital systems.

The book as a whole aims to be an engineering treatise and not a mathematical one. Indeed, as far as possible the supporting mathematics have been omitted, although it is hoped that the reader may readily find such support from the given chapter references. Therefore engineering-oriented postgraduate students starting their research and system designers in industry can hopefully find some information of interest and use in this book, which may in its turn speed on the undoubted new developments and innovations which will be forthcoming in the world of digital signals. If this book does in some small way help in promoting this interest and development it will have amply served its modest purpose.

S.L.H.

Acknowledgments

The author wishes to acknowledge that the research work done by him on threshold logic and spectral techniques during the past few years was supported in part by the Science Research Council of Great Britain. Their support of several postgraduate students working on differing aspects of logic design at the University of Bath is also acknowledged. Lastly, but by no means least, he must acknowledge a great debt to Dr. C. R. Edwards, Research Fellow with the School of Electrical Engineering at the University of Bath, whose enthusiasm for new logic concepts and initiative in pursuing so many of them provides such a stimulating background to many of these associated concepts.

List of Symbols Used

$f(x)$ the output of a binary network or system, taking the logic value 0 or 1

n the number of independent input variables in a network or system

$x_i, i = 1, 2, \ldots, n$ the independent binary inputs into a network or system, each input taking the logic value 0 or 1

$+$ Boolean summation or the logical OR operator, unless otherwise stated (see particularly threshold-logic equations below)

$\cdot$ Boolean multiplication or the logical AND operator, unless otherwise stated (see also threshold logic below)

$^{-}$ the logical negation (or complementation or inversion) operator, e.g. $\bar{x}_1, \overline{f(x)}$

$=$ mathematically equals (in a mathematical expression) or "is realized by" (in a logical expression)

$\neq$ does not equal, or is not realized by.

$\equiv$ equals in all respects

$\triangleq$ equals by definition

$\oplus$ Modulo-2 addition with any carries discarded, or the logical exclusive-OR (or non-equivalence) operator

$\odot$ Modulo-2 addition $+ 1$, with all carries discarded, or the logical exclusive-NOR (or equivalence) operator

$f_{00\ldots0}(x), f_{00\ldots1}(x)$, etc. the value of any binary function $f(x)$ when expanded down to minterm level, i.e. $f_{00\ldots0}(x) =$ value of $f(x)$ at minterm $00\ldots0$

Σ normal arithmetic or Boolean summation

$\Sigma\!\!\!\!/$ Modulo-2 summation with any carries discarded

To differentiate between Boolean expressions and threshold functions, the following notations are employed.

Boolean expressions employ [ ] for any outer brackets and ( ) for any necessary inner brackets. Within these defining brackets $+$ and $\cdot$ take the normal Boolean meaning as defined above.

Threshold functions employ ⟨ ⟩ for outer brackets and { } for any necessary inner brackets. Within these defining brackets the normal arithmetic rules of multiplication and addition hold.

$a_i, i = 1, 2, \ldots, n$ the independent real-number coefficients or "weights" associated with the binary inputs $x_i, i = 1, 2, \ldots, n$ respectively of a binary threshold-logic function or gate (unless otherwise defined in the text)

m total input summation $\sum_{i=1}^{n} a_i$ of a binary threshold-logic function or gate

List of Symbols Used

t real-number output threshold value of a binary single-output threshold-logic function or gate, $1 \leq t \leq m$; often referred to simply as the "gate threshold"

t_j individual gate thresholds of a multi-output binary threshold-logic gate, $1 \leq t_j \leq m$, $t_j < t_{j+1}$

R the radix of a numbering system, or the number of values in a multi-valued system ($R = 2$ in the binary case, 3 in the ternary case, 4 in the quaternary case, and so on)

$f(X)$ the output of a multi-valued network or system, $R > 2$; generally three-valued (ternary) unless otherwise being considered, with logic values generally taken as 0, 1 and 2 (cf. $f(x)$ above for the $R = 2$ case)

X_i, $i = 1, 2, \ldots, n$ the independent multi-valued inputs into a network or system, $R > 2$; three valued unless otherwise being considered, with logic values generally taken as 0, 1 and 2 (cf. x_i above for the $R = 2$ case)

X_i^0, X_i^1, X_i^2, the value 0, 1 and 2 respectively of a ternary variable X_i

& the "minimum-value-of" in a multi-valued algebraic expression (cf. AND in the binary case)

v the "maximum-value-of" in a multi-valued algebraic expression (cf. OR in the binary case)

$f_{00\ldots0}(X), f_{00\ldots1}(X)$, etc. the value of a multi-valued function $f(X)$ when expanded down to minterm level, i.e. $f_{00\ldots0}(X)$ = value of $f(X)$ at minterm $00\ldots0$

A, B, C, etc. fuzzy sets, each a member of some overall classification Z (used in Chapter 1, §1.10 only)

a, b, c, etc. grade of membership of any sample set z in the fuzzy sets A, B, C, etc., respectively, or fuzzy input variables in a fuzzy logic situation (used in Chapter 1, §1.10 only)

y_i, $i = 1, 2, \ldots, n$ the independent binary inputs into a network or system, each input taking the logic value -1 or $+1$ instead of 0 or 1 as in the x_i case; numerically $y_i = (2x_i - 1)$

$f(y)$ the output of a binary function or network, taking the logic value -1 or $+1$; numerically $f(y) = (2f(x) - 1)$

z_i, $i = 1, 2, \ldots, n$ the independent binary inputs into a network or system, each input taking the logic values of $+1$ or -1 instead of 0 or 1 as in the x_i case; numerically $z_i = -(2x_i - 1)$

$f(z)$ the output of a binary function or network taking the logic values $+1$ or -1; numerically $f(z) = -2(f(X)-1)$

a_0 an additional real-number coefficient to augment the n input weights a_i, $i = 1 \ldots n$, previously defined

y_0 an additional input variable to augment the n input variables y_i, $i = 1$ to n, where $y_0 \triangleq$ always $+1$

x_0 the additional input variable in the 0, 1 domain corresponding to y_0, where $x_0 \triangleq$ always 1

p general symbol for any of the 2^n binary minterms

List of Symbols Used

b_i, $i = 0, 1, 2, \ldots, n$ the arithmetic summation over all 2^n minterms p of the product $\{f(y) \cdot y_i\}$ (see Chapter 2, equation (2.7)); also known as the Chow parameters

A_1, etc. a vector quantity, for example $A_1 = a_{11} + a_{12} + a_{13}$

M, **H**, etc. a matrix, normally a square matrix of size $2^n \times 2^n$ in our particular application

R_i, $i = 0, 1, 2, \ldots, n$ the Rademacher functions, $R_i \in \{+1, -1\}$

R_i, $i = 0, 1, \ldots, n, 12, 13, \ldots, 12..n$ the Rademacher–Walsh functions, generated from the set of n Rademacher functions, $R_i \in \{+1, -1\}$

R'_i, $i = 0, 1, \ldots, n, 12, 13, \ldots, 12..n$ the Rademacher–Walsh functions as above, but renumbered 0 for $+1$ and 1 for -1, i.e. $R'_i \in \{0, 1\}$

r_i, $i = 0, 1, \ldots, n, 12, 13, \ldots, 12..n$ the Rademacher–Walsh coefficients of an n-variable binary function $f(x)$

D_k the density (or ratio) of the true to false minterms in a binary Karnaugh map construction, particularly for linearly separable functions (see Appendix D)

$=_0$, $=_1$, $=_2$ "equals 0 (or 1, or 2) when"; used in ternary algebraic expressions (see Chapter 6) when numerical equality of the left-hand and right-hand sides of the written expression is not necessarily present

A_i, $i = 1, 2, \ldots, n$ independent real-number coefficients, or "weights" associated with the ternary inputs X_i, $i = 1, 2, \ldots, n$, respectively of a ternary threshold-logic function (used in Chapter 6, §6.7 only)

T_1, T_2 real-number threshold values of a ternary single-output threshold-logic function (used in Chapter 6, §6.7 only)

$f_0(X)$ the output of a network or subsystem which detects the 0-valued minterms of a ternary function $f(X)$; note that the output $f_0(X)$ is not necessarily 0 when $f(X) = 0$, nor is it necessarily a ternary output signal

$f_1(X)$ as above for the $f(X) = 1$ minterms

$f_2(X)$ as above for the $f(X) = 2$ minterms

$f(X)_{02}$ the output of a network or subsystem used in a ternary realization with output values 0 and 2 only, 0 being the quiescent value

$f(X)_{01}$ as above with output values 0 and 1 only, 0 being the quiescent value

$f(X)_{20}$ as above with output values 2 and 0 only, 2 being the quiescent value

$f(X)_{21}$ as above with output values 2 and 1 only, 2 being the quiescent value

Chapter 1

A Survey of Binary and Higher-Valued Logic

Introduction

In this book we shall be looking at many aspects of digital signals, and how they may be handled and processed during the analysis or synthesis of a digital system. It is assumed that the reader will already be familiar with the basic conventional tools of the logic designer, such as binary AND, OR, NAND, and NOR gates, binary counters, the essentials of Boolean algebra, the principles of combinatorial minimization, and the use of the Karnaugh map for displaying simple binary functions or sequences. In other words he will be familiar with the fundamental ideas taught in most academic courses on logic and used by most industrial logic designers.

These aspects will not be detailed in this book, but instead they will be extended to cover further ideas, some of which have yet to find a place in the teaching or engineering world. Indeed some of the topics expounded here may never find a prominent place in the armory of the logic designer, but different aspects and developments of a subject help towards a more perfect understanding of the subject as a whole.

The Karnaugh map will be found used in this book in many applications. In real life the serious limitation of this map construction is of course that it cannot conveniently be used for binary situations containing more than, say, four independent input variables. However, for up to four variables its two-dimensional construction is unsurpassed for visually displaying the input–output relationships or other features of a binary system, and therefore its use in this book will be largely for geometrically displaying the principles of whatever process or situation is being discussed. Having grasped the principles from some simple example, the reader may then more readily apply such principles to higher-ordered situations which the map cannot clearly represent, for example during the formulation of computer programs for CAD studies.

1.1. The use and limitation of conventional Boolean ("vertex") gates in logic design

Present-day logic systems are almost universally built up around the basic Boolean-logic building blocks illustrated in Fig. 1.1. To these must of course be added the bistable-circuit assembly which forms the necessary memory element for all sequential applications. Of all the basic building blocks illustrated in Fig. 1.1, possibly the NAND gate is presently the most widely used, with the type JK bistable-circuit assembly as the most popular type of bistable circuit.

The choice of AND and OR as basic logic gates follows directly from Boolean algebra. In the algebraic formulation of a combinatorial problem the problem requirements are frequently first expressed in a simple *sum-of-product form*, for example

$$f(x) = [\bar{x}_1\bar{x}_2x_3 + x_1\bar{x}_2\bar{x}_3 + \bar{x}_1x_2\bar{x}_3]$$

where $f(x)$ represents the required network output of value 0 or 1, and where x_1, x_2, x_3 represent the three independent binary input variables to the network. Now whilst this expression is an exact, rigorous, mathematically correct equation, the value 0 or 1 for $f(x)$ on the left-hand side of the equals sign always being given by the resultant Boolean multiplication and addition of 0's and 1's on the right-hand side of the equation[1], nevertheless such an expression is more usually taken as a logic expression rather than a Boolean equation. Each of the product terms of the right-hand side is considered to be an AND function, with the three AND terms connected by an OR, giving that this function $f(x)$ may be realized by three three-input AND gates, the outputs of which are combined together by one three-input OR gate. A product-of-sums expression, for example

$$f(x) = [(x_1 + x_2 + \bar{x}_3)(\bar{x}_1 + x_2 + x_3)(x_1 + \bar{x}_2 + \bar{x}_3)]$$

is likewise an exact numerical equation, although again it would most usually be read as a logic expression which may be realized by three three-input OR gates followed by one three-input AND gate.

Thus AND and OR logic gates correlate directly with classical Boolean algebraic expressions. NAND and NOR gates may be re-

---

[1] Boolean addition and multiplication is as follows:
$0 + 0 = 0$ $0 \times 0 = 0$
$0 + 1 = 1$ $0 \times 1 = 0$
$1 + 0 = 1$ $1 \times 0 = 0$
$1 + 1 = 1$ $1 \times 1 = 1$

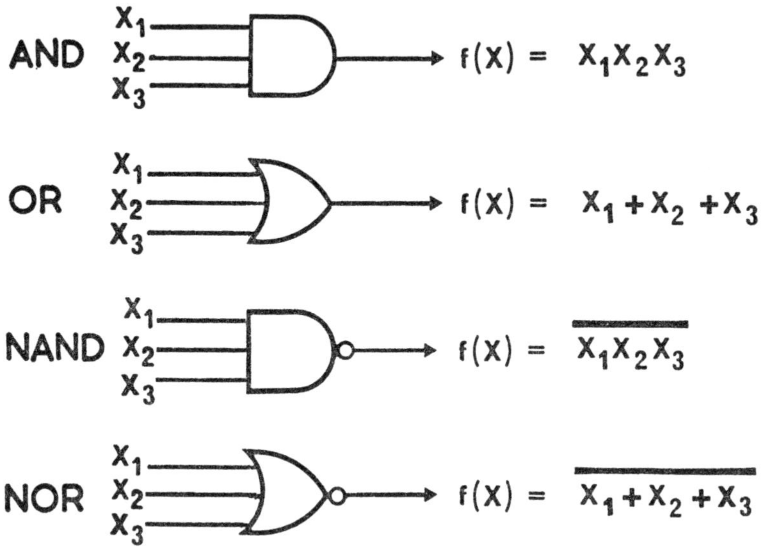

AND $\begin{matrix}X_1\\X_2\\X_3\end{matrix}$ ⟶ $f(X) = X_1 X_2 X_3$

OR $\begin{matrix}X_1\\X_2\\X_3\end{matrix}$ ⟶ $f(X) = X_1 + X_2 + X_3$

NAND $\begin{matrix}X_1\\X_2\\X_3\end{matrix}$ ⟶ $f(X) = \overline{X_1 X_2 X_3}$

NOR $\begin{matrix}X_1\\X_2\\X_3\end{matrix}$ ⟶ $f(X) = \overline{X_1 + X_2 + X_3}$

GATE INPUTS			GATE OUTPUT $f(X)$			
X_1	X_2	X_3	AND	OR	NAND	NOR
0	0	0	0	0	1	1
0	0	1	0	1	1	0
0	1	0	0	1	1	0
0	1	1	0	1	1	0
1	0	0	0	1	1	0
1	0	1	0	1	1	0
1	1	0	0	1	1	0
1	1	1	1	1	0	0

Fig. 1.1. The basic Boolean logic gates and their truth tables. (Note that the single-input INVERTER gate may be formed from a multi-input NAND or NOR gate; three inputs are shown on all multi-input gates purely for illustrative purposes).

garded as merely variations of the AND and OR, respectively, with simple inversion of each output but with no more or no less logic discrimination capability. The various rules of Boolean algebra may be used to convert such simple sum-of-products or product-of-sums expressions into an all-NAND or all-NOR form when necessary. For instance our first sum-of-products example may be manipulated into an all-NAND form by the simple algebraic steps

$$f(x) = [\bar{x}_1\bar{x}_2 x_3 + x_1\bar{x}_2\bar{x}_3 + \bar{x}_1 x_2 \bar{x}_3]$$
$$\overline{f(x)} = [\overline{\bar{x}_1\bar{x}_2 x_3 + x_1\bar{x}_2\bar{x}_3 + \bar{x}_1 x_2 \bar{x}_3}]$$
$$= \overline{(\bar{x}_1\bar{x}_2 x_3)} \cdot \overline{(x_1\bar{x}_2\bar{x}_3)} \cdot \overline{(\bar{x}_1 x_2 \bar{x}_3)}$$

whence

$$f(x) = \left[\overline{\overline{(\bar{x}_1\bar{x}_2 x_3)} \cdot \overline{(x_1\bar{x}_2\bar{x}_3)} \cdot \overline{(\bar{x}_1 x_2 \bar{x}_3)}}\right].$$

Details of the standard manipulations possible will be found in all the usual textbooks on logic design[1-4].

It will also be remembered that functional completeness, that is, the ability to be able to construct all possible logic functions from the x_i inputs including the extreme trivial cases of $f(x)$ always equal to 0 or 1, is provided by the gates shown in Fig. 1.1. Proof that this is so springs from the mathematical work of Post[5-7], which shows that any arbitrary Boolean function $f(x)$ may be expressed by the canonic (or "ordered") expansion along any r of its n input variables, $0 \leq r \leq n$, in the form

$$f(x) = \sum^{2^r}\{\dot{x}_1 \dot{x}_2 \ldots \dot{x}_r \ f_{\phi_1,\phi_2,\ldots \phi_r}(x)\}$$

where

$$\dot{x}_i = x_i \text{ or } \bar{x}_i$$
$$\phi_i = 1 \text{ if } \dot{x}_i = x_i$$
$$\quad = 0 \text{ if } \dot{x}_i = \bar{x}_i$$

$f_{\phi_1,\phi_2,\ldots,\phi_r}(x)$ = the expansion of the function $f(x)$ when

$$x_i = \phi_i, i = 1 \ldots r,$$

and where $\sum$ denotes the Boolean sum operation, as defined on p. 2.

For an expansion into *minterm form*[2], that is, when $r = n$, we therefore have the summation of the 2^n terms:

---

[2] "Minterms" are AND terms each containing all the input variables $x_1, x_2, \ldots, x_n$ in either true or complemented form. They therefore correspond to the 2^n individual input combinations of a truth table for a function with n inputs. Notice that although any given function $f(x)$ may be expressed in a sum-of-products minterm form as above, where it is possible to merge (minimize) two or more same-valued minterms into a single prime-implicant term[1-3] a sum-of-products prime-implicant form is clearly more compact.

A Survey of Binary and Higher-Valued Logic

$$f(x) = [\bar{x}_1\bar{x}_2\ldots\bar{x}_n \underline{f_{00\ldots0}(x)} + \bar{x}_1\bar{x}_2\ldots x_n \underline{f_{00\ldots1}(x)} + \ldots + x_1 x_2\ldots x_n \underline{f_{11\ldots1}(x)}]$$

where $\underline{f_{00\ldots0}(x)}$, $\underline{f_{00\ldots1}(x)}$ etc. now take the value 0 or 1 according to whether the associated minterm is of value 0 or 1 in the required function $f(x)$. Omitting the numerically zero terms in this expansion we have the normal sum-of-products minterm expression for $f(x)$.

Thus functional completeness is guaranteed by the logical AND plus the logical OR operators, together with the NOT or INVERT operator. However, AND plus INVERT, or OR plus INVERT, are themselves complete, as the OR operator may be made from AND plus INVERT, and the AND operator from OR plus INVERT. Hence NAND operators, or NOR operators, are each individually complete, enabling any arbitary function $f(x)$ to be constructed using only one or other of these types of gate, a result which of course is widely appreciated and employed.

There is therefore a powerful incentive to use and to continue to use the logic gates shown in Fig. 1.1. The classic Boolean algebra enables us to manipulate the expressions for a given logic problem into alternative equivalent forms or to use other design techniques which stem from this algebra, from which we may finally realize the problem with an appropriate net of the directly corresponding logic gates.

However, are these the only logic gates which can be used? Are they the most efficient or logically powerful types of gate? Are there completely alternative ways of expressing a logic requirement other than by a classic Boolean expression? These are some of the questions which we shall now start to consider.

The conventional AND, OR, NAND, and NOR Boolean gates shown in Fig. 1.1 may be collectively referred to as "vertex" gates, which distinguishes them from all other possible types of binary gate which may be proposed. The reason given below for this terminology will also show that such gates have a very poor logic discrimination capability; in fact they are the least efficient type of logic gate that can be proposed. The circuit arrangements of such gates, however, may be simpler than the circuits necessary to produce more complex gates such as those considered later on, and therefore it is understandable why such gates were first designed and brought into extensive use.

Let us consider the case of a simple network with three input variables x_1, x_2, x_3, somewhere within which the signals $x_1, x_2, \bar{x}_3$ are applied to (i) a three-input AND gate, (ii) a three-input OR gate, (iii) a three-input NAND gate, and (iv) a three-input NOR gate. Table 1.1 lists the output conditions of these four gates as the input variables sequence through all the possible 2^3 input minterms.

Table 1.1. The full truth table for three-input AND, OR, NAND, and NOR gates with signal inputs x_1, x_2 and $\bar{x}_3$

Input Variables	Input minterm reference number	Actual gate input signals	Gate outputs			
			AND gate	OR gate	NAND gate	NOR gate
x_1 x_2 x_3		x_1 x_2 $\bar{x}_3$				
0 0 0	0	0 0 1	0	1	1	0
0 0 1	1	0 0 0	0	0	1	1
0 1 0	2	0 1 1	0	1	1	0
0 1 1	3	0 1 0	0	1	1	0
1 0 0	4	1 0 1	0	1	1	0
1 0 1	5	1 0 0	0	1	1	0
1 1 0	6	1 1 1	1	1	0	0
1 1 1	7	1 1 0	0	1	1	0

Now in all cases the gate output is unchanged for seven out of the eight input minterms, and only changes on the appropriate eighth input combination. The change on receipt of the appropriate eighth minterm may be from a 0 to a 1 output (see the AND and NOR gates) or from a 1 to a 0 output (see the OR and NAND gates). All four types of gates, however, are by themselves unable to distinguish in any way between seven of the eight input minterms. Looking at a gate output when any one of these seven input conditions is present gives us no information whatsoever on which of the seven is present. Further, the outputs given by the AND and NOR gates, and by the OR and NAND gates, are identical for six of the eight input conditions.

An alternative way of illustrating this lack of discrimination, and which illustrates the terminology "vertex gates", is the *hypercube construction* shown in Fig. 1.2. The hypercube construction is a Euclidean geometry multi-dimensional construction, each corner or node uniquely representing one input minterm condition. Like the Karnaugh map this hypercube construction rapidly becomes increasingly difficult to draw and interpret clearly as the number of input variables increases. However, for $n = 3$ the picture is still clear and concise.

A Survey of Binary and Higher-Valued Logic

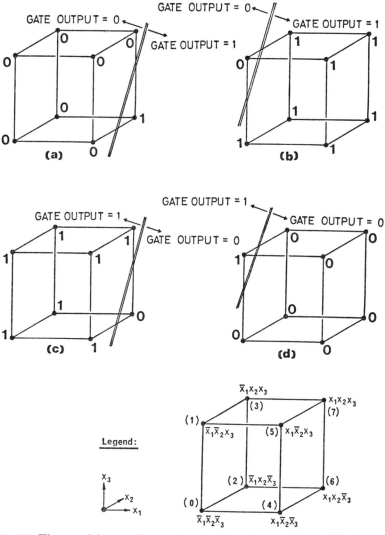

Fig. 1.2. The $n = 3$ hypercube construction showing the AND, OR, NAND, and NOR functions: (a) $x_1 x_2 \bar{x}_3$, (b) $x_1 + x_2 + \bar{x}_3$, (c) $\overline{x_1 x_2 \bar{x}_3}$, and (d) $\overline{x_1 + x_2 + \bar{x}_3}$, respectively.

It will be obvious from Fig. 1.2 that there is a clear separation between seven input minterms which give one particular gate output signal, and the eighth minterm which gives the other output signal.

Each gate thus uniquely detects merely one node or "vertex" of the total, which one it detects depending of course on the type of gate and which input signals are applied to the gate.

Hence whichever way the situation is considered, our commonly used vertex gates are logically not very powerful. To extend the specific $n = 3$ case here considered to the general case, we may state that any n-input vertex gate is unable to distinguish between $2^n - 1$ of the 2^n input minterms which may be applied to it, and unambiguously detects only one out of this total of 2^n. This of course is why it is frequently necessary to use a very large number of vertex gates for what initially might appear to be a simple design requirement.

1.2 The Exclusive-OR Gate and Its Derivatives

The exclusive-OR gate is not a vertex gate as it is not responsive to just one input minterm condition. Further, as the usually listed rules of Boolean algebra do not normally include the exclusive-OR operator, it is frequently not classed as a Boolean logic gate in the same sense as the AND, OR, NAND, NOR family. The fact that it is somewhat different to and possibly more difficult to utilize than normal vertex gates is demonstrated by the fact that it is not referred to at all in some logic textbooks. Yet, as will be shown in several future sections of this book, it and its variations are extremely powerful logic building blocks.

The basic exclusive-OR gate is generally considered to be a two-input gate, which with inputs x_1 and x_2 obeys the following rules:

$$\text{output } f(x) = 1 \text{ iff } x_1 \neq x_2$$
$$= 0 \text{ iff } x_1 = x_2$$

Algebraically we may re-express this action in the more usual way:

$$f(x) = [x_1 \odot x_2]$$

where the symbol $\odot$ is the exclusive-OR operator. The alternative terminology "non-equivalent" may also be found used, with the operator symbol $\neq$, but this terminology is gradually falling out of favor.

The relationship between the exclusive-OR operator and the normal Boolean operators is given by:

$$[x_1 \odot x_2] \triangleq [\bar{x}_1 \cdot x_2 + x_1 \cdot \bar{x}_2]$$
$$= [(x_1 + x_2) \cdot (\bar{x}_1 + \bar{x}_2)]$$

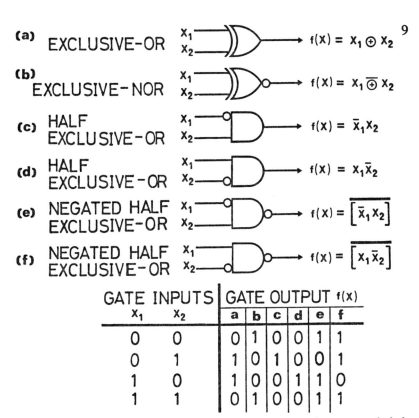

Fig. 1.3. Two-input exclusive-OR and exclusive-NOR logic gates and their possible derivatives.

This relationship illustrates how an exclusive-OR logic assembly may be constructed from normal vertex gates, although as we shall see in a later chapter it is far preferable to consider the circuit design of an exclusive-OR gate from first principles and not as an assembly of existing vertex gates.

Still dealing for the moment with two-input gates, variations on the exclusive-OR gate so far considered can be proposed. The one most widely known is the exclusive-NOR gate, which having an output complementary to the exclusive-OR relationship is also sometimes referred to as the "equivalent" gate. Algebraically we may express its action by

$$f(x) = [x_1 \overline{\oplus} x_2]$$

where $\overline{\oplus}$ is the exclusive-NOR operator. The alternative operator symbol $\equiv$ may be found when the terminology "equivalent" is in use.

However, the full range of interesting two-input "exclusive"-type functions may be extended further, as illustrated in Fig. 1.3. The last four of the six gates listed in Fig. 1.3 have reverted to being vertex gates, being effectively two-input AND or NAND gates but with one of their two inputs negated. The terminology "half exclusive" has been suggested for these gates, as logically they each realize one of the two Boolean terms of the full exclusive-OR or exclusive-NOR relationship. There is a further reason for this terminology in that the circuit realization for these latter gates may conceivably be half the circuitry of a full exclusive-OR or exclusive-NOR gate design, and therefore directly available on a full exclusive gate. The terminology "inhibit" and "imply" may also be found for these "half-exclusive" gates, but such terminology does not find great favor in engineering circles. Finally, notice that it is not possible to formulate any other useful two-variable non-vertex function other than the exclusive-OR and exclusive-NOR functions, as Table 1.2 will confirm.

Table 1.2. All possible 2^{2^n} functions of two input variables x_1 and x_2. Note that functions f_0 and f_{15} are trivial, f_1, f_7, f_8, and f_{14} are the two-input AND, OR, NOR, and NAND functions, f_3, f_5, f_{10}, and f_{12} are single-input functions, leaving the remaining six functions those detailed in Fig. 1.3.

x_1	x_2	f_0	f_1	f_2	f_3	f_4	f_5	f_6	f_7	f_8	f_9	f_{10}	f_{11}	f_{12}	f_{13}	f_{14}	f_{15}
0	0	0	0	0	0	0	0	0	0	1	1	1	1	1	1	1	1
0	1	0	0	0	0	1	1	1	1	0	0	0	0	1	1	1	1
1	0	0	0	1	1	0	0	1	1	0	0	1	1	0	0	1	1
1	1	0	1	0	1	0	1	0	1	0	1	0	1	0	1	0	1

Because the exclusive-OR and exclusive-NOR operators are not so widely known as the more common Boolean operators, it may be useful to record here some of the algebraic relationships which hold with both these operators. They are as follows:

(a) Laws involving one binary variable and/or a constant only:

$[x_1 \oplus 0] = x_1$ $[x_1 \overline{\oplus} 0] = \bar{x}_1$
$[x_1 \oplus 1] = \bar{x}_1$ $[x_1 \overline{\oplus} 1] = x_1$
$[x_1 \oplus x_1] = 0$ $[x_1 \overline{\oplus} x_1] = 1$

(continued)

$$[\bar{x}_1 \oplus x_1] = 1 \qquad [\bar{x}_1 \overline{\oplus} x_1] = 0$$
$$[\bar{x}_1 \oplus \bar{x}_1] = 0 \qquad [\bar{x}_1 \overline{\oplus} \bar{x}_1] = 1$$
$$[0 \oplus 0] = 0 \qquad [0 \overline{\oplus} 0] = 1$$
$$[0 \oplus 1] = 1 \qquad [0 \overline{\oplus} 1] = 0$$
$$[1 \oplus 1] = 0 \qquad [1 \overline{\oplus} 1] = 1$$

(b) Laws involving more than one binary variable:

$$[x_1 \oplus (x_2 \oplus x_3)] = [(x_1 \oplus x_2) \oplus x_3] = [x_1 \oplus x_2 \oplus x_3]$$
$$[x_1 \overline{\oplus} (x_2 \overline{\oplus} x_3)] = [(x_1 \overline{\oplus} x_2) \overline{\oplus} x_3] = [x_1 \overline{\oplus} x_2 \overline{\oplus} x_3]$$

(the associative laws)

$$[x_1 \oplus x_2] = [x_2 \oplus x_1]$$
$$[x_1 \overline{\oplus} x_2] = [x_2 \overline{\oplus} x_1]$$

(the commutative laws)

$$[x_1(x_2 \oplus x_3)] = [x_1 x_2 \oplus x_1 x_3]$$
$$[x_1(x_2 \overline{\oplus} x_3)] = [x_1 x_2 \overline{\oplus} x_1 x_3]$$

(the distributive laws)

$$\overline{[x_1 \oplus x_2]} = [\bar{x}_1 \oplus x_2] = [x_1 \oplus \bar{x}_2] \equiv [x_1 \overline{\oplus} x_2]$$
$$\overline{[x_1 \overline{\oplus} x_2]} = [\bar{x}_1 \overline{\oplus} x_2] = [x_1 \overline{\oplus} \bar{x}_2] \equiv [x_1 \oplus x_2]$$
$$\overline{[x_1 \oplus x_2 \oplus \ldots x_i]} = [x_1 \overline{\oplus} x_2 \overline{\oplus} \ldots x_i] \left. \begin{array}{l} \\ \end{array} \right\} \text{ for } i \text{ even}$$
$$\overline{[x_1 \overline{\oplus} x_2 \overline{\oplus} \ldots x_i]} = [x_1 \oplus x_2 \oplus \ldots x_i] \quad \text{numbered}$$

(the complementation laws)

(c) Other useful relationships:

$$[x_1 \oplus x_2 \oplus \ldots x_i] = [x_1 \overline{\oplus} x_2 \overline{\oplus} \ldots x_i] \text{ for } i \text{ odd numbered}$$

If $x_i = [x_1 \oplus x_2]$, then $x_1 = [x_i \oplus x_2]$ and $x_2 = [x_i \oplus x_1]$

If $x_i = [x_1 \overline{\oplus} x_2]$, then $x_1 = [x_i \overline{\oplus} x_2]$ and $x_2 = [x_i \overline{\oplus} x_1]$

$$[x_i f_1(x) \oplus \bar{x}_i f_2(x)] = [x_i f_1(x) + \bar{x}_i f_2(x)]$$

where $f_1(x)$ and $f_2(x)$ are any two identical or dissimilar Boolean functions

$$[x_1(x_1 \oplus x_2)] = [x_1 \bar{x}_2]$$
$$[x_1(x_1 \overline{\oplus} x_2)] = [x_1 x_2]$$
$$[x_1 + (x_1 \oplus x_2)] = [x_1 + x_2]$$
$$[x_1 + (x_1 \overline{\oplus} x_2)] = [x_1 + \bar{x}_2]$$

(continued)

$$[x_1 \oplus (x_1 x_2)] = [x_1 \bar{x}_2]$$
$$[x_1 \overline{\oplus} (x_1 x_2)] = [x_1 x_2]$$
$$[x_1 \oplus (x_1 + x_2)] = [\bar{x}_1 x_2]$$
$$[x_1 \overline{\oplus} (x_1 + x_2)] = x_1$$

It is possible to use these and further like relationships to minimize algebraic expressions involving these operators in a manner similar to that which is employed with more usual Boolean equations. As a passing example, consider the following function:

$$\begin{aligned} f(x) &= [x_1 \oplus \bar{x}_3 \oplus x_1 x_3 \oplus x_2 x_4 \oplus x_1 x_3 x_4] \\ &= [x_1 \oplus (x_1 + \bar{x}_3) \oplus x_2 x_4 \oplus x_1 x_3 x_4] \\ &= [\bar{x}_1 \bar{x}_3 \oplus x_2 x_4 \oplus x_1 x_3 x_4] \\ &= [(\bar{x}_1 \bar{x}_3 + x_1 x_3 x_4) \oplus x_2 x_4]. \end{aligned}$$

Currently the use of the exclusive-OR and NOR operators and hence the application of exclusive-OR and NOR gates is not as widespread and familiar as Boolean operators and vertex gates. Nevertheless considerable economy in logic network design may be produced if such gates are employed[8–11]. This will be referred to more fully later on, particularly in Chapters 3, 5, and 7.

The exclusive-OR (and NOR) operator by itself does not provide functional completeness. However, the AND plus exclusive-OR operators form a functionally complete set of operators but require to be augmented with steady logic 1 signals in order to provide full functional completeness (see below). This latter point is of no practical disadvantage in an engineering situation, as steady logic 0 or 1 signals may always be obtained from the logic supply rails. The "half-exclusive" operators, on the other hand, are merely particular cases of two-input AND and NAND operators, and therefore obey the same functional completeness qualifications as normal AND and NAND gates. However, there is no merit whatsoever in considering these "half-exclusive" relationships by themselves as especially useful[12].

Proof of exclusive-OR and AND completeness comes from the Reed–Muller canonical expansion[7,13,14], which states that any arbitrary function $f(x)$ may be mathematically expressed by

$$f(x) = \sum_{i=0}^{2^n - 1} a_i \{ x_j x_k \ldots x_m \}$$

where
a_i, $0 \leq i \leq 2^n - 1$, = constant 0 or 1

$x_j x_k \ldots x_m$ = all possible different subsets of the variables x_i taken from the complete set x_1 to x_n, taken r at a time, $0 \leq r \leq n$, including the empty set

and where Σ denotes the exclusive-OR summation, that is, modulo-2 addition with all carries disregarded. For a three-variable function ($n = 3$) the Reed–Muller expansion is therefore

$$f(x) = [a_0 \oplus a_1 x_1 \oplus a_2 x_2 \oplus a_3 x_3 \oplus a_4 x_1 x_2 \oplus a_5 x_1 x_3 \oplus a_6 x_2 x_3 \oplus a_7 x_1 x_2 x_3]$$

where the expansion coefficients a_0, a_1, etc. are each 0 or 1 as required to realize $f(x)$. Notice that the a_0 term in this expression may need to be 1; therefore unless 1 is available in addition to the input variables x_i, $i = 1 \ldots n$, the expansion cannot be realized in practice. Notice also two other interesting points in this expansion for $f(x)$: firstly there are *no complements* of any of the x_i inputs involved, as there are in the Boolean minterm expansion for $f(x)$ previously considered in §1.1, and secondly the a_0 term serves to complement the resultant value of the remaining terms of the expansion, depending upon whether a_0 is 1 or 0.

A simple method for determining the a_i coefficients for any given function $f(x)$ in this uncomplemented ("positive") Reed–Muller expansion is as follows. Assume that $f(x)$ is given in a minimized prime-implicant form. Examine the prime-implicant product terms to see if they are mutually exclusive. This may be done by mapping, or by expanding each prime-implicant term into its component minterms, looking for any minterms which appear in two (or more) prime implicants. If any such overlap is found, then divide the prime implicants such that they finally contain no common minterm factors; that is they are mutually exclusive (disjoint).

When mutually exclusive prime implicants are present, then the OR operator which connects the product terms can be replaced by the exclusive-OR operator without altering the function in any way. To make the final expansion complement-free, the three identities

$$\bar{x}_i = x_i \oplus 1$$
$$x_{ij} \oplus x_{ij} = 0$$

and $\quad x_i (x_j \oplus x_k) = x_i x_j \oplus x_i x_k$

then must be applied, from which the final 1-valued terms of an uncomplemented Reed–Muller expansion result. All other terms in the general expansion must be zero valued.

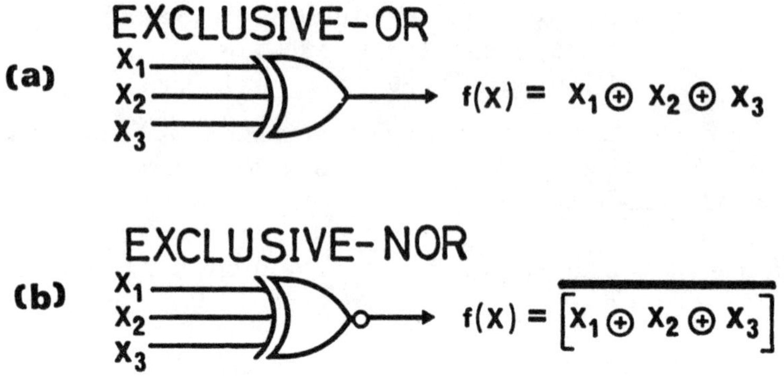

EXCLUSIVE-OR

(a) $X_1, X_2, X_3 \rightarrow f(x) = X_1 \oplus X_2 \oplus X_3$

EXCLUSIVE-NOR

(b) $X_1, X_2, X_3 \rightarrow f(x) = \overline{[X_1 \oplus X_2 \oplus X_3]}$

GATE INPUTS			GATE OUTPUT	
X_1	X_2	X_3	(a)	(b)
0	0	0	0	1
0	0	1	1	0
0	1	0	1	0
0	1	1	0	1
1	0	0	1	0
1	0	1	0	1
1	1	0	0	1
1	1	1	1	0

Fig. 1.4. Three-input exclusive-OR and exclusive-NOR logic gates with their truth table.

As a simple example, consider the three-variable function

$$f(x) = [\bar{x}_1 x_2 + x_1 \bar{x}_2 + x_1 x_3].$$

A check will show that the last two product terms are not mutually exclusive as they share minterm $x_1 \bar{x}_2 x_3$ between them. To make them mutually exclusive let us alter the third prime implicant term to $x_1 x_2 x_3$ making them now disjoint, giving us

A Survey of Binary and Higher-Valued Logic

$$f(x) = [\bar{x}_1 x_2 + x_1 \bar{x}_2 + x_1 x_2 x_3]$$
$$\equiv [\bar{x}_1 x_2 \oplus x_1 \bar{x}_2 \oplus x_1 x_2 x_3].$$

Replacing the complemented variables and rearranging we obtain

$$f(x) = [x_2(x_1 \oplus 1) \oplus x_1(x_2 \oplus 1) \oplus x_1 x_2 x_3]$$
$$= [x_1 x_2 \oplus x_2 \oplus x_1 x_2 \oplus x_1 \oplus x_1 x_2 x_3]$$
$$= [x_1 \oplus x_2 \oplus x_1 x_2 x_3].$$

The full Reed–Muller expansion coefficients therefore are

$$f(x) = [0 \oplus 1 \cdot x_1 \oplus 1 \cdot x_2 \oplus 0 \cdot x_3 \oplus 0 \cdot x_1 x_2 \oplus 0 \cdot x_1 x_3 \oplus 0 \cdot x_2 x_3 \oplus 1 \cdot x_1 x_2 x_3].$$

Notice that if we select a different set of disjoint prime-implicant terms as our starting point for developing the canonic Reed–Muller expansion for any given function $f(x)$, the steps in our development would be dissimilar but the final positive canonic expression would be the same. This may be illustrated using the above example function. Thus there may be a number of possible exclusive-OR equations for a given function $f(x)$ involving one or more complemented variables, but only one positive canonic Reed–Muller expansion—cf. prime-implicant equations and minterm expansions for any given $f(x)$.

A modified Reed–Muller expansion incorporating the exclusive-NOR operator in place of the exclusive-OR may be proposed, but has no particular mathematical significance or merit. Both cases require the availability of the AND operator together with logic 1 as well as the exclusive operator in order to provide functional completeness. Finally, if all x_i's, $i=1\ldots n$, are replaced by their complement by the substitution $\bar{x}_i = [x_i \oplus 1]$, then we obtain the "negative" canonic Reed–Muller expansion for $f(x)$. Again this has no particular significance, except to emphasize that Reed–Muller expansions for any $f(x)$ can be obtained in all-positive (uncomplemented) or in all-negative (complemented) form, unlike prime-implicant or minterm expressions which may contain mixtures of true and complemented input variables.

"Exclusive" functions with more than two inputs x_1 and x_2 may be proposed. Figure 1.4 illustrates three-input exclusive-OR and exclusive-NOR gates and their corresponding truth tables. It will be noted that because of the associative rules which hold with the exclusive operators, namely

$$[x_1 \oplus x_2 \oplus x_3] = [(x_1 \oplus x_2) \oplus x_3] = [x_1 \oplus (x_2 \oplus x_3)], \text{ etc.}$$

the three-input exclusive-OR gate may also be regarded as an odd-parity gate, that is a gate whose output signal is 1 when an odd number of its input signals is 1. Likewise the 3-input exclusive-NOR gate may be regarded as an even-parity gate, for the same train of reasoning.

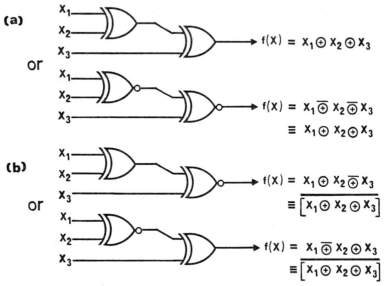

Fig. 1.5. Three-input exclusive-OR/NOR assemblies: (a) exclusive-OR made from two-input exclusive gates; (b) exclusive-NOR made from two-input exclusive gates.

Three-input exclusive-OR gates are logically equivalent to two two-input exclusive-OR or two two-input exclusive-NOR gates in cascade, as shown in Fig. 1.5a. Figure 1.5b shows the corresponding three-input exclusive-NOR equivalents. However, in the same manner that one should not regard a two-input exclusive-OR or NOR gate as necessarily an assembly of separate vertex gates, neither should these three-input versions be necessarily regarded as a cascade of separate two-input exclusive gates. A useful place will be shown later for these higher-input exclusive gates, particularly in association with threshold-logic gates (see Chapter 4) but currently little practical work has been pursued on the optimum circuit design for such gates.

To summarize this section, the exclusive-OR and NOR relationships and the corresponding logic gates are potentially very useful to supplement other logic relationships and their logic gates. A great deal more will be heard of them in subsequent chapters of this book.

1.3. MAJORITY Gates

The majority gate is a multi-input–single-output gate whose output is the same value as that of the majority of its inputs. To avoid possible confusion over what constitutes a "majority", it is universally taken that the number of gate inputs is an odd number, that is $n = 3$ or 5 etc., from which the gate output may be expressed as

$$f(x) = 1 \text{ iff } \sum_{i=1}^{n} x_i \geq n/2$$
$$= 0 \text{ otherwise}$$

where normal arithmetic summation, not Boolean, is involved in summing the number of logic 0 or 1 input signals. However, the terminology "majority" is normally sufficiently explicit to convey the gate operation. Figure 1.6 illustrates a three-input and a five-input majority gate.

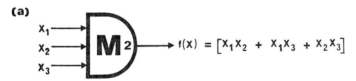

Fig. 1.6. Majority logic gates: (a) three inputs; (b) five inputs. Note that the output $f(x)$ will switch from 0 to 1 when a majority of its inputs are a logic 1, that is 2 or more for (a), and 3 or more for (b). The numbers 2 and 3 inside the gate symbols thus represent the minimum number of inputs that have to be at 1 to make $f(x) = 1$.

If the output of a majority gate is negated, then we have a minority gate. The gate output now takes the signal value corresponding to the minority of its signal inputs, which in the extreme case is output 1 when all inputs are 0 and *vice versa*. However, it is true to say that no particular use has yet been found to warrant considering the minority gate in its own right and therefore we shall here continue to talk mainly about majority relationships.

If one or more inputs of a majority gate are energized by steady 0 or 1 logic signals, the function realized by the remaining variable inputs obeys a modified relationship. Take the simple case of a three-input majority gate. If one input is steadily energized by, say, a 0 logic signal, then the two remaining inputs have to sum to 2 ($\geq n/2$, where $n = 3$) in order to change the gate output from 0 to 1. Thus the gate has become a simple two-input AND gate under these circumstances. Conversely if one input is held at 1, the gate becomes a simple two-input OR gate to the two variable inputs. With a five-input majority gate, selectively energizing one or two inputs with steady 0 or 1 logic signals provides four different gate operations for the remaining three inputs.

Algebraically the gate action with all inputs freely energized, or with one or more inputs steadily energized with 0 or 1 logic signals, is given by the simple Boolean algebraic development following. For illustration take the case of the five-input gate shown in Fig. 1.6 *b* whose output may be expressed by

$$f(x) = [x_1x_2x_3 + x_1x_2x_4 + x_1x_2x_5 + x_1x_3x_4 + x_1x_3x_5 + x_1x_4x_5$$
$$x_2x_3x_4 + x_2x_3x_5 + x_2x_4x_5 + x_3x_4x_5].$$

With one input, say x_5, always 0 we have

$$f(x) = [x_1x_2x_3 + x_1x_2x_4 + x_1x_3x_4 + x_2x_3x_4] = \text{a ``3-out-of-4'' gate.}$$

With two inputs, say x_4 and x_5, both always 0 we have

$$f(x) = [x_1x_2x_3] = \text{a ``3-out-of-3'' or AND gate.}$$

With one input, say x_5, always 1 we have

$$f(x) = [x_1x_2x_3 + x_1x_2x_4 + x_1x_2 + x_1x_3x_4 + x_1x_3 + x_1x_4 + x_2x_3x_4$$
$$+ x_2x_3 + x_2x_4 + x_3x_4]$$
$$= [x_1x_2 + x_1x_3 + x_1x_4 + x_2x_3 + x_2x_4 + x_3x_4]$$
$$= \text{a ``2-out-of-4'' gate.}$$

With two inputs, say x_4 and x_5, both always 1 we have

$$f(x) = [x_1x_2x_3 + x_1x_2 + x_1x_2 + x_1x_3 + x_1x_3 + x_1 + x_2x_3 + x_2x_3$$
$$+ x_2 + x_3]$$
$$= [x_1 + x_2 + x_3]$$
$$= \text{a ``1-out-of-3'' or OR gate.}$$

Notice that although the Boolean algebra for all these situations is absolutely explicit, it is becoming clumsy in expressing input–output

A Survey of Binary and Higher-Valued Logic

relationships more readily grasped by the concept of *at least p-out-of-q inputs energized*, where $1 \leq q \leq n$, $p \leq q$.

This simple development shows that AND and OR gates may be considered as special cases of majority gates, provided the latter have a sufficiently large number of inputs available which may be used for steady logic input signals in addition to the variable x_i inputs. If it is desired to use a majority gate with which to make an n-input AND or an n-input OR gate, then the gate must be equipped with $n - 1$ additional inputs to which the steady 0 or 1 preconditioning logic signals may be applied. The total number of gate inputs is now $2n - 1$, the gate requiring that n of them shall be energized to give $f(x) = 1$. Figure 1.7 illustrates that with both true and complemented outputs on such a gate, then it is possible to generate all the vertex functions shown in Fig. 1.1 in addition to the remaining "p-out-of-q" functions and their complements.

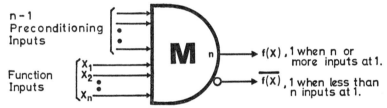

Fig. 1.7. The possible use of additional inputs on a majority type gate. Note that with all preconditioning inputs at 1, $f(x)$ and $\overline{f(x)}$ are the OR or NOR respectively of inputs x_1 to x_n, whilst with all such inputs at 0 $f(x)$ and $\overline{f(x)}$ realize AND and NAND.

Majority gates by themselves do not form a functionally complete set, as the invert operation is unavailable. Neither is it possible to generate from the variable x_i inputs the steady logic signals of 0 and 1. However, majority gates (with a sufficient number of inputs) plus inverter gates are functionally complete, as such pairing could always generate AND plus Invert, or OR plus Invert, with the INVERTER gate generating steady 0 or 1 inputs if and where necessary. Minority gates alone would form a functionally complete set, as they incorporate the Invert operation, but to date they have not been considered to be very useful candidates for general-purpose use.

The use of both these types of gates as general-purpose building blocks is complicated because there is no convenient algebraic expansion for any arbitrary function $f(x)$ in terms of majority or minority operators. One expansion which has been suggested is in terms of three-input majority operators, which expands $f(x)$ as follows:

$$f(x) = [1 \# x_1 f_1(x) \# x_1 f_0(x)]$$

where # # represents the three-input majority operator,

$$f_1(x) = f(1, x_2, \ldots, x_n),$$

and

$$f_0(x) = f(0, x_2, \ldots, x_n)$$

that is $f_1(x)$ and $f_0(x)$ are the two functions that result when x_1 in the given function $f(x)$ is replaced by 1 and 0 respectively. Notice also that this initial expansion is identical to

$$f(x) = [x_1 f_1(x) + \bar{x}_1 f_0(x)]$$

that is, the OR decomposition about x_1 of the given function.

Each term $f_1(x)$ and $f_0(x)$ may now be expanded in a similar manner, giving a second-level expansion of

$$x_1 f_1(x) = [x_1 \# x_2 f_{11}(x) \# \bar{x}_2 f_{10}(x)]$$

and

$$\bar{x}_1 f_0(x) = [\bar{x}_1 \# x_2 f_{01}(x) \# \bar{x}_2 f_{00}(x)].$$

This expansion continues until we reach the stage where we have three-input majority functions with x_n and $\bar{x}_n$ inputs, that is an n-level expansion down to the minterm level. Figure 1.8 illustrates a final expansion of a four-variable function $f(x)$.

Clearly this is not an easy or necessarily minimum expansion to use. Notice that as the majority operator we are using is by definition a three-input operator, associative decomposition of, say,

$$\{p \# q \# r\} \text{ into } \{(p \# q) \# r\} \text{ or } \{p \# (q \# r)\}$$

has no meaning. It is therefore difficult to manipulate algebraically such expressions, and the algebraic realization of $f(x)$ using this majority expansion can in fact be less minimal than that which may be formed by intuitive or mapping techniques[15-23]. It is also difficult to conceive these expressions as arithmetic equations giving the correct value 0 or 1 for $f(x)$. This is because the majority operator has to obey the unusual arithmetic summation

$$0+0+0=0, \quad 0+0+1=0, \quad 0+1+1=1, \quad 1+0+0=0,$$
$$1+0+1=1, \quad 1+1+0=1, \text{ and } 1+1+1=1.$$

A Survey of Binary and Higher-Valued Logic

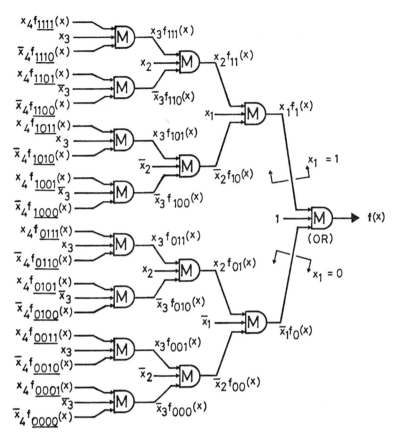

Fig. 1.8. A network realization of a four-variable function $f(x)$ using three-input majority gates. Note that $f_{0000}(x)$, etc. represent the minterm values of the given function $f(x)$ and that the upper and lower inputs into each gate are always mutually exclusive.

The logic functions which are directly realized by majority and minority gates are in the class of *symmetric functions*. All symmetric functions are characterized by some "symmetry" in their input variables, whereby the interchanging of two (or more) inputs causes no change to result in the output function $f(x)$[12,20,23]. Mathematically a Boolean function is said to be symmetric in any two variables x_i, x_j, $i \neq j$, if

$$f(x_1, \ldots, x_i, x_j, \ldots x_n) \equiv f(x_1, \ldots, x_j, x_i, \ldots x_n)$$

that is, the function remains completely unchanged ("invariant") if wherever x_i occurs x_j is substituted, and wherever x_j occurs x_i is substituted. If this invariance holds for all possible pairs of x_i, x_j, $1 \leq i, j \leq n$, then the function is said to be *totally* or *completely symmetric*.

Our basic AND/OR vertex gates are clearly completely symmetric in their inputs, as it makes no difference whatsoever in which order the inputs x_1 to x_n are connected to a single gate. Similarly all the majority and p-out-of-q-type functions are completely symmetric. A function such as

$$f(x) = [x_1x_3 + x_2x_3 + \bar{x}_4]$$

however, is not completely symmetric but is *partially symmetric*, having a symmetry in the input variables x_1 and x_2.

Symmetric functions form an important class of Boolean functions, and recognition of the symmetry may aid in producing an efficient realization of the function. Completely symmetric functions were extensively investigated in the days when relay contacts formed the heart of switching networks, and very elegant relay-switching solutions were devised[24]. However, because the concept of zero or infinite transmission and the very powerful availability of the change-over contact are not directly realizable by solid-state switching means, much of this earlier development has tended to be overlooked in recent years. This is a pity, but continuance of the earlier developments and the application of symmetries in logic design is gradually reappearing[12-21]; we shall certainly refer to symmetries again in a later chapter of this book.

This general consideration of majority gates has, however, introduced for the first time in this book two features, as follows.

(i) It is possible to specify a logic gate more powerful than the simple vertex gates but from which it is possible to generalize the basic AND/OR functions. This may also be viewed from the other approach, namely that our more common vertex gates are but extreme cases of more general logic functions. We shall come back to this point with even greater emphasis in the section immediately following. Notice, however, that the exclusive-OR/NOR gates are not in this category; they are something special and do not lie in this category of p-out-of-q functions.

(ii) The functions realized individually by both vertex and majority gates are completely symmetric functions, as are also the exclusive-

A Survey of Binary and Higher-Valued Logic 23

OR/NOR functions. Of course in appropriate combinations they may be used to realize other functions, whether symmetric in any sense or not. However, in the section following logic gates are considered which individually may not be completely symmetric; they therefore represent an even more powerful type of logic gate than we have so far considered.

1.4. THRESHOLD-LOGIC gates and a comparison with previous types

The threshold-logic gates which we shall consider in this section have binary-valued input and output signals exactly the same as all the previous classes of gate we have so far considered. However, unlike all previous gates the gate inputs need not each have the same "importance" in determining the 0 or 1 gate output state. We shall see shortly that both vertex gates and majority gates may be considered as particular and relatively simple cases of the general class of threshold-logic gates; the latter may therefore be regarded as logically the most fundamental and powerful of our binary gates, and hence the associated algebraic and mathematical considerations of threshold-logic relationships will form a fundamental subject area in several subsequent chapters of this book.

At this point it may be worth pausing to note that ease of circuit realization of general threshold-logic gates has been the major drawback to their adoption to date. However, with the furtherance of integrated circuit technologies and expertise more complex circuit configurations no longer constitute such a severe practical limitation, and increased "logic power per gate input connection" may be a much more useful circuit parameter to consider. (It is difficult satisfactorily to define and quantify this latter parameter, but the general concept is intuitively clear!) Chapter 8 of this book will be looking at the possible circuit configurations for these and other non-vertex gates.

The threshold-logic gates which we shall here consider are multi-binary-input gates, with one or more binary output. Unlike vertex gates, however, they do not obey simple Boolean input–output relationships, but rather each output $f(x)$ obeys an *arithmetic summation* relationship as follows:

$f(x) = 1$ if $\langle a_1 x_1 + a_2 x_2 + \ldots a_n x_n \rangle \geq$ gate threshold value t,

$ = 0$ if $\langle a_1 x_1 + a_2 x_2 + \ldots a_n x_n \rangle <$ gate threshold value t

where a_1, a_2, etc. are the input weighting factors, or more simply the "weights," associated with the respective binary inputs, and where normal arithmetic multiplication and addition, not Boolean or modulo-2, are involved in between the $\langle \rangle$ brackets. This basic gate-summation action and equating against the threshold value t may be more compactly expressed by

$$f(x) = 1 \text{ if } \sum_{i=1}^{n} a_i x_i \geq t$$
$$= 0 \text{ if } \sum_{i=1}^{n} a_i x_i < t.$$

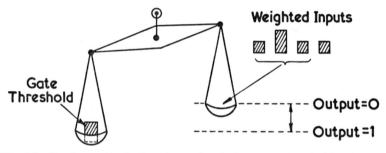

Fig. 1.9. Simple mechanical analogy of a single-output threshold-logic gate.

A simple mechanical analogy of the circuit action of such a gate is shown in Fig. 1.9. Remembering that each input may take the value 0 or 1, so a specific input summation $\sum_{i=1}^{n} a_i x_i$ is present for each input minterm, which summation must be equal or greater than t for the gate output to switch from 0 to 1.

The threshold-logic gate is thus seen to be a logic circuit which can by some means "weight" its various binary inputs, so as to give some more "importance" than others, sum the resultant weighted products, and give a gate output 1 or 0 if this weighted sum is above or below a chosen threshold value. Theoretically the individual weights a_i and the threshold value t can be any real numbers, positive or negative, integer or non-integer. In practice it is usually more convenient to restrict this freedom to the real positive integers 1, 2, 3, etc. This is certainly so as far as gate specifications and system design using such gates is concerned.

Figure 1.10a illustrates the general symbol used for a threshold-logic gate, together with the general expression for the gate output $f(x)$.

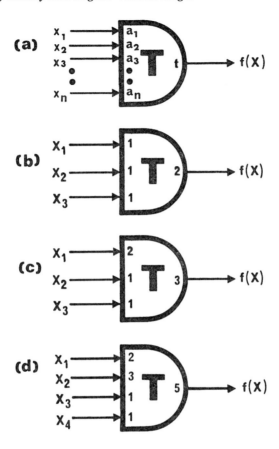

Fig. 1.10 Symbols for threshold-logic gates:

(a) general gate symbol
$$f(x) = \langle a_1 x_1 + a_2 x_2 + \ldots + a_n x_n \rangle_t$$

(b) three-input gate
$$f(x) = \langle x_1 + x_2 + x_3 \rangle_2, \text{ threshold,}$$
$$= [x_1 x_2 + x_1 x_3 + x_2 x_3], \text{ Boolean}$$

(c) three-input gate
$$f(x) = \langle 2x_1 + x_2 + x_3 \rangle_3, \text{ threshold}$$
$$= [x_1 x_2 + x_1 x_3], \text{ Boolean}$$

(d) four-input gate
$$f(x) = \langle 2x_1 + 3x_2 + x_3 + x_4 \rangle_5, \text{ threshold}$$
$$= [x_1 x_2 + x_2 x_3 x_4], \text{ Boolean.}$$

Figure 1.10b to d illustrate some specific threshold-logic gates, together with the threshold expression and corresponding Boolean expression for each. Examination of these very simple examples shows the potential advantages offered by the use of threshold-logic gates, as in each case one threshold-logic gate is realizing a logic requirement

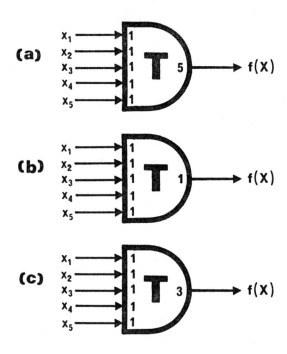

Fig. 1.11 AND, OR, and majority functions realized by a threshold-logic gate:
(a) five-input AND gate, $f(x) = [x_1 x_2 x_3 x_4 x_5]$
(b) five-input OR gate, $f(x) = [x_1 + x_2 + x_3 + x_4 + x_5]$
(c) five-input majority gate,
$f(x) = [x_1 x_2 x_3 + x_1 x_2 x_4 + \ldots + x_3 x_4 x_5]$

A Survey of Binary and Higher-Valued Logic

that would entail the use of three or more conventional AND and OR vertex gates. This reduction in the number of gates per system, and equally if not more important the number of gate interconnections, is the principal attraction of threshold-logic realizations.

Normal AND and OR vertex gates may be regarded as extreme cases of threshold-logic gates. The AND relationship is obtained when the gate threshold t is set at $\sum_{i=1}^{n} a_i$, whilst the OR relationship is obtained when t is set at unity. The input weights a_i may all be set to unity, as no one input has to be allocated a greater "importance" than any other in controlling the gate output. Similarly our majority gate may be considered as another special case, this time the gate threshold of course being set to $\frac{1}{2}(\sum_{i=1}^{n} a_i + 1)$, assuming n to be an odd number as is usual with a majority gate. These specific cases of threshold-logic relationships are illustrated in Fig. 1.11. Notice also that if it were possible to adjust the threshold value of a gate, as is now a feasible proposition, then a range of simple logic relationships ranging from OR through to AND may be obtained from the one gate.

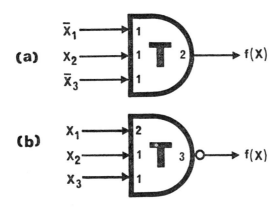

Fig. 1.12 Complemented inputs and outputs on threshold-logic gates:
(a) $f(x) = \langle \bar{x}_1 + x_2 + \bar{x}_3 \rangle_2$, threshold
$= [\bar{x}_1 x_2 + \bar{x}_1 \bar{x}_3 + x_2 \bar{x}_3]$, Boolean
(b) $f(x) = \langle 2x_1 + x_2 + x_3 \rangle_3$, threshold
$= [x_1 x_2 + x_1 x_3]$, Boolean

The complement $\bar{x}_i$ of any one or more variable x_i may be applied to a threshold-logic gate, exactly as with any other type of gate. Such an input signal $\bar{x}_i$ adds to the input summation when the signal value of $\bar{x}_i$ is 1 and contributes nothing to the summation when $\bar{x}_i = 0$, that is when the input variable $x_i = 1$. Similarly the output from a threshold-logic gate may be inverted to give an overall NOT relationship, as in conventional NAND/NOR gates. These variations are illustrated in Fig. 1.12.

Finally, to conclude this initial non-mathematical introduction to threshold logic, it is possible to consider threshold-logic gates with a given set of input weights but with two or more parallel outputs, each separate output corresponding to a specific gate-output threshold value t_j, where the value of each t_j may range from unity, the OR function, to a maximum useful value of $\sum_{i=1}^{n} a_i$, the AND function. One possible multi-output gate with three outputs is illustrated in Fig. 1.13. Negation of one or more of these parallel outputs is also possible.

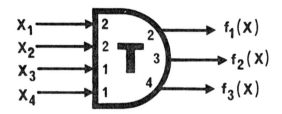

Fig. 1.13 Multiple outputs on a threshold-logic gate:

$f_1(x) = \langle 2x_1 + 2x_2 + x_3 + x_4 \rangle_2$, threshold
$= [x_1 + x_2 + x_3 x_4]$, Boolean

$f_2(x) = \langle 2x_1 + 2x_2 + x_3 + x_4 \rangle_3$, threshold
$= [x_1(x_2 + x_3 + x_4) + x_2(x_3 + x_4)]$, Boolean

$f_3(x) = \langle 2x_1 + 2x_2 + x_3 + x_4 \rangle_4$, threshold
$= [x_1(x_2 + x_3 x_4) + x_2 x_3 x_4]$, Boolean

So far this discussion should have shown why threshold logic has been considered for digital system realization as an alternative to or to supplement the more restricted but simpler Boolean logic functions, but clearly the design procedures necessary to utilize fully the appreciable power and flexibility of threshold relationships, as well as the

A Survey of Binary and Higher-Valued Logic

circuit design of the gates themselves, are likely to be more complex than our more familiar world of Boolean functions and vertex gates, but overall considerable advantages may accrue[17,21,23,25–34]

However, before we leave this area and come back to it again in more formal detail in later chapters, let us just survey the range of binary gates which we have now looked at and see how they share or differ in certain mathematical or geometric features.

1.5. A survey so far

1.5.1. Functional Completeness and Canonical Equations for any Given Boolean Function $f(x)$

The AND, OR, NAND, NOR family together form a functionally complete set, enabling any given Boolean function $f(x)$ to be synthesized. The NAND and the NOR gates are themselves each functionally complete. The Post canonic expansion for any given $f(x)$ is the sum-of-products expression involving Boolean multiplication and addition, which if expanded into its most fundamental form, becomes the sum-of-minterms expression

$$f(x) = \left[m_0 f_{00\ldots0}(x) + m_1 f_{00\ldots1}(x) + \ldots + m_{2^n-1} f_{11\ldots1}(x) \right]$$

where $m_0 = $ minterm $\bar{x}_1 \bar{x}_2 \ldots \bar{x}_n$, $m_1 = \bar{x}_1 \bar{x}_2 \ldots x_n$, etc. up to $m_{2^n-1} = x_1 x_2 \ldots x_n$, and where $f_{00\ldots0}(x)$, $f_{00\ldots1}(x)$ etc. each take the value 0 or 1 as required to realize $f(x)$.

The exclusive-OR/NOR gates basically require the addition of the AND operator and the logic-1 signal to ensure functional completeness. The canonic expansion is now the modulo-2 Reed–Muller expression, which states that any function $f(x)$ is given by

$$f(x) = \left[a_0 \oplus a_1 x_1 \oplus \ldots \oplus a_n x_n \oplus a_{n+1} x_1 x_2 \oplus \ldots \oplus a_{2^n-1} x_1 x_2 \ldots x_n \right]$$

where the a's each take the value 0 or 1 as required to realize $f(x)$. Notice that there are 2^n terms in this Reed–Muller equation as there are in the sum-of-minterms equation, since the number of different combinations of n inputs is always $2^n - 1$ for any value of n. Some of the terms in both expressions may of course be zero valued, and therefore neglected as far as the evaluation or realization of $f(x)$ is concerned. Also notice that in both the sum-of-products and the Reed–Muller expansion, the associative laws enable the corresponding circuit reali-

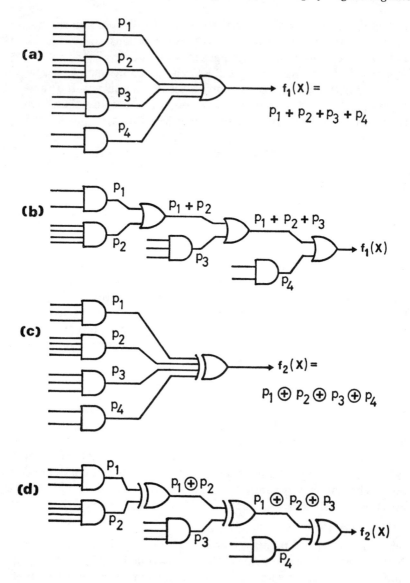

Fig. 1.14. Network realizations of Boolean and exclusive-OR canonical expansions for typical functions: (a) two-level AND/OR; (b) corresponding four-level AND/OR; (c) two-level AND/exclusive-OR; (d) corresponding four-level AND/exclusive-OR.

zation to be made up as a two-level network, which provides the fastest network response time, or as a greater-than-two-level network, which minimizes the fan-in necessary on the OR gates. This is illustrated in Fig. 1.14.

Majority gates require the Invert operator to be made available in order to form a functionally complete set. Minority gates may be considered as majority plus inverter gates combined, and therefore are themselves functionally complete. There is, however, no convenient general expansion for $f(x)$ using the majority operator which is of great practical significance, and none which enables a choice of two-level or higher-level networks to be realized, corresponding to the illustrations of Fig. 1.14.

Finally let us consider the threshold-logic gates. Like majority gates they do not form a functionally complete set by themselves, unless the Invert operator is also available on the gate or as a separate INVERTER gate. Also no algebraic expansion for a given function $f(x)$ in terms of threshold operators is available. Indeed when the fundamental specification of the threshold operation is considered, it is seen to be basically an arithmetic summation and discrimination, not a Boolean or modulo-2 summation. The problem which then arises is that there is no linear operator which can relate this arithmetic summation and discrimination with the discrete two-valued Boolean domain. There is a completely non-linear relationship between these two mathematical domains.

Thus there are no algebraic expressions which may be used to specify a given Boolean function $f(x)$ in threshold-logic form, apart of course from the case where the threshold-logic operations have all degenerated into vertex or simple majority functions. Therefore logic network synthesis using threshold-logic gates is not algebraically based; instead other techniques, particularly *spectral-domain* techniques, may be employed, a topic which will be introduced in Chapter 2.

1.5.2. Symmetry of the Various Functions

The functions realized by simple vertex gates are completely symmetric, as the gates do not distinguish between their n independent inputs in determining the gate output state. Similarly majority gates realize completely symmetric functions. We may consider OR and NOR gates as "at-least-1-out-of-n" gates, AND and NAND gates as "n-out-of-n" gates, and majority gates as "at-least-p-out-of-n" gates, where $p = (n + 1)/2$, n odd. In the latter case the p-out-of-n concept emphasizes both the logical function and the symmetric property more directly than the corresponding Boolean algebra.

Exclusive-OR/NOR gates, whether two input ($n = 2$) or more, also realize completely symmetric functions. The exclusive-OR gate effectively realizes "exactly-p-out-of-n," where p is odd numbered, $1 \leq p \leq n$, whilst the exclusive-NOR gate effectively realizes "exactly-p-out-of-n," p even, $0 \leq p \leq n$. Again the symmetry of all these functions is emphasized by this non-Boolean terminology.

The two input half-exclusive-OR/NOR gates, however, are not symmetric. Interchange of the input variables does not result in invariance of the gate output. Figure 1.15 illustrates geometrically the sym-

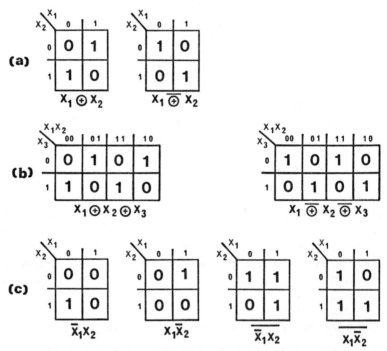

Fig. 1.15 Karnaugh-map plots of simple exclusive functions showing symmetries or lack of symmetries in the input variables: (*a*) two-input exclusive-OR/NOR functions, completely symmetric in the two inputs x_1 and x_2; (*b*) three-input exclusive-OR/NOR functions, completely symmetric in the three inputs x_1, x_2, and x_3; (*c*) the two-input half-exclusive functions of Fig. 1.3c to f inclusive, not symmetric in their inputs x_1 and x_2.

A Survey of Binary and Higher-Valued Logic

metry of exclusive-OR/NOR gates and the lack of symmetry of the half-exclusive-OR/NOR variations.

From previous developments it should be clear that threshold-logic gates with equal weighting on all gate inputs realize completely symmetric functions, as all inputs have precisely the same "importance" in determining the gate output. Hence interchange of any two or more input variables will result in no change in the output function.

However, this is a particular and restricted case of the general possibilities for such gates. If we consider a more general case, such as the gate $\langle 5, 4, 3, 2, 1 \rangle_7$, shown in Fig. 1.16a, then the function realized by this gate is completely unsymmetric, no two inputs having the same "importance" in determining the gate output. However, the gate $\langle 3, 2, 2, 1 \rangle_4$ shown in Fig. 1.16b is partially symmetric, having a symmetry in the inputs x_2 and $\bar{x}_3$ but no symmetry in any other pairs of variables.

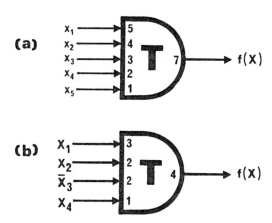

Fig. 1.16 Symmetries in threshold-logic gates:
(a) gate $\langle 5x_1 + 4x_2 + 3x_3 + 2x_4 + x_5 \rangle_7$
$= [x_1(x_2 + x_3 + x_4) + x_2(x_3 + x_4x_5)]$,
no symmetry in any pair of input variables.
(b) gate $\langle 3x_1 + 2x_2 + 2\bar{x}_3 + x_4 \rangle_4$
$= [x_1(x_2 + \bar{x}_3 + x_4) + x_2\bar{x}_3]$,
= symmetric in x_2 and $\bar{x}_3$ only.

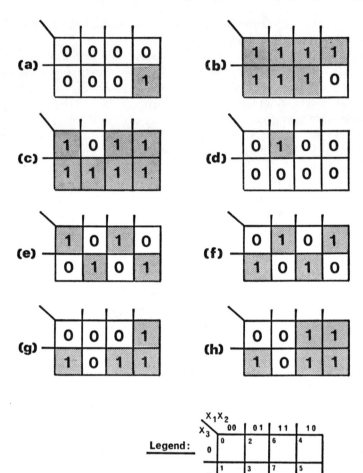

Fig. 1.17 Logic coverage of various simple gates:
(a) AND, $f(x) = [x_1 \bar{x}_2 x_3]$
(b) NAND, $f(x) = \overline{[x_1 \bar{x}_2 x_3]}$
(c) OR, $f(x) = [x_1 + \bar{x}_2 + x_3]$
(d) NOR, $f(x) = \overline{[x_1 + \bar{x}_2 + x_3]}$
(e) exclusive, $f(x) = [x_1 \oplus \bar{x}_2 \oplus x_3]$
(f) exclusive, $f(x) = \overline{[x_1 \oplus \bar{x}_2 \oplus x_3]}$
(g) threshold, $f(x) = \langle x_1 + \bar{x}_2 + x_3 \rangle_2$, = majority
(h) threshold, $f(x) = \langle 2x_1 + \bar{x}_2 + x_3 \rangle_2$

A Survey of Binary and Higher-Valued Logic

Thus the input weighting values of a threshold-logic gate give a direct indication of symmetries in the function realized by the gate. However, notice that, although all functions realized by a threshold-logic gate with equal input weightings are completely symmetric, the converse is not true. The exclusive-OR gate is a good counter-example of a completely symmetric function, but one which fundamentally cannot be realized by any possible threshold-logic relationship.

1.5.3. Logic Coverage, Hamming Distances

If we consider the logic coverage provided by all the types of binary gate we have considered by plotting typical functions of each on Karnaugh-map layouts, an overall situation as shown in Fig. 1.17 is illustrated. Three-input functions $f(x_1,x_2,x_3)$ are taken in each case purely for convenience. The vertex gates, by definition, merely differentiate between one input minterm condition and all remaining input conditions. The three-input exclusive-OR and NOR gates, however, differentiate between every adjacent input condition. The two threshold-logic gates, on the other hand, identify a group, or *pattern*, of input minterms. Notice that the map patterns of the two chosen threshold-logic functions (g) and (h) cannot be minimized to a one- or two-input function; all three inputs x_1, x_2, and x_3 are required in order to specify $f(x)$ fully in these cases.

The concept of *Hamming distances* may also be considered in this context[35]. A Hamming distance of 1 refers to any two minterms which differ in the value of one variable only, for example $\bar{x}_1 x_2 \bar{x}_3 \bar{x}_4$ and $\bar{x}_1 x_2 \bar{x}_3 x_4$. Similarly, a Hamming distance of 2 relates to two minterms which differ in the value of two variables only, for example $\bar{x}_1 x_2 \bar{x}_3 \bar{x}_4$ and $\bar{x}_1 x_2 x_3 x_4$, and so on.

On a Karnaugh-map construction, therefore, Hamming distances of 1 are associated with any two horizontally or vertically adjacent squares; Hamming distances of 2 are associated with any two squares which involve two horizontal and/or vertical moves to traverse one from the other, and so on. (Notice that diagonal moves are not allowed.) The counterpart of this Karnaugh-map picture of Hamming distances is the hypercube construction, as used in Fig. 1.2, wherein Hamming distances of 1 involve the traverse of any one side member of the hypercube, Hamming distances of 2 involve the traverse of any two joined side members, and so on for higher-value Hamming distances.

Now conventional Boolean minimization involves the recognition of Hamming distances of 1 in the logic-0 or the logic-1-valued minterms. Such a Hamming distance is therefore involved in the minimized realization of any function $f(x)$ using vertex gates. Similarly, a Hamming distance of 1 is also present in the coverage given by any threshold-logic gate, of which of course the vertex and majority gates are special cases. However, looking at Fig. 1.17e and f, we see that Hamming distances of 2 and not 1 are involved. So exclusive-OR/NOR gates are again shown to be fundamentally dissimilar to all other types of gate, and because they involve a Hamming distance of 2 rather than 1 the associated algebra and minimization is in general unfamiliar.

1.5.4. Linear Separability

Finally let us glance at another characteristic of binary functions, one which we shall refer to again in greater detail in the following chapter. This is linear separability.

Looking back at Fig. 1.2 for a moment, it will be seen that in this hypercube construction it is possible to construct a single plane through the hypercube which will separate the true (1) nodes from the false (0) nodes of all vertex gates. In these first examples the separating plane has merely to separate one node from all remaining $2^n - 1$ nodes.

If we apply this concept to other binary functions, we shall find that all threshold-logic functions, and hence all majority and vertex functions, are characterized by being able to specify such a separating plane to divide the true from the false nodes. In fact this is a necessary and sufficient condition for any given function $f(x)$ to be realizable with a single threshold-logic gate, and all possible threshold-logic functions may be visualized as those functions which result as a separating plane is swung generally across the hypercube construction through every possible solid angle.

Threshold-logic functions are therefore often termed *"linearly separable"* functions. In contrast the two or more input exclusive-OR and NOR gates are not linearly separable. They are in practice as far from a linearly separable function as it is possible to specify, as Fig. 1.18 attempts to illustrate. This feature is of course illustrated in another manner in the maps of Fig. 1.17. However, it must be noted that there are a vast number of functions which are neither linearly separable nor exclusive functions—in fact the vast majority of possible functions where the number of input variables becomes large, say $n \geq 5$, are non-linearly separable functions which have in practice to be realized by some appropriate assembly of logic gates of one or more types.

A Survey of Binary and Higher-Valued Logic

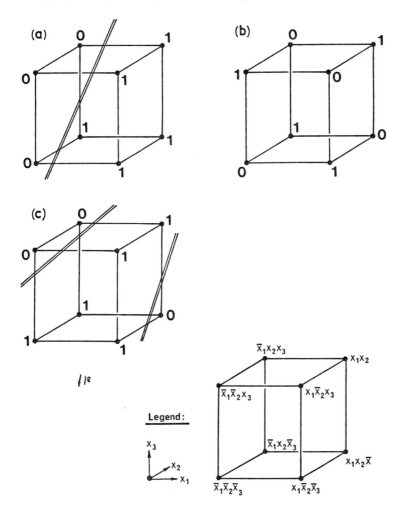

Fig. 1.18 The hypercube construction and linear separability:

(a) a linearly separable function
$$f(x) = [x_1 + x_2\bar{x}_3]$$
$$= \langle 2x_1 + x_2 + \bar{x}_3 \rangle_2$$

(b) the three-input exclusive-OR function
$$f(x) = [x_1 \oplus x_2 \oplus x_3], \text{ completely non-linearly séparable}$$

(c) a further non-linearly separable function
$$f(x) = [x_1 x_3 + \bar{x}_1 \bar{x}_3 + \bar{x}_2 \bar{x}_3], \text{ requiring two distinct separating planes to divide to 0- and 1-valued minterms.}$$

The final result of this initial survey into binary gates is summarized in Table 1.3. Very briefly, threshold-logic gates can be degenerated into both vertex and majority gates, but definitely not into any form of exclusive gate. Threshold-logic gates and exclusive gates may therefore in some sense be considered as "opposite" to each other, belonging to a different genus; if we can optimally use them both in network synthesis then we shall have the most powerful combination available. Boolean algebra, however, will be inadequate for this purpose, and hence in subsequent chapters we shall see other ways of handling binary data which may be more relevant for this purpose.

For a more comprehensive treatment of the supporting mathematics of the subject matter we have so far considered in this chapter, reference may be made to more specialized texts, for example Dietmeyer[12] for the binary work and Muroga[21] for further coverage of threshold-logic concepts.

Table 1.3. A survey of the principal characteristics of binary gates

Class of gate	Functionally complete	Canonic expansion for any $f(x)$ available	Logic coverage	Associated Hamming distances	Gate-input symmetry	Linearly separable
Vertex	AND+INVERT OR+INVERT NAND alone NOR alone	Yes; easy to manipulate	Very poor	1	Completely symmetric	Yes
Majority	With INVERT availability	Yes, but very cumbersome	Better	1	Completely symmetric	Yes
Threshold	With INVERT availability	No	Good	1	May be completely symmetric, partially symmetric, or asymmetric	Yes
Exclusive OR/NOR	With AND and logic 1 availability	Yes, but cumbersome	Good	2	Symmetric	No

1.6. The theoretical advantages of higher-valued logic

The present status and application of two-valued logic has developed largely because of the ready availability of efficient two-state devices with which to make the hardware realization. The necessary mathematical background starting with the work of George Boole[36,37]

and the subsequent more recent developments of Boolean algebra and associated concepts, of which the work of Shannon[38] is the classic starting point, aided this expansion.

However, leaving aside for the moment the problem of hardware realization, the question which should be asked is whether two-valued logic is an optimum choice. In many practical areas there is no clear binary yes/no requirement; situations such as yes/no/not defined, or up/down/stop, or left/right/straight ahead abound in the engineering world, and thus it could be argued that three-valued (radix 3) digital realizations would be more appropriate than binary. Equally in the mathematical area, most people would feel a greater affinity with (and possibly affection for!) a machine which worked in a 10-valued (radix 10 or denary) mode if technology could but provide efficient 10-state digital mechanisms.

Considering the question slightly more scientifically, the greatest significance of higher-radix working is the potential ability of being able to *reduce the number of interconnections* per system or subsystem. In our normal binary situation the information carried by each connection is minimal, being only one or other of two possibilities. Higher-radix working increases the information-carrying capacity of each connection, and hence must, if practical, reduce the number of total interconnections per system, except where the system itself is a trivial entirely two-state system. It is also interesting to note that one of the biggest problems now present for microelectronics manufacturers in large scale integration is that of interconnections, both on-chip and inter-chip.

However, is there an optimum radix for general-purpose system realization? Consider the arguments as follows.

In any system involving numbers the smaller the radix R that is chosen the greater will be the number of digits necessary to express the range of numbers present. Similarly, in a control-type system the smaller the number of logic levels available then the greater will be the number of circuit elements and interconnections necessary to cover the given number of differing input conditions.

Now the number of digits d necessary to express a range of N numbers is given by $N = R^d$. If we now assume that the number and/or cost of the basic hardware components C is proportional to the "digit capacity" $R \times d$, then we have that

$$C = k(Rd)$$
$$= k\left(R \frac{\log N}{\log R}\right)$$

where k is some constant. Differentiating this cost equation with respect to the radix R and equating to zero gives that R should equal e (= 2.718) for minimum cost. As in practice R must be an integer, this suggests that radix $R = 3$ should be more economical than the binary radix $R = 2$.

As an alternative supposition, if we consider that devices or circuits are available which provide two, three, four, or more stable digital signals without any increase in individual costs for the higher-valued radices, then in such ideal circumstances total cost C would be proportional to d. Hence

$$C = kd = k \frac{\log N}{\log R}$$

which is a gradually decreasing total cost C with increasing R.

The results of both these derivations are illustrated in Table 1.4. Both assumptions will be seen to indicate that radix 2 is not the most economical radix to choose, but whether the assumptions upon which these results are based are valid or not is the subject for the hardware component designer to prove or disprove. To date it would be fair to say that these assumptions have not been realized in semiconductor realizations, but this may be because so little development work has been put into higher-radix circuit realizations compared with the vast amount of research and development which has taken place in the binary field.

Table 1.4. Theoretical variation of total system cost C assuming radix 2 (binary) system costs of 100

Radix R	Cost C assuming cost proportional to Rd	Cost C assuming cost independent of R
1	∞	∞
2	100	100
2.718	94.5	69.9
3	95	63.1
4	100	50.0
5	107.9	43.1
10	150.5	30.1

There is clearly considerable scope and potential in pursuing research and development in the higher-radix digital field, and the extensive know-how available in microelectronic technology may finally

A Survey of Binary and Higher-Valued Logic

give us efficient multi-valued logic circuits. A useful review of the current status and hopes in this area, together with useful bibliographies, has recently been published [39-42].

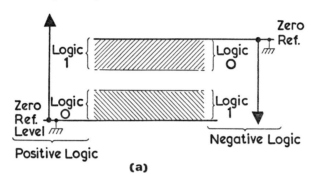

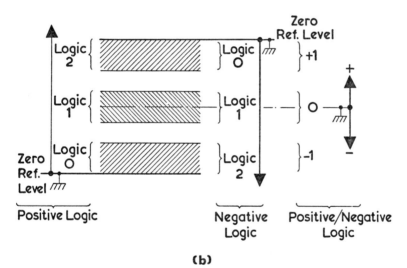

Fig. 1.19 Possible different conventions for both binary and ternary systems, where voltage levels are used to represent the logic values: (*a*) binary; (*b*) ternary. (Note that the areas between the shaded bands are the areas of ambiguity in which the actual logic voltages should never lie.)

1.7. Radix 3 (ternary) considerations

Next in sequence to our normal binary systems are of course *ternary* (radix 3) digital systems and it is therefore not surprising that three-state systems have received most consideration when radices higher than binary are being investigated. This is apart from any optimization which theoretically may accrue from the adoption of radix 3 rather than any other radix.

The three digital values of a ternary system conventionally may be identified as

$$0 \quad 1 \quad 2$$

or as

$$-1 \quad 0 \quad +1.$$

It is irrelevant to the possible circuit realizations which of these conventions is used, although there is sometimes confusion if the electrical signals representing these three values are quoted as a negative potential, a zero potential, and a positive potential with respect to the so-called system zero. For example three voltages listed as, say, -5 V, 0 V, and $+5$ V may equally be regarded as 0 V, $+5$ V, and $+10$ V respectively with respect to the most negative supply potential. Figure 1.19 illustrates possible binary and ternary conventions when voltages are used to represent each logic state.

There is, however, a difference in the way these ternary symbols may be used in *number representation*. Table 1.5 lists how simple denary numbers may be represented in two possible ternary numbering systems. Notice that in both cases the denary number D is given by the summation

$$D = \{T_n 3^n + T_{n-1} 3^{n-1} + \ldots + T_1 3^1 + T_0 3^0\}$$

where

T = ternary digit 0, 1, or 2, or $-1, 0, +1$

n = significance of the ternary digit, T_0 = least significant

T_n = most significant.

However, the $-1, 0, +1$ numbering system has a unique advantage that any number can be changed from a positive value to the corresponding negative value by merely changing all -1's to $+1$'s and *vice versa*, leaving all zeros unchanged (i.e. sign reversals only), whilst any given number is shown to be a positive or a negative number by examination of the sign of its most significant non-zero digit only. Addition and subtraction are therefore possible with the same hardware, with suitable sign changes of the addend or subtrahend as required.

A Survey of Binary and Higher-Valued Logic

Now to efficiently design any ternary (or other) digital system, two aspects require to have been investigated. The first of these is an algebra or other mathematical means whereby the given data may be expressed and possibly manipulated and minimized; the second of course is the circuits with which the system may be constructed.

Since the existing information in neither of these two ternary areas is as well known as that in the binary field, we shall consider them both in this book. In the immediately following pages we shall look at ternary algebras and functional completeness criteria, whilst in a later chapter we shall return to the state of the art of ternary circuits and ternary system design.

Table 1.5. Two possible notations to represent denary numbers in a ternary notation

Denary number D	Ternary notation using numbers 0, 1, and 2	Ternary notation using the numbers $-1, 0, +1$ (written $-, 0, +$ for convenience)
0	0000	0 0 0 0
1	0001	0 0 0 +
2	0002	0 0 + −
3	0010	0 0 + 0
4	0011	0 0 + +
5	0012	0 + − −
6	0020	0 + − 0
7	0021	0 + − +
8	0022	0 + 0 −
9	0100	0 + 0 0
10	0101	0 + 0 +
11	0102	0 + + −
etc.		

1.8. Multi-valued algebras and functional completeness

If we consider first a system with n multi-valued inputs X_1 to X_n and one multi-valued output $f(X)$, generally as shown in Fig. 1.20a, then the number of different signal input combinations is R^n and the number of different functions $f(X)$ of the n inputs is $R^{(R)^n}$. For a ternary system as shown in Fig. 1.20b there are therefore 3^n input combinations and $3^{(3)^n}$ possible ternary functions of these input variables. The latter of

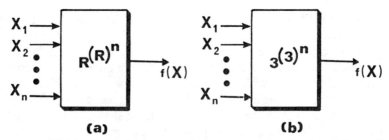

Fig. 1.20. Multi-valued single-output system: (*a*) R-valued, $R^{(R)^n}$ possible functions $f(X)$; (*b*) three-valued (ternary), $3^{(3)^n}$ possible functions $f(X)$.

course includes the trivial cases of $f(X)$ = constant, plus all the degenerate functions of fewer than n inputs. Notice how the number of possible functions of n input variables rapidly increases from the two-valued binary case in the ternary and higher-valued cases. With merely one input X_i there are 27 possible ternary functions of $f(X_i)$, compared with only four for the binary situation, although many of these single-variable functions are logically trivial as Table 1.6 shows.

Table 1.6. All possible functions of a single variable for the binary and the ternary case

Binary		Ternary	
x_i	$f(x_i)$	X_i	$f(X_i)$
0	0 0 1 1	0	0 0 0 0 0 0 0 0 0 1 1 1 1 1 1 1 1 1 2 2 2 2 2 2 2 2 2
1	0 1 0 1	1	0 0 0 1 1 1 2 2 2 0 0 0 1 1 1 2 2 2 0 0 0 1 1 1 2 2 2
		2	0 1 2 0 1 2 0 1 2 0 1 2 0 1 2 0 1 2 0 1 2 0 1 2 0 1 2

We have already considered the concept of *functional completeness* in the binary case. In the same manner full functional completeness for the ternary case involves the ability to be able to synthesize all possible $3^{(3)^n}$ functions of $f(X)$. If the algebra or available circuit elements enable us to synthesize all such functions except the constant-value ones, then such *operational completeness* may be sufficient for all practical purposes.

Several tests exist for determining whether any given algebra is functionally complete or not. One such test is to test directly whether it is possible to synthesize all possible functions of two variables ($n=2$). If this is possible, then the algebra or circuit elements are functionally complete for any n[43,44]. Another possible test is that of Slupecki[45,46] which states the following:

(i) It must be possible to generate all possible single-variable functions $f(X_i)$

(ii) There exists at least one two-variable function $f(X_1, X_2)$ which is not equivalent to a single-variable function $f(X_1)$ and which when appropriate truth values from the set $0, 1, \ldots, R-1$ are substituted gives the function value $f(X_1, X_2) = i$, where i represents all the possible truth values $0, 1, \ldots, R-1$ (cf. the binary AND function in which $f(0,0) = 0$ and $f(1,1) = 1$).

However, the algebras by which a functionally complete multi-valued system of $R>2$ may be described are dissimilar from our familiar Boolean albegra. It may be shown that Boolean algebra is complete only for the two-valued case[46] and therefore alternative algebras become necessary for the higher-valued (non-binary) cases. Various algebras to provide functional completeness for $R \geqslant 2$ have been investigated. These include the following:

(a) An algebra based upon the original work of Post (1921) generally known as *Post algebra*.

(b) A modulo-sum and modulo-product algebra based upon the original work of Bernstein (1924) and generally known as *modular algebras*.

(c) The single-operator algebra of Webb (1935) which itself is a generalization of the binary *Sheffer–Stroke operator*.

(d) Various *augmented-Boolean* and *device-orientated algebras* which attempt to correlate directly with specific logic circuit configurations, the majority of which are concerned specifically with the ternary case.

It is instructive to look at these various algebras, and to see how in many cases they reduce to our more familiar binary operators when applied to the $R = 2$ case.

1.8.1. Post Algebra

The original contribution of Post[5] was to show that two primitive operators only are necessary to express any multi-valued system.

These two operators are an *addition modulo-R operator*, $\sim$, with any carries disregarded, and the *"maximum-value" operator*, V_R, where

$$\tilde{X}_i \triangleq \{1 + X_i\}_{\text{mod}_M}$$

and

$$X_i \ V_R \ X_j \triangleq \text{maximum } X_i, X_j$$

$$X_i, X_j = 0, 1, \ldots, R-1.\text{[3]}$$

Notice that in the binary case these two operators become the NOT and the OR operator, respectively. For the ternary case, these two operators behave as detailed in Table 1.7.

Table 1.7. The two Post operators which provide functional completeness in a ternary ($R = 3$) system

X_i	$\tilde{X}_i$	X_i	X_j	$X_i \ V_R \ X_j$
0	1	0	0	0
1	2	0	1	1
2	0	0	2	2
		1	0	1
		⋮	⋮	⋮
		2	1	2
		2	2	2

The use of these two primitive operators only, however, results in very unwieldy canonical equations for realization purposes[5,43].

Repeated use of the $\sim$ operator is required in very lengthy expressions of the form:

$$f(X) = \{\tilde{X}_i^{-r}(\tilde{X}_j^{-s} \ldots V_R \ldots \tilde{X}_k^{-t})V_R \ldots\}^p$$

[3] The symbols used here for the Post operators and for certain subsequent operators are dissimilar to the original authors' symbols. Also we shall continue to use for the truth values the set of real numbers $0, 1, 2, \ldots, R-1$ instead of numbers or subscripts $1, 2, 3, \ldots, R$ adopted by Post and others, the choice here being made to maintain a certain consistency in all sections.

A Survey of Binary and Higher-Valued Logic

where $\sim^p, \sim^q, \ldots$ represent the possibly repeated application of the $\tilde{X}$ operator $p, q, \ldots$ times, $0 \leq p, q, \ldots \leq R - 1$. Notice that this corresponds in the binary ($R = 2$) case to having to express all operations with the two Boolean operators of NOT and OR only; for example the simple AND function of two variables would require to be expressed as $\overline{[\bar{x}_1 + \bar{x}_2]}$ and so on.

1.8.2. The Modular Algebras

The primitive Post operators, therefore, are not convenient propositions for $R > 2$ and preferably require to be augmented with further operators for convenience. The modular algebras, on the other hand, based on the first publication of Berstein[47] and employing the natural numbers 0, 1, 2, upwards, provide a more compact expansion for a given function $f(X)$ than the basic Post expansion. The two fundamental operators involved are *modulo-R arithmetic summation*, $+\mathrm{mod}_M$, and *modulo-R arithmetic multiplication*, $\times\mathrm{mod}_M$, with any carries disregarded. For the binary and ternary cases these two operators behave as shown in Table 1.8.

Table 1.8. The modular algebra operators $+\mathrm{mod}_R$ and $\times\mathrm{mod}_R$ in the binary and ternary cases

x_1	x_2	$+\mathrm{mod}_2(x_1,x_2)$	$\times\mathrm{mod}_2(x_1,x_2)$	X_1	X_2	$+\mathrm{mod}_3(X_1,X_2)$	$\times\mathrm{mod}_3(X_1,X_2)$
0	0	0	0	0	0	0	0
0	1	1	0	0	1	1	0
1	0	1	0	0	2	2	0
1	1	0	1	1	0	1	0
				$\vdots$	$\vdots$	$\vdots$	$\vdots$
				2	1	0	2
				2	2	1	1

It may be noticed that in the development of this modular algebra, the truth table format of Table 1.8 may be replaced by a matrix format, as shown in Table 1.9[46,47]. A certain degree of normal matrix manipulation of the real numbers which occur in these formats is possible, although no overriding advantages accrue.

Table 1.9. The modular algebra operators $+\mathrm{mod}_R$ and $\times\mathrm{mod}_R$ defined in matrix form: (a) binary, (b) ternary, and (c) quarternary

(a)

$+\mathrm{mod}_2$	0	1
0	0	1
1	1	0

$\times\mathrm{mod}_2$	0	1
0	0	0
1	0	1

(b)

$+\mathrm{mod}_3$	0	1	2
0	0	1	2
1	1	2	0
2	2	0	1

$\times\mathrm{mod}_3$	0	1	2
0	0	0	0
1	0	1	2
2	0	2	1

(c)

$+\mathrm{mod}_4$	0	1	2	3
0	0	1	2	3
1	1	2	3	0
2	2	3	0	1
3	3	0	1	2

$\times\mathrm{mod}_4$	0	1	2	3
0	0	0	0	0
1	0	1	2	3
2	0	2	0	2
3	0	3	2	1

These two operators together with the logic value 1 are functionally complete for any prime number R. For $R = 2$ the operators become the more familiar exclusive-OR and AND operators, considered in §1.2.

Various expansions have been proposed using these modular operators[43,46,48], but the expansion due to Tamari is the most direct. This states that any R-valued function $f(X)$ may be expressed as

$$f(X) = \sum_{}^{R^n} \{a_\alpha X_\beta\}_{\mathrm{mod}\,R}$$

where X_β are all possible different combinations of powers of the n input variables X_1 to X_n, ranging from $X_1, X_1^2, \ldots, X_1^{R-1}$, up to $X_1^{R-2}X_2^{R-1}\ldots X_n^{R-1}, X_1^{R-1}X_2^{R-1}\ldots X_n^{R-1}$, and a_α are real-number coefficients associated with each X_β, taking values in the range $0 \leq a_\alpha \leq R-1$ as required to realize $f(X)$,

and where each of the R^n product terms and the overall summation are modulo-R multiplications and additions, respectively.

For a ternary function of two variables X_1 and X_2 this expansion becomes

$$f(X) = \{a_{00} + a_{10}X_1 + a_{20}X_1^2 + a_{01}X_2 + a_{11}X_1X_2 + a_{21}X_1^2X_2 \\ + a_{02}X_2^2 + a_{12}X_1X_2^2 + a_{22}X_1^2X_2^2\}_{\mathrm{mod}\,3}$$

whilst for the binary case it reduces to the Reed–Muller canonical form previously encountered, namely

$$f(x) = [a_0 \oplus a_1x_1 \oplus a_2x_2 \oplus \ldots \oplus a_{2^n-1}x_1x_2\ldots x_n].$$

A Survey of Binary and Higher-Valued Logic

This Tamari expansion may therefore sometimes be referred to as a generalized Reed–Muller expansion[49]. It also provides a direct means of function realization in terms of the two basic operators, as may be demonstrated by the following ternary example:

Function $f(X)$ truth table

X_1	X_2	$f(X)$
0	0	0
0	1	0
0	2	0
1	0	0
1	1	0
1	2	2
2	0	2
2	1	1
2	2	2

The values for the required coefficients a_{00}, a_{01}, etc. are now evaluated by solving the simultaneous equations which occur in the tabulation following. Notice that all operations are modulo-3 in this example.

Input values X_1 X_2	Coefficients of the non-zero terms in the mod_3 expansion	Required value of the mod_3 additions i.e. $f(X)$
0 0	a_{00}	0
0 1	$a_{00} + a_{01} + a_{02}$	0
0 2	$a_{00} + 2a_{01} + a_{02}$	0
1 0	$a_{00} + a_{10} + a_{20}$	0
1 1	$a_{00} + a_{01} + a_{02} + a_{10} + a_{11} + a_{12} + a_{20} + a_{21} + a_{22}$	0
1 2	$a_{00} + 2a_{01} + a_{02} + a_{10} + 2a_{11} + a_{12} + a_{20} + 2a_{21} + a_{22}$	2
2 0	$a_{00} + 2a_{10} + a_{20}$	2
2 1	$a_{00} + a_{01} + a_{02} + 2a_{10} + 2a_{11} + 2a_{12} + a_{20} + a_{21} + a_{22}$	1
2 2	$a_{00} + 2a_{01} + a_{02} + 2a_{10} + a_{11} + 2a_{12} + a_{20} + 2a_{21} + a_{22}$	2

Solution of these nine simultaneous equations rapidly gives

$$a_{00} = 0, \quad a_{01} = 0, \quad a_{02} = 0, \quad a_{10} = 2, \quad a_{11} = 2$$
$$a_{12} = 0, \quad a_{20} = 1, \quad a_{21} = 0, \quad a_{22} = 1$$

Hence $f(X) = \{2X_1 + 2X_1X_2 + X_1^2 + X_1^2X_2^2\}_{\text{mod}3}$
$= \{2X_1(1+X_2) + X_1^2(1+X_2^2)\}_{\text{mod}3}$

which, if gates were available to realize the modulo-3 operators, would yield the circuit realization shown in Fig. 1.21.

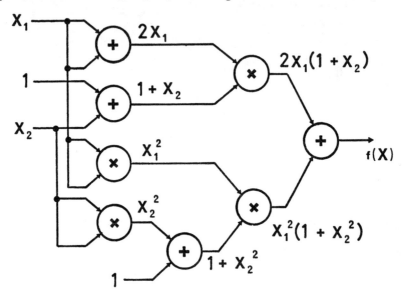

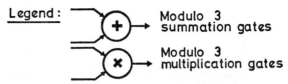

Legend:
+ → Modulo 3 summation gates
× → Modulo 3 multiplication gates

Fig. 1.21. Realization of a given ternary function $f(X)$ from the modular algebraic expression

$f(X) = \{2X_1(1 + X_2) + X_1^2(1 + X_2^2)\}_{\text{mod}3}$

It is interesting to consider that, in the interpretation of the ternary network shown in Fig. 1.21, the conventional logic concepts of AND and OR are inadequate to impart any meaning to the operation of this network. This is so even if we reconsider the AND operator as "the minimum value of" and the OR operator as "the maximum value of."

A Survey of Binary and Higher-Valued Logic 51

Instead we are having to consider networks as mathematical networks, the digital signals of value 0, 1, or 2 obeying specific mathematic rules, in this particular case the modulo-3 rules of multiplication and addition.

1.8.3. The Single-Operator Multi-Valued Algebra of Webb

Less successful than the modular algebras when applied to ternary and higher-valued systems is the single functionally complete operator proposed by Webb and by Martin[51,52]. This operator, based upon the *Sheffer–Stroke operator* of Sheffer[50], is defined as follows for any R-valued system:

$$X_1 | X_2 \triangleq 0 \text{ if } X_1 \neq X_2$$
$$= (1 + X_1)_{\text{mod} R} \text{ if } X_1 = X_2.$$

For $R = 2$ (binary) and $R = 3$ (ternary) this gives the truth table relationships shown in Table 1.10. Notice that in the binary case the operator is the simple 0-level NAND or 1-level NOR operator, which of course is functionally complete for the two-valued case.

Webb proved that any multi-valued function $f(X)$ may be synthesized using this single operator. Martin details its application specifically to the three-valued case. However, in general inordinately long and cumbersome expansions result, and as a consequence no practical convenience or attraction has been found for adopting just this one functionally complete operator.

Table 1.10. The functionally complete Webb operator: (*a*) binary case; (*b*) ternary case

(a)

X_1	X_2	$X_1 \| X_2$
0	0	1
0	1	0
1	0	0
1	1	0

(b)

X_1	X_2	$X_1 \| X_2$
0	0	1
0	1	0
0	2	0
1	0	0
1	1	2
1	2	0
2	0	0
2	1	0
2	2	0

1.8.4. Augmented Boolean-Type Operators for Ternary Systems

Leaving general multi-valued algebras and considering now algebras developed specifically for the three-valued case, we find a number of suggestions for augmenting Boolean-type operators in order to obtain functional completeness for the $R = 3$ situation. Now the ternary operators which in theory are the simplest to realize in hardware form are the two-variable conjunction operator

$$f(X_1, X_2) = \text{minimum } (X_1, X_2), = X_1 \,\&\, X_2$$

and the two-variable disjunction operator

$$f(X_1, X_2) = \text{maximum } (X_1, X_2), = X_1 \vee X_2.$$

Notice that in the binary case these two operators are the AND and OR operators respectively. Hence they are often referred to as AND and OR for the higher-valued case also.

These two operators in the three-valued case continue to possess commutative and associative properties and are therefore amenable to algebraic manipulation. However, as in the binary case, they are not by themselves functionally complete[53]. The addition of appropriate single-variable operators is necessary in order to achieve functional completeness; exactly as in the binary case the NOT operator has to be made available. However, in the ternary case several choices of single-variable operators are possible, and indeed a non-minimal set of single-variable operators may be useful to reduce the overall algebraic expansion.

Table 1.11. The basic ternary AND and OR operators used in augmented algebras for three-valued systems

X_1	X_2	$X_1 \,\&\, X_2$	$X_1 \vee X_2$
0	0	0	0
0	1	0	1
0	2	0	2
1	0	0	1
1	1	1	1
1	2	1	2
2	0	0	2
2	1	1	2
2	2	2	2

A Survey of Binary and Higher-Valued Logic

In all these augmented Boolean-type algebras the two-variable AND and OR operators remain as previously defined[4], which gives the truth-table relationships detailed in Table 1.11.

Several sets of single-valued operators have been suggested which with the above AND and OR operators give functional completeness. The most commonly quoted ones are as tabulated in Table 1.12 [54-58]. Notice that these operators provide a means of converting a three-valued signal into a binary-valued signal, which frequently allows a Boolean-type expansion for any given function $f(X)$ to be developed.

Table 1.12. Single-variable functions which have been proposed to achieve functional completeness of the Boolean-type algebras

(a)

X_i	$\overline{X_i}$	$\overset{\vee}{X_i}$	$\overset{\wedge}{X_i}$
0	2	0	0
1	1	2	0
2	0	2	2

(b)

X_i	$\overline{X_i}$	$\equiv_0 X_i$	$\equiv_1 X_i$	$\equiv_2 X_i$
0	2	2	0	0
1	1	0	2	0
2	0	0	0	2

(c)

X_i	Inv X_i	Opp X_i	Con X_i
0	2	1	2
1	1	0	0
2	0	0	0

(d)

X_i	$j_0 X_i$	$j_1 X_i$	$j_2 X_i$
0	2	0	0
1	0	2	0
2	0	0	2

(a) Rohleder's "Not," "Cup," and "Cap" operators.
(b) Goto's "Not" and "Equivalence" operators.
(c) Vacca's "Inverse," "Opposite," and "Converse" operators.
(d) Mühldorf's j_k, $k = 0,1,2$, operators, the same as Goto's "Equivalence" operators.

Notice the set of single-variable operators tabulated in Table 1.12a, b, c give functional completeness when employed with the AND and OR operators, but Mühldorf's j_k operators have to be supplemented with logic 1 to achieve functional completeness. Also the three operators of Vacca can be combined to produce further g and h two-valued

---

[4] Some authors have defined AND and OR opposite to as here defined, but this in the main occurs owing to considering the truth-value symbols 0,1,2 in the opposite order to normal arithmetic significance.

operators which may be of more direct use in the algebraic expansions for $f(X)$.

The expansions for any given function using any of these sets of operators is relatively straightforward[46,59], although not as satisfying as the Tamari modular algebra expansion or the Chen and Lee development which will be covered shortly. In general all involve an expansion of the given function $f(X)$ to minterm level, with the maximum and minimum operators effectively selecting one minterm value at a time determined by the particular minterm input. Some rearrangement and simplification of the complete expansion, however, is usually possible.

As an example, the Mühldorf expansion takes the initial form:

$$f(X) = \left\{ \left(k_0 \ \& \ f_{00\ldots0}(X)\right) \lor \left(k_1 \ \& \ f_{00\ldots1}(X)\right) \lor \ldots \lor \left(k_{3^n-1} \ \& \ f_{n-1\ldots n-1}(X)\right) \right\}$$

where k_i, $i = 0$ to 3^n-1, are functions formed from the basic set whose value is 0 except on the respective i^{th}-input minterm when they take the value 2, and where $f_i(X)$, $i = 00\ldots0$ to $n-1\ldots n-1$ is the output value of the function $f(X)$ on the respective i^{th} minterm. Hence all brackets except one are zero-valued on every input minterm, the exception being the one bracket addressed by the particular input minterm combination. This one bracket takes the value given by $f_i(X)$, which is set to the required function output value $f(X)$.

As an example, the two-variable function synthesized in §1.8.2 by the modular algebra approach is as follows:

$$f(X) = \{(k_0 \ \& \ 0) \lor (k_1 \ \& \ 0) \lor (k_2 \ \& \ 0) \lor (k_3 \ \& \ 0) \lor (k_4 \ \& \ 0) \\ \lor (k_5 \ \& \ 2) \lor (k_6 \ \& \ 2) \lor (k_7 \ \& \ 1) \lor (k_8 \ \& \ 2)\}$$

which, as it is unnecessary to realize any of the $(k_i \ \& \ 0)$ terms and because every $(k_i \ \& \ 2)$ term $\equiv k_i$, becomes

$$f(X) = \{k_5 \lor k_6 \lor (k_7 \ \& \ 1) \lor k_8\}.$$

Now the k_i functions are formed from the basic set by simple rules and for this particular example yield the final expansion

$$f(X) = \{(j_1 X_2 \ \& \ j_2 X_1) \lor (j_2 X_2 \ \& \ j_0 X_1) \lor (j_2 X_2 \ \& \ j_1 X_1 \ \& \ 1) \\ \lor (j_2 X_2 \ \& \ j_2 X_1)\}.$$

The close analogy of this expansion with our more familiar Boolean sum-of-products minterm expression will be noticed. In both cases zero-valued terms may be dropped, but in the binary case it is unnecessary to define the value of the remaining non-zero terms. Here in the ternary case, however, it is necessary to distinguish between the non-zero truth values 1 and 2, as illustrated by the (k_7 & 1) term in the above expansion. Without this 1-value the k_7 term would give the truth value 2 for $f(X)$ at this point.

However, algebraically there appears to be little to choose between all these Boolean-type algebras; from the practical circuit realization of the operators Mühldorf's set may be the simplest, but even so appears less promising than other alternatives not based upon the AND and OR minimum and maximum operators.

1.8.5. The Three-Valued Algebra of Lee and Chen

In contrast to the previous section where a multiplicity of basic function operators was present, the work of Lee and Chen[60] illustrates how an increased simplicity of expansion may be achieved by using logically more powerful operators in the chosen basic set.

Lee and Chen's proposals involve one basic ternary operator only, with four ternary inputs, which together with the three truth values 0, 1, and 2 provide a functionally complete set. The ternary operator proposed is termed the "*T*-gate" operator and is defined as follows:

$$T(X_1, X_2, X_3; X_4) = X_1 \quad \text{if} \quad X_4 = 0$$
$$= X_2 \quad \text{if} \quad X_4 = 1$$
$$= X_3 \quad \text{if} \quad X_4 = 2.$$

Thus the *T*-gate effectively selects as its output the value of input X_1 or X_2 or X_3, depending solely upon whether X_4 is 0, 1, or 2, respectively. Figure 1.22*a* illustrates the basic action of this *T*-gate operator in a mechanical switching form.

Now the expansion for any two-variable ternary function $f(X_1, X_2)$ is directly given by

$$f(X_1, X_2) = T\{f(X_1, 0), f(X_1, 1), f(X_1, 2); X_2\};$$

that is, the value of X_2 selects in turn each of the reduced functions $f(X_1, 0)$, $f(X_1, 1)$, and $f(X_1, 2)$ as X_2 takes the values 0, 1, and 2, respectively. Further, each of the reduced functions are functions of the single variable X_1 only, each of which may be expanded as

$$f(X_1) = T\{f(0), f(1), f(2); X_1\}$$

where $f(0)$, $f(1)$, and $f(2)$ are the appropriate function truth values when $X_1 = 0, 1,$ and 2, respectively.

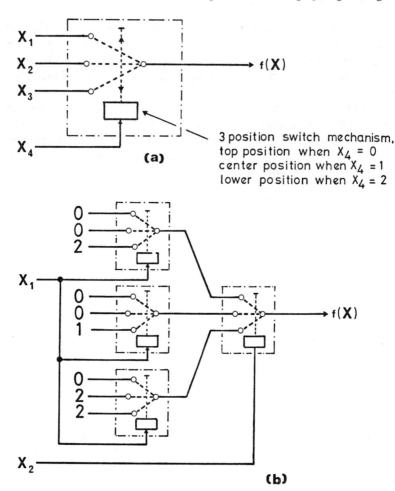

Fig. 1.22. The Lee and Chen functionally complete ternary T-gate operator, which multiplexes three of its inputs under the control of the fourth input:
 (a) mechanical interpretation of the operator;
 (b) mechanical interpretation of the expansion of a two-variable ternary function

$$f(X_1, X_2) = 0,0,0,0,0,2,2,1,2$$

Hence a complete expansion for the two-variable function $f(X_1, X_2)$ is directly obtained, as illustrated in mechanical switching terms in Fig. 1.22 b. The ternary function first synthesized in §1.8.2. with the truth-

A Survey of Binary and Higher-Valued Logic

table output values 0,0,0,0,0,2,2,1,2, becomes as follows in this expansion:

$$f(X_1, X_2) = T\{T(0,0,2;X_1), T(0,0,1;X_1), T(0,2,2;X_1); X_2\}$$

The T-gate operator is a beautifully direct operator to use with three-position mechanical switches. The general expansion, however, gives an n-level realization for an n-variable function $f(X)$, although some informal and intuitive minimization can be made by grouping together input combinations with the same output function value. It would be very interesting to have this operator available in modern microelectronic form.

1.8.6. Other Device-Oriented Proposals

Because the present state of electronic circuit realization favors the use of binary switching elements working in an on–off mode, a number of possible circuit arrangements for ternary applications have been suggested which are basically binary in nature but which are combined so as to provide an overall three-level result[46,61–63]. Figure 1.23 illustrates the general arrangement of many such proposals.

The algebra associated with such proposals may generally be arranged in the following form:

$$f(X) = \text{maximum } \{f(X)_{01}, f(X)_{02}\}$$
or
$$\text{minimum } \{f(X)_{20}, f(X)_{21}\}$$

where $f(X)_{01}$, $f(X)_{02}$, $f(X)_{20}$ and $f(X)_{21}$ represent the two-valued signals of 0 or 1, 0 or 2, 2 or 0, and 2 or 1, respectively, from the separate binary decision networks.

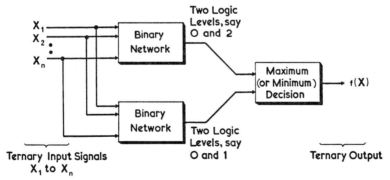

Fig. 1.23. Ternary circuit composition using binary subsystems.

Within each binary decision network the ternary input signals first require to be grouped into two signal levels, which may be done by appropriate single-input "conversion" functions. Following this, appropriate binary conjunction (AND) and disjunction (OR) operators may be employed. For example, Table 1.13 details one non-minimal basic set which has been proposed[59,61].

Table 1.13. The device-oriented non-minimal set of ternary functions

(a)

X_i	$C_{01}X_i$	$C_{02}X_i$	$C_{10}X_i$	$C_{12}X_i$	$C_{20}X_i$	$C_{21}X_i$
0	1	2	2	0	2	2
1	0	0	0	2	2	2
2	0	0	2	0	0	1

(b)

X_1	X_2	0 AND	1 AND	2 AND	0 OR	1 OR	2 OR
0	0	2	2	2	2	0	2
0	1	0	2	2	2	2	2
0	2	0	2	2	2	0	0
1	0	0	2	2	2	2	2
1	1	0	0	2	0	2	2
1	2	0	2	2	0	2	0
2	0	0	2	2	2	0	0
2	1	0	2	2	0	2	0
2	2	0	2	0	0	0	0

(a) Single-variable "conversion" functions.
(b) Multi-input logical-AND and logical-OR functions (two or more inputs).

Synthesis techniques follow directly from the truth-table of the required function, using whichever operators are most convenient in the network synthesis. Taking our previous example once again, the synthesis procedure would build up as illustrated in Fig. 1.24. The comparable nature of this synthesis with normal binary realization will be recognized, including the possibilities of being able to apply certain minimization measures similar to normal binary minimization techniques[59,64].

However, more powerful multi-variable operators than the examples detailed in Table 1.13b would be advantageous, exactly as in the

A Survey of Binary and Higher-Valued Logic

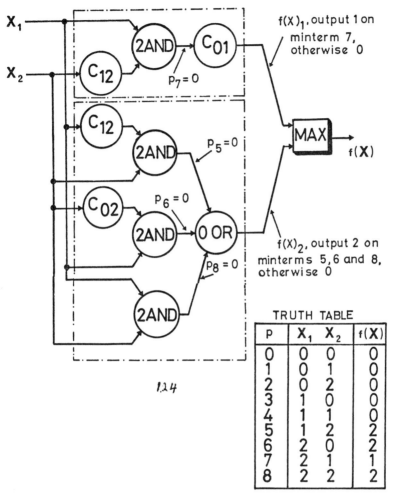

Fig. 1.24. Synthesis of $f(X)$ using device-oriented functions (not a minimized solution).

binary case where the weakness of AND/OR vertex gates has already been encountered.

1.9. Summary of this Higher-Valued Logic

In spite of the considerable research in the underlying algebras of higher-valued systems, advancement of this area clearly is hindered by the lack of readily available circuit elements. In view of the capital involvement in binary systems it is not surprising that the digital field is at present binary dominated. However, ternary logic, being the next-higher-value logic to binary, will undoubtedly exercise the attention of more and more authorities in the future as the demand for more powerful and compact digital systems increases.

The multi-valued algebras which we have briefly considered will form a basis for this development. As we have seen they may be more complex than our more familiar Boolean algebra, but all of them can be shown to bear a likeness to expansions which are encountered in the binary field. It is, however, not clear at this time what form of algebra and what form of multi-valued circuit realization will finally be adopted for engineering applications; this will to a large extent depend upon whatever microelectronic circuit configurations are developed and marketed. For further general reading in this area with the supporting mathematics reference may be made to several published texts[65-67].

Basic research, however, is not standing still. Already research on yet a further area which we do not propose to cover here in any detail is under way, this area being *multi-valued threshold logic*[68-70]. Precisely as in the binary case where binary threshold-logic gates are more powerful than vertex gates, so in the $R > 2$ case multi-valued threshold-logic gates will be considerably more powerful than the non-threshold types of gates such as we have looked at in preceding pages. The total volume of research needed fully to exploit this area, including possible concepts of symmetry and other aspects, however, has hardly been started.

1.10 Fuzzy Logic

In all the sections we have considered so far, the area with which we have been concerned is the area of *deterministic logic*. By this we mean that we are dealing with a digital network or system, binary or higher valued, wherein each required function is fully defined, there being no uncertainty or probability or chance in the specification of the system. Notice that "don't care" conditions are still part of a fully deterministic system; for example "don't care" conditions in a binary system still have the truth value 0 or 1 and do not introduce any additional truth values or any probability considerations[5]. There are, however, two

---

[5] For the purpose of logic design, however, it may be useful in some synthesis techniques to consider "don't care" conditions as a further "value" which the techniques will subsequently optimize to 0 or 1 as required. There is no such "value" occurring in the actual input and output signals, however.

A Survey of Binary and Higher-Valued Logic

main logic areas which lie outside this fully deterministic area. The first of these is *stochastic* (or *probabilistic*) logic, whilst the other is what has been termed *continuously variable* or *fuzzy* logic.

Stochastic logic deals with the logic area wherein the digital input variables X_i, $i = 1 \ldots n$, are still statistically independent, but where the feature of importance is the probability of chosen input signal values, say 1 in a binary system, occurring within a given period of time. Such probabilities of course may be related to the total number of occurrences within a given period of time. Currently the principal proposed application of stochastic logic is where such probabilities are used to represent analog quantities which, when appropriate simple logic operations are performed, give rise to the concept of digital stochastic computing machines[71-73]. However, wider fields may develop as stochastic logic signals may possibly represent the effect of logic signals buried in noise, which may in turn model the action of neurons in the human brain and which collectively have to process information in the presence of considerable background noise. We shall not, however, have space to consider any of these newer areas in this book.

Fuzzy logic, however, is basically dissimilar to stochastic logic, but nevertheless possesses several features in common. By definition fuzzy logic is a logic wherein the logic values do not possess definite numerical values. Instead they may range over the whole interval, which is normally taken as from 0 to $1^{(74-76)}$. This must be contrasted with, say, normal binary working, where the logic values are exclusively 0 or 1, with no meaningful signal-level existence between these two extremes. In effect fuzzy logic is a kind of multi-valued logic, but where in theory there are an infinite number of values between the two extreme values. Figure 1.25 attempts to illustrate this distinction.

It may be argued that, if the logic values freely range over the whole interval 0 to 1, we are no longer dealing with a digital system, which by definition must involve discrete values or levels. However, as we shall briefly review in a moment, the present ideas for dealing with a fuzzy logic situation involve (i) the algebra of and the logical manipulation of equations for *fuzzy sets* and (ii) the subdivision of the continuous logic value into *ranges,* which may then be handled by the same algebra as developed for the fuzzy sets. However, as probability considerations are involved in fuzzy set theory, some authorities consider fuzzy logic to be a form of probabilistic logic, both fuzzy logic and stochastic logic therefore arising from an underlying general area of probabilities and decisions[77].

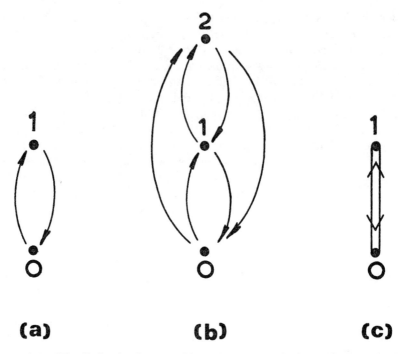

Fig. 1.25. The distinction between binary and ternary logic and fuzzy logic: (*a*) binary, values 0 and 1 only, including any "don't cares"; (*b*) ternary, values 0, 1 and 2, including any "don't cares"; (*c*) fuzzy, any values between 0 and 1 theoretically possible.

The concept of fuzzy sets is very relevant for many real-life situations, where hard-and-fast classification of members of a set of items or signals under consideration cannot be made into precisely defined categories. The "set of all intelligent women" or the "set of all thin men" cannot be precisely defined, and hence membership of such sets is not a clear-cut yes/no binary decision. In the engineering world, such as in pattern-recognition and artificial intelligence applications, many such situations arise which require to be handled in some logical manner. A robot mechanism in space exploration, for example, may have to make a number of intelligent decisions from "fuzzy" data collected and processed by its various sensors.

It would clearly be advantageous to be able to combine probability theory with symbolic logic. The current ideas in this area are largely based upon the concept of *fuzzy set theory* first proposed by Zadeh in

1965[74,78], although certain earlier work may be found referenced[79]. The "logic" involved in the current techniques involves algebraically expressing and manipulating combinations of input signals, each of which may take a range of signal values. Also as these values have a proper arithmetic significance, involving "greater than" and "less than" considerations, the standard techniques of truth tables, Karnaugh maps, and the like are inappropriate in this area. However, the algebra which may be used has a strong likeness to our normal logic algebra.

The basic fuzzy set theory of Zadeh which leads on to fuzzy switching circuit possibilities gives three major definitions. However, first let us define the symbols which we shall be using. In order to distinguish between our previous system for which we have been using the symbols x_i and X_i for the various independent digital inputs, let us now introduce the upper-case symbols A, B, C, etc., to represent *sets*, where A, B, C, etc. are all subsets of some overall class Z, i.e. A, B, C, etc. $\in Z$. Further, let us use the lower-case symbols a, b, c, etc. to represent the *grade of membership* of any sample set z, $z \in Z$, in the fuzzy sets A, B, C, etc., respectively, where the values of a, b, c, etc. may freely range between the values of 0 and 1. The nearer the value of, say, a to unity, then the higher the membership of the sample set z in the fuzzy set A.

The three definitions which we shall mention here are as follows:

(1) The *union* of two fuzzy sets A and B is a fuzzy set C, written as $C = A + B$, the grade of membership in C of any sample set z being given by $c =$ maximum (a, b).

(2) The *intersection* of two fuzzy sets A and B is a fuzzy set C, written as $C = A \cdot B$, the grade of membership in C of any sample set z being given by $c =$ minimum (a, b).

(3) The *complement* of a fuzzy set A is the fuzzy set A', where the grade of membership in A' of any sample set z is given by $a' = (1 - a)$.

Based upon these definitions an algebra similar in many but not all respects to Boolean algebra may be built up. The "grade of membership" may now be replaced by a more appropriate parameter for engineering purposes, namely *fuzzy variable*, where the numerical value of each fuzzy variable may at any particular time be anywhere in the range 0 to 1. This is illustrated in Fig. 1.26.

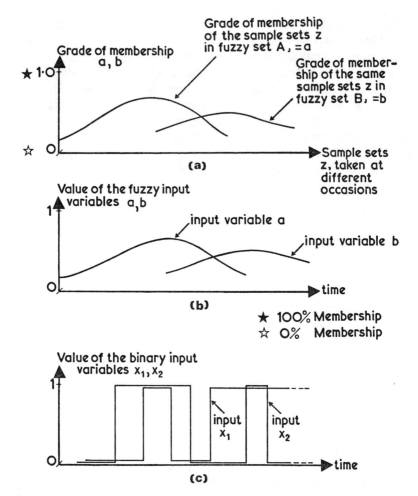

Fig. 1.26. Fuzzy sets and fuzzy variables: (*a*) possible distribution of the grade of membership of random sample sets z in the two fuzzy sets A and B, A,B,z all being members of the overall classification Z; (*b*) two fuzzy variables a and b of a system, the value of a and b each being anywhere in the range 0 to 1; (*c*) the Boolean situation corresponding to (*b*), with two inputs x_1 and x_2 each taking the values 0 or 1 exclusively.

At this stage one major difference between this "soft algebra" and normal Boolean algebra must be noted, this difference concerning the complements of the terms. If we consider the union (sometimes termed

A Survey of Binary and Higher-Valued Logic

the "addition") of a grade of membership or a fuzzy variable a and its complement a' we have

$$a + a' = \text{maximum}(a, a')$$
$$= \text{maximum}(a, 1-a)$$

which is not unity as it would be in the Boolean case. Similarly if the intersection (sometimes termed "multiplication") of a and a' is considered, we have:

$$a \cdot a' = \text{minimum}(a, a')$$
$$= \text{minimum}(a, 1-a)$$

which is not zero as it would be in the Boolean case. The lack of these two Boolean simplifications strongly reduces the minimization which is possible in more complex expressions. For example,

$$a \cdot b + a \cdot b' = a \cdot (b + b')$$
$$= \text{minimum}\left(a, (\text{maximum } b, b')\right)$$

which is not necessarily a as it always would be in the Boolean case. On the other hand

$$a \cdot a' \cdot b + a \cdot a' \cdot b' = a \cdot a' \cdot (b + b')$$
$$= a \cdot a'$$

as $a \cdot a'$ must always be less than $(b + b')$, irrespective of the values which a and b may take. So in the first example the term $(b + b')$ is not redundant, but in the second example the same term is redundant. These results may be quickly demonstrated by putting any value between 0 and 1 for the terms a and b, remembering that $a' = 1-a$ and $b' = 1-b$.

However, if we know that for some reason the value of certain variables is always equal or less than the value of other variables, then simplification of complex expressions is frequently possible[80]. Take the expression

$$f(a,b,c) = a \cdot (b' + a' \cdot c) + b \cdot (a + a' \cdot c').$$

If now a is known to be always $\leq b$, this expression simplifies to

$$f(a,b,c) = a + a' \cdot b \cdot c'.$$

Such simplifications, which in effect are allowing certain terms to be discounted in comparison with more significant terms, are certainly not as straightforward to handle as Boolean or multi-valued expressions, but nevertheless are collectively obeying rigid "minimum of" and "maximum of" algebraic rules.

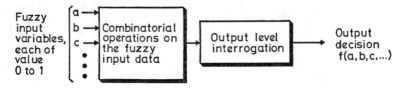

Fig. 1.27. The schematic arrangement of a fuzzy logic system.

Now a system with fuzzy inputs may be considered as shown in Fig. 1.27. Each fuzzy input may in theory take a value which at any particular time lies anywhere between the limits of 0 and 1. The system has to accept these inputs, make some logical operations upon their values, and give a final output decision. The fuzzy input variables themselves and the combinational expressions which combine them do not themselves contain any "decision" information; instead we have to define appropriate conditions or limits for our system which must be met in order to give a definite output decision. To do this the continuous range of the value of each fuzzy input variable a, b, etc., requires to be grouped into a *finite number of ranges r*, defined as follows:

range r: $\quad R_{r-1} \leqslant a \leqslant 1$

⋮

range 2: $\quad R_1 \leqslant a < R_2$

range 1: $\quad 0 \leqslant a < R_1$

where $0 < R_1 < R_2 < \ldots < R_{r-1} < 1$.

Now in considering a fuzzy system, one is interested in the input conditions which must be present in order that the function output $f(a,b,c, \ldots)$ falls into some appropriate chosen range. For example, we may possibly wish the value of the function to be in the range

$$R_{r-2} \leqslant f(a,b,c, \ldots) < R_{r-1}$$

for the system to give the desired output decision. In general we may say that the function value must be

$$R_\ell \leqslant f(a, b, c, \ldots) < R_u$$

A Survey of Binary and Higher-Valued Logic

where R_ℓ, R_u are the lower and upper limits, respectively, of the selected output range. Two relationships, therefore, have to be satisfied, namely

Group 1 Relationship: $f(a, b, c \ldots) \geq R_\ell$

and

Group 2 Relationship: $f(a, b, c \ldots) < R_u$.

The two procedures of system *analysis* and system *synthesis* may now be built up, having specified the ranges of the input variable values and chosen the required output decision range which the function must meet.

For the analysis procedure the function specification and output range is given. As an example, let the given function be

$$f(a, b, c) = ab' + a'bc'.$$

In the *analysis* we are interested in determining the input signal values which the inputs a, b, and c must take in order to satisfy the requirement that $f(a, b, c)$ belongs to the chosen output range. Therefore

$$R_\ell \leq ab' + a'bc' < R_u$$

which gives the following relationships

Group 1: $\left\{ \text{and} \begin{array}{l} a \geq R_\ell \\ b' \geq R_\ell \end{array} \right\}$ or $\left\{ \begin{array}{l} \text{and} \; a' \geq R_\ell \\ \text{and} \; b \geq R_\ell \\ \phantom{\text{and}} \; c' \geq R_\ell \end{array} \right\}$

$= \left\{ \text{and} \begin{array}{l} a \geq R_\ell \\ b \leq 1 - R_\ell \end{array} \right\}$ or $\left\{ \begin{array}{l} \text{and} \; a \leq 1 - R_\ell \\ \text{and} \; b \geq R_\ell \\ \phantom{\text{and}} \; c \leq 1 - R_\ell \end{array} \right\}$

Group 2: $\left\{ \text{or} \begin{array}{l} a < R_u \\ b' < R_u \end{array} \right\}$ and $\left\{ \begin{array}{l} \text{or} \; a' < R_u \\ \text{or} \; b < R_u \\ \phantom{\text{or}} \; c' < R_u \end{array} \right\}$

$= \left\{ \text{or} \begin{array}{l} a < R_u \\ b > 1 - R_u \end{array} \right\}$ and $\left\{ \begin{array}{l} \text{or} \; a > 1 - R_u \\ \text{or} \; b < R_u \\ \phantom{\text{or}} \; c > 1 - R_u \end{array} \right\}$

It will be noticed that in such simple cases these two relationships are dual in nature, as Group 2 may be obtained from Group 1 by interchanging all the "and's" and "or's", reversing the inequality signs, and interchanging the upper and lower range limits. Should the given function specification be in a "products-of-sums" form, that is the "minimum of several maximum" terms, then a similar pair of group relationships as above may be formulated, but with appropriate "and" and "or" reversals in each group relationship. Various rules may be established which help to handle and possibly simplify the above procedures, references to which will be found published[80,81].

For system *synthesis* a set of requirements must first be formulated which the system has to meet. This may, for example, start with a collection of statement such as "the output must be switched on when a is greater than a certain value, or when b and c are both less than a" and so on. From such verbal specifications two initial groups of inequalities may be generated, representing the upper and lower bands of the output decision, such as the following.

Group 1: $\left\{\text{and } \begin{array}{c} a \geqslant R_1 \\ b \leqslant R_2 \end{array}\right\}$ or $\left\{\begin{array}{c} \text{and } a \leqslant R_3 \\ \text{and } b \geqslant R_4 \\ c \leqslant R_5 \end{array}\right\}$

Group 2: $\left\{\text{or } \begin{array}{c} a < R_6 \\ b > R_7 \end{array}\right\}$ and $\left\{\begin{array}{c} a > R_8 \\ b < R_9 \\ \text{or } c > R_{10} \end{array}\right\}$

Notice that any range of the fuzzy input signals may be present in the initial group conditions. However, by suitable individual scaling, that is recognizing one range and on recognition generating another approparite signal range, we may scale all the inputs such that we finally achieve group equations with common lower and upper range values R_ℓ and R_u, as follows.

Group 1: $\left\{\text{and } \begin{array}{c} a \geqslant R_\ell \\ b' \geqslant R_\ell \end{array}\right\}$ or $\left\{\begin{array}{c} \text{and } a' \geqslant R_\ell \\ \text{and } b \geqslant R_\ell \\ c' \geqslant R_\ell \end{array}\right\}$

Group 2: $\left\{\text{or } \begin{array}{c} a < R_u \\ b' < R_u \end{array}\right\}$ and $\left\{\begin{array}{c} \text{and } a' < R_u \\ \text{and } b < R_u \\ c' < R_u \end{array}\right\}$

This particular example will be recognized as the example we previously considered in the analysis considerations, and represents the fuzzy function $f(a, b, c) = ab' + a'bc'$. However, in a more general situation the function equation $f(a, b, c, ...)$ obtained from the scaled group equations may not initially be in a minimized form, as is the simple example we have taken. Hence minimization using the appro-

A Survey of Binary and Higher-Valued Logic

priate rules of the fuzzy algebra should be attempted before realization of the system equation into a network is made.

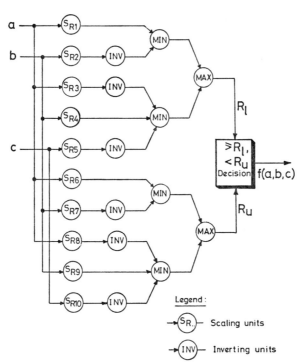

Fig. 1.28 Schematic circuit realization of the fuzzy function $f(a,b,c) = ab' + a'bc'$

The schematic realization of the particular example used here is shown in Fig. 1.28. The system as a whole may finally be considered as a form of multi-valued logic system, although it is hard to discount the fact that effectively analog signals are being handled by the scaling units at the input end of the complete assembly.

We shall not pursue this particular topic any further in this book. It has, however, been included in this first chapter for the sake of completeness in this review of the breadth and scope of present digital logic considerations. For further reading around this particular area reference may be made to several sources referenced at the end of this chapter, in particular 65–67, 76, 80.

1.11 Chapter Summary

This first chapter has ranged fairly widely but not in great depth over the field of digital signals and logic configurations, and has not confined itself to our more familiar binary pastures. As a result this chapter is somewhat more lengthy than some of the following chapters will be. Conventional Boolean techniques have not been detailed, as it is assumed the reader will already have this information available to him from prior sources.

Our broad field therefore is as shown in Table 1.14, with present-day digital logic engineering occupying basically the upper left-hand section only. In our subsequent chapters we shall attempt to extend this section to embrace newer concepts of handling both binary and higher-valued logic signals, and to cover new ideas of possible circuit configurations for logic gates.

Table 1.14. The broad field of digital signals and digital algebra with subsequent chapters covering the newer areas of development

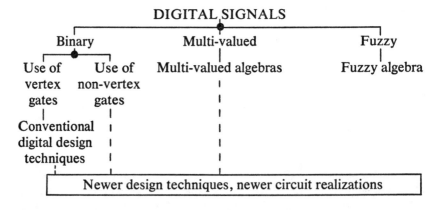

References

1. Lewin, D. *Logical Design of Switching Circuits*. Elsevier/North-Holland, New York; Nelson, London, 1974.

2. Barna, A., and Porat, D.I. *Integrated Circuits in Digital Electronics*. John Wiley Interscience, New York, 1973.

3. Bannister, B.R., and Whitehead, D.G. *Fundamentals of Digital Systems.* McGraw-Hill, Maidenhead, U.K., 1973.

4. Zissos, D. *Logic Design Algorithms.* Oxford University Press, Oxford, U.K., 1972.

5. Post, E.L. Introduction to a general theory of elementary propositions. *Am. J. Math.* **43**, 163–85, 1921.

6. Shannon, C.E. The synthesis of two-terminal switching circuits. *Bell Syst. Tech. J.* **28** (1), 59–98, 1949.

7. Mukhopadhyay, A. Complete sets of logic primitives. In *Recent Development in Switching Theory.* Academic Press, New York, 1971.

8. Akers, S.B. On a theory of Boolean functions. *J. Soc. Ind. Math.* **7** (4), 487–98, Dec. 1959.

9. Meissler, M.H. Use of exclusive-OR gates for Boolean minimization. *Proc. IEE* **119**, 1269–72, 1972.

10. Edwards, C.R. The manipulations of Boolean expressions containing the exclusive-OR operator. *Int. J. Electron.* **39** (6) 687–90, 1975.

11. Mukhopadhyay, A., and Schmitt, G. Minimization of exclusive-OR and logical equivalence switching circuits. *Trans. IEEE* **C19**, 132–40, Feb. 1970.

12. Dietmeyer, D.L. *Logical Design of Digital Systems* Allyn & Bacon, Boston, 1971.

13. Reed, I.S. A class of multiple-error-correcting codes and decoding scheme. *Trans. IRE* **IT4**, 38–49, 1954.

14. Muller, D.E. Application of Boolean algebra to switching circuit design and error correction, *Trans. IRE* **EC3**, 6–12, Sept. 1954.

15. Akers, S.B. On the algebraic manipulation of majority logic. *Trans. IRE,* **EC10,** 779, Dec. 1961.

16. Akers, S.B. Synthesis of cominational logic using three-input majority gates. *Proc. Symp. on Switching Theory and Logical Design,* 149–57. American Institute of Electrical Engineers, 1962.

17. Lewis, P.M., and Coates, C.L. *Threshold Logic.* John Wiley, New York, 1966.

18. Lindaman, R. A theorem for deriving majority-logic networks with an augmented Boolean algebra. *Trans. IRE* **EC9**, 338–42, Sept. 1960.

19. Miller, H.S., and Winder, R.O. Majority logic synthesis by geometric methods. *Trans. IRE* **EC11,** 89–90, Feb. 1962.

20. Muroga, G. Functional forms of majority functions and a necessary and sufficient condition for their realizability. *Trans. IEEE Commun. Electron.* **83**, 474–86, Sept. 1964.

21. Muroga, S. *Threshold Logic and its Application.* John Wiley Interscience, New York, 1971.

22. Negrin, A.E. Synthesis of practical three-input majority logic networks. *Trans. IEEE* **C17**, 978–85, Oct. 1968.

23. Sheng, C.L. *Threshold Logic.* Academic Press, New York, 1969.

24. Caldwell, S.H. *Switching Circuits and Logical Design.* John Wiley, New York, 1958.

25. Coates, C.L., and Lewis, P.M. DONUT—a threshold-gate computer. *Trans IEEE* **C18**, 240–7, June 1964.

26. Dertouzos, M.L. *Threshold-Logic: A Synthesis Approach.* MIT Press, Cambridge, Mass., 1965.

27. Hu, S.T. *Threshold Logic* University of California Press, Berkeley, Calif., 1965.

28. Lewis, P.M. Practical guide to threshold logic. *Electron. Des.* **15**, 66–88, Oct. 1967.

29. Winder, R.O. Threshold logic will cut costs. *Electronics* **41**, 94–103, 27 May 1968.

30. Winder, R.O.: The status of threshold logic. *RCA Rev.* **30**, 62–84, March 1969.

31. Hurst, S.L. An introduction to threshold logic: a survey of present theory and practice. *Radio Electron. Eng.* **37**, 339–51, June 1969.

32. Hampel, D. Beinart, J. and Prost, K. *Threshold Logic Implementation of a Modular Computer System Design.* NASA Rep. No. NASA-CR-1668, Washington, D.C., Oct. 1970.

33. Hurst, S.L. *Threshold Logic: An Engineering Survey* Mills & Boon, London, 1971.

34. Hurst, S.L. Threshold logic can cut gate count. *Electron. Des.* **22**, 106–18, 25 Oct. 1974.

35. Hamming, R.W. Error detecting and error correcting codes. *B.S.T J.*, **29**, 147–60, Apr. 1950.

36. Boole, G. *An Investigation of the Laws of Thought.* 1849. Reprinted by Dover Publications, New York, 1954.

37. Boole, G. *The Mathematical Analysis of Logic.* Reprinted by Blackwell, Oxford, U.K., 1948.
38. Shannon, C.E. Symbolic analysis of relay and switching circuits. *Trans. AIEE* **57,** 713–23, 1938.
39. Rine, D.C. Multiple-valued logic and computer science in the 20th century. *IEEE Computer* **7,** 18, 19, Sept. 1975.
40. Epstein, G., Frieder, G., and Rine, D.C. The development of multiple-valued logic as related to computer science. *IEEE Computer* **7,** 20–33, Sept. 1975.
41. Vranesic, Z. G., and Smith, K. C. Engineering aspects of multi-valued logic system. *IEEE Computer* **7,** 34–41, Sept. 1975.
42. Abraham, G. Variable radix multi-stable integrated circuits. *IEEE Computer* **7,** 42–59, Sept. 1975.
43. Berlin, R.D., Synthesis of N-valued switching circuits. *Trans. IRE* **EC7,** 52–6, March 1958.
44. Webb, D.L. Generation of any N-valued logic by one binary operator. *Proc. Natn. Acad. Sci.* **21** 252–4, 1953.
45. Slupeki, J. A criterion of fullness of many-valued systems of propositional logics. *C. R. Séanc. Soc. Sci. Lett. Varsovie* (Classe III) **32,** 102–9, 1939.
46. Lowenschuss, O. Non-binary switching theory. *IRE Natn. Conv. Rec.* **6** (4), 305–17, 1958.
47. Bernstein, B.A. Modular representations of finite algebras. *Proc. 7th Int. Congr. on Mathematics, 1924,* Vol. 1, pp. 207–16. University of Toronto Press, Toronto, Canada, 1928.
48. Tamari, D. Some mutual applications of logic and mathematics. *Proc. 2nd Int. Colloq. of Mathematical Logic, August 1952,* pp. 89, 90.
49. Pradham, D.K. A multi-valued switching algebra based on finite-fields. *IEEE Proc Int. Symp. on Multivalued Logic, May 1974,* pp. 95–111.
50. Sheffer, H.M. A set of five independent posulates for Boolean algebra with application to logical constants. *Trans. Am. Math. Soc.* **14,** 481–8, 1913.
51. Webb, D.L. Generation of any N-valued logic by one binary operation. *Proc. Natn. Acad. Sci.* **21,** 252–4, May 1935.
52. Martin, N.M. The Sheffer functions of 3-valued logic. *J. Symbol. Logic* **19,** 45–51, March 1954.

53. Rosenbloom, P.C. Post algebras: postulates and a general theory. *Am. J. Math.*, **64**, 167–88, 1942.

54. Rohleder, H. Three-valued calculus of theoretical logics and its application to the description of switching circuits which consist of elements of two states. *Z. Angew. Math. Mech.* **34**, 308–11, 1954.

55. Goto, M. *Theory and structure of the automatic relay computer ETL Mark II*. Industrial Science Research of the Electrotechnical Laboratory, Japan, No. 556, 1956.

56. Vacca, R. *A Three-Valued System of Logic and its Application to Base Three Digital Circuits*. Rep. No. UNESCO/NS/ICIP, G.2.14, 1957.

57. Mühldorf, E. Ternare schaltalgebra. *Arch. Elektrish. Urbertrag.* **12**, 138–48, March 1958.

58. Muehldorf, E.I. Multi-valued switching algebras and their application in digital systems. *Proc. Natn. Electron. Conf.* **15**, 467–80, Oct. 1959.

59. Hurst, S.L. *An investigation into the Realisation and Algebra of Electronic Ternary Switching Functions*. M.Sc. Thesis, University of London, 1966.

60. Lee, C.Y., and Chen, W.H. Several-valued switching circuits, *Trans. AIEE* **25**, (1), 278–83, July 1957.

61. Hurst, S.L. Semiconductor circuits for 3-state logic applications. *Electron. Eng.* **40**, 197–202, Apr. 1968; 256–9, May 1968.

62. Birk, J.E., and Farmer, D.E. Design of multivalued switching circuits using principally binary components. *Proc. Int. Symp. on Multiple-Valued Logic, May 1974*, pp. 115–33.

63. Mouftah, H.T., and Jordan, I.B. Integrated circuits for ternary logic. *Proc. Int. Symp. on Multiple-Valued Logic,* May 1974, pp. 285–302.

64. Hurst, S.L. An extension of binary minimization techniques to ternary equations. *Computer J.* **11**, 277–86, Nov. 1968.

65. Rosser, J.B., and Turquette, A.R. *Many Value Logics*. North Holland, Amsterdam, 1952.

66. Ackerman, R. *Introduction to Many Valued Logics*. Dover Publications, New York, 1967.

67. Rescher, N. Many-Valued Logic. McGraw-Hill, New York, 1969.

68. Nazarala, J., and Moraga, C.R. Minimal realization of ternary threshold functions. *Proc. Int. Symp. on Multiple-Valued Logic, May 1974,* pp. 347–59.

69. Druzeta, A., Vranesic, Z.G., and Sedra, A.S. Application of multi-threshold elements in the realization of many-valued logic networks. *Trans. IEEE* **C23**, 194–98, Nov. 1974.
70. Abara, T., and Akagi, M. Enumeration of ternary threshold functions of three variables. *Trans. IEEE* **C21**, 402–7, Apr. 1972.
71. Gaines, B.R. Stochastic computing. *AFIPS, Spring Joint Computer Conf.* **30**, 149–56, 1967.
72. Popplebaum, W.J., Afuso, C., and Esch, J.W. Stochastic computing elements and systems. *AFIPS, Fall Joint Computer Conf.* **31**, 635–44, 1967.
73. Miller, A.J., Brown, A.W., and Mars, P. Adaptive logic circuits for digital stochastic computers. *Electron. Lett.* **9**, 500–2, Oct. 1973.
74. Zadeh, L.A. Fuzzy sets. *Informa. Control* **8**, 338–53, 1965.
75. Marinos, P.N. Fuzzy logic and its application to switching systems. *Trans. IEEE* **C18**, 343–8, Apr. 1969.
76. Lee, R.C.T. Fuzzy logic and the resolution principle. *J. Ass. Computing Mach.* **19**, 109–19, Jan. 1972.
77. Gaines, B.R. Stochastic and fuzzy logics. *Electron. Lett.* **11**, 188–9, May 1975.
78. Zadeh, L.A. Fuzzy sets and systems. *Symp. on System Theory.* Polytechnic Institute of Brooklyn, Brooklyn, N.Y. 1965.
79. Preparata, F.B., and Yeh, R.T. On a theory of continuously valued logic. *Conf. Record Symp. on the Theory and Application of Multiple-valued Logic Design, Buffalo, N.Y., May 1971*, pp. 124–32.
80. Marinos, P.N. Fuzzy logic and its application to switching systems. *Trans. IEEE* **C18**, 343–8, Apr. 1969.
81. Tien, P.S. The analysis and synthesis of fuzzy switching circuits. *Conf. Record Symp. on the Theory and Application of Multiple-valued Logic Design, Buffalo, N.Y., May 1971*. pp. 165–72.

Chapter 2

The Classification of Logic Functions

2.1. The object of classifying functions

The number of different functions of n binary inputs, including all degenerate and trivial ones[1], is 2^{2^n}. Clearly this becomes a very large number for n greater than, say, two or three, and certainly too large to enable us to tabulate comfortably. Even worse, of course, is the situation for higher-valued functions, $R > 2$, where the number of functions now becomes R^{R^n}.

However, different functions may possess certain likenesses or structures which enable us to classify all the possible different functions into a small number of classes. One objective of such a classification procedure may be to list more compactly all the possible 2^{2^n} functions, but a second more practical objective may be to enable us to specify a certain small set of "standard functions," or "prototype functions," from which all the individual 2^{2^n} functions may be made. Further, if such "standard functions" are fully investigated and documented, then standard fault diagnosis procedures may be more readily available for networks incorporating such standard functions.

---

[1] Degenerate functions of n variables are functions which do not depend upon all n inputs to determine the function output, that is one or more inputs are redundant. Trivial functions are the extreme cases where the function output is constant, that is $f(x) = 0$ or 1 in a binary system.

The Classification of Logic Functions

As simple illustrations of functions which readily may be classed as "similar", consider the two functions

$$f_1(x) = [x_1x_2 + x_2\bar{x}_3]$$

and

$$f_2(x) = [x_1x_3 + \bar{x}_2x_3].$$

These two functions are similar in that both are realized with precisely the same logic network, but with inputs x_2 and x_3 interchanged in $f_2(x)$ in comparison with their connection in $f_1(x)$.

Similarly the two functions

$$f_3(x) = [x_1\bar{x}_2 + x_2x_3]$$

and

$$f_4(x) = [x_1x_2 + \bar{x}_2x_3]$$

are similar in the sense that both are realized with the same logic network, but with the complement of x_2 fed into $f_4(x)$ in comparison with $f_3(x)$.

The classification procedures for binary functions have received much more effort than for higher-valued functions, as would be expected. Therefore in general we shall be dealing with binary-function classification in the following pages, showing how such classification procedures have progressed through various stages, each stage representing some more powerful and compact characterizing procedure in comparison with its predecessors.

It will also be seen that linearly separable functions, that is, functions which may be realized by the use of one threshold-logic gate, have received the greatest attention, and therefore a great deal of this chapter will be spent in covering the characteristics and characterizing parameters of threshold-logic functions. This will introduce certain numerical considerations as distinct from algebraic means which may be used for function classification, such numerical classification techniques, however, not being restricted to the linearly separable case alone.

2.2. The PN, NPN, and SD algebraic classifications for binary functions

The three main algebraic classifications which have been developed for binary functions are the *PN classification,* the *NPN classification,* and the *SD classification*. Each is progressively more compact in classifying the 2^{2^n} possible binary functions, as we shall now show.

2.2.1. The PN classification

The earliest published method for the classification of binary functions was the method used in producing the Harvard switching function tables[1,2]. This method has subsequently been termed the PN classification (standing for Permutation/Negation) by several authorities, in order to provide a ready means of distinguishing this earliest method from subsequent more compact methods.

The original Harvard table was produced by the Harvard Computation Laboratory in the early days of digital computers and listed vacuum-tube realizations of all possible combinational functions of up to four variables. The total number of possible functions of not more than four variables, that is 65536, was compressed into 402 classified representatives by this method, this reduction being achieved by considering all functions which differ only by some permutation of the input variables and/or by negation (complementation) of one or more of the input variables as being in the same classification entry. The listing in this and in all subsequent classification methods was done in a form which involved all input variables written in a preferred order and uncomplemented where possible, a form which may be termed a "positive canonic" form. The compactness of this first classification is indicated in Table 2.1.

Table 2.1. The PN classification statistics for binary functions of $n \leq 6$ based on the original Harvard classification for $n \leq 4$

Number of input variables n	1	2	3	4	5	6
Total number of functions of $\leq n$ variables $= 2^{2^n}$	4	16	256	65536	$\approx 4.3 \times 10^9$	$\approx 1.8 \times 10^9$
Total number of functions of exactly n variables	2	10	218	64594	$\approx 4.3 \times 10^9$	$\approx 1.8 \times 10^{19}$
PN classification of functions of $\leq n$ variables	3	6	22	402	1228158	$\approx 4 \times 10^{14}$
PN classification of functions of exactly n variables	1	3	16	380	1227756	$\approx 4 \times 10^{14}$

The Classification of Logic Functions

For illustration, the full PN classification of functions for $n \leq 2$ is as follows:

$f(x) = 0$	(one function per entry)
$f(x) = 1$	(one function per entry)
$f(x) = x_1$	(four functions per entry)
$f(x) = x_1 x_2$	(four functions per entry)
$f(x) = x_1 + x_2$	(four functions per entry)
$f(x) = x_1 \oplus x_2$	(two functions per entry)

total: 16 functions of $n \leq 2$.

2.2.2 The NPN Classification

In the NPN classification, negation (complementation) of the overall function as well as each input variable is allowed in each class[3-5]. This extra operation allows functions which are related by the application of De Morgan's theorem to be classed in the same entry, for example the positive canonic PN entries of $x_1 x_2$ and $x_1 + x_2$ are both in the same positive canonic NPN entry.

The application of this more concise classification reduces the number of classification entries to almost half those listed in the PN classification method. For example, for $n \leq 2$ the six PN entries detailed in §2.2.1. become four NPN entries, as follows:

$f(x) = 1$	(two functions per entry)
$f(x) = x_1$	(four functions per entry)
$f(x) = x_1 + x_2$	(eight functions per entry)
$f(x) = x_1 \oplus x_2$	(two functions per entry).

2.2.3. The SD Classification

The third and most compact of the algebraic classification techniques is the SD *(self-dual)* classification of Goto and Takahasi[4,6,7]. This classification embraces the previous PN and NPN classifications, but in addition includes a stronger operation to reduce further the classification entries for any function value n. Although the SD classification is particularly suitable for linearly separable (threshold) functions, and will be referred to again in subsequent sections, it is applicable to all binary functions whether linearly separable or not. However, we must first define the terms which will be encountered in this area.

Firstly, the dual $f^d(x)$ of any given binary function $f(x)$ is defined as follows:

$$f^d(x) \triangleq \overline{f(\bar{x})}$$

where the bar within the parentheses indicates that all the variables x_i of the given function $f(x)$ are individually complemented. For example the dual of the function $f(x) = [x_1\bar{x}_2 + \bar{x}_3 x_4]$ is $[\bar{x}_1 x_2 + x_3 \bar{x}_4]$.

If now $f^d(x) \equiv f(x)$, then the given function $f(x)$ is said to be *self-dual*. As an example of a self-dual function, consider the function

$$f(x) = [x_1 x_2 + x_2 x_3 + x_1 x_3].$$

Then

$$\begin{aligned}
f^d(x) &= \overline{[\bar{x}_1 \bar{x}_2 + \bar{x}_2 \bar{x}_3 + \bar{x}_1 \bar{x}_3]} \\
&= [(x_1 + x_2)(x_2 + x_3)(x_1 + x_3)] \\
&= [x_1 x_2 + x_2 x_3 + x_1 x_3] \\
&\equiv f(x).
\end{aligned}$$

Now associated with any arbitrary binary function $f(x)$ of n input variables $x_1, \ldots, x_n$ is a self-dual *hyperfunction* $f^h(x)$ of $n+1$ variables[2] where

$$f^h(x) \triangleq [x_{n+1} f(x) + \bar{x}_{n+1} f^d(x)]$$

and where x_{n+1} is an additional binary variable not included in $f(x)$. All such hyperfunctions $f^h(x)$ are self-dual functions, as may be proved by evaluating the dual of $f^h(x)$ as follows:

$$\begin{aligned}
f^d(f^h(x)) &\triangleq \overline{[\bar{x}_{n+1} f(\bar{x}) + x_{n+1} f^d(\bar{x})]} \\
&= \left[\overline{(\bar{x}_{n+1} f(\bar{x}))} \cdot \overline{(x_{n+1} f^d(\bar{x}))}\right] \\
&= \left[(x_{n+1} + \overline{f(\bar{x})})(\bar{x}_{n+1} + \overline{f^d(\bar{x})})\right] \\
&= \left[(x_{n+1} + f^d(x))(\bar{x}_{n+1} + f(x))\right] \\
&= [x_{n+1} f(x) + \bar{x}_{n+1} f^d(x) + f(x) f^d(x)] \\
&= [x_{n+1} f(x) + \bar{x}_{n+1} f^d(x)] \triangleq f^h(x).
\end{aligned}$$

[2] The terminology "self-dualized" function was used by Goto and Takahasi for the $(n+1)$-variable function which we have here termed a hyperfunction. The original terminology has not been retained as confusion with a naturally self-dual function of $\leq n$ variables may arise. Goto and Takahasi's original paper will illustrate this difficulty.

The Classification of Logic Functions

Therefore

$$f^h(x) \text{ is self-dual.}$$

Notice also that the hyperfunction $f^h(x)$ of any self-dual function $f(x)$ is merely the function $f(x)$, as

$$[x_{n+1}f(x) + \bar{x}_{n+1}f^d(x)] = f(x)[x_{n+1} + \bar{x}_{n+1}]$$
$$= f(x).$$

Now in the previous NPN classification the dual of any function $f(x)$ is included in the same classification entry by virtue of the two negation operations in the NPN classification. The hyperfunction of $f(x)$, however, certainly does not fall into the same NPN classification, unless $f(x)$ happens to be a self-dual function, in which special case the hyperfunction of $f(x)$ remains merely $f(x)$. However, if we consider any hyperfunction of $n + 1$ variables $f_1(x_1, x_2, \ldots, x_n, x_{n+1})$, it may be decomposed into *two non-self-dual functions of n variables* about any single variable. For example, decomposing about x_{n+1} gives the two functions

$$f_1(x_1, x_2, \ldots, x_n), \quad x_{n+1} = 1$$

and

$$f_2(x_1, x_2, \ldots, x_n), \quad x_{n+1} = 0.$$

What the SD classification method now states is that both the original self-dual function of $n + 1$ variables and all the possible decompositions into functions of n variables may be classed in the same SD classification entry.

As an example, consider the hyperfunction, given in a positive canonic form,

$$f(x_1, x_2, x_3, x_4) = [(x_1 + x_2 + x_3)x_4 + x_1x_2x_3].$$

Decomposing this function about, say, variable x_4, we obtain the two three-variable functions

$$f_1(x_1, x_2, x_3) = [x_1 + x_2 + x_3 + x_1x_2x_3]$$
$$= [x_1 + x_2 + x_3] \quad \Big\} \quad x_4 = 1$$

and

$$f_2(x_1, x_2, x_3) = [x_1x_2x_3] \quad \Big\} \quad x_4 = 0.$$

Decomposing about x_3, however, gives

$$f_3(x_1, x_2, x_4) = [x_1x_4 + x_2x_4 + x_4 + x_1x_2]$$
$$= [x_4 + x_1x_2] \quad\quad\quad\quad\} \; x_3 = 1$$

and

$$f_4(x_1, x_2, x_4) = [x_1x_4 + x_2x_4]$$
$$= [x_4(x_1 + x_2)] \quad\quad\quad\quad\} \; x_3 = 0.$$

Decompositions about x_2 and x_1 yield similar functions to $f_3(x)$ and $f_4(x)$ above which under appropriate permutations give the standard canonic forms

$$f_3(x) = [x_1 + x_2x_3]$$

and

$$f_4(x) = [x_1(x_2 + x_3)].$$

Hence the original four-variable hyperfunction and the subsequent four three-variable functions are all classed under the same SD entry. Notice that full NPN permutations and negations are still appropriate within each SD classification.

For $n \leq 3$, the number of entries in the SD classification is seven, compared with 14 and 22 in the NPN and PN classifications respectively. For illustration, the first four of these seven SD classifications are shown in Table 2.2; the remainder, which tend to be increasingly cumbersome to detail, will be found in published tables[4,7].

Table 2.2. The initial PN, NPN, and SD classification of binary functions showing four of the seven SD classifications for $n \leq 3$

Type of Classification		
PN representative functions	NPN representative functions	SD representative functions
0 1	$\Big\} \, 1$	$\Big\} \, x_1$
x_1	x_1	
$x_1 x_2$ $x_1 + x_2$	$\Big\} \, x_1 + x_2$	$x_1x_2 + x_2x_3 + x_1x_3$
$x_1 \oplus x_2$	$x_1 \oplus x_2$	$x_1 \oplus x_2 \oplus x_3$

(continued)

The Classification of Logic Functions

Type of Classification *(Table 2.2 continued)*		
PN representative functions	NPN representative functions	SD representative functions
$x_1x_2 + x_2x_3 + x_1x_3$	$x_1x_2 + x_2x_3 + x_1x_3$	Already listed above
$x_1 \oplus x_2 \oplus x_3$	$x_1 \oplus x_2 \oplus x_3$	Already listed above
$x_1x_2x_3$ $x_1 + x_2 + x_3$	$\left.\begin{array}{c}\\ \\\end{array}\right\} x_1 + x_2 + x_3$	$\left.\begin{array}{c}\\ \\\end{array}\right\} (x_1 + x_2 + x_3)x_4 + x_1x_2x_3$
$x_1(x_2 + x_3)$ $x_1 + x_2x_3$	$\left.\begin{array}{c}\\ \\\end{array}\right\} x_1(x_2 + x_3)$	
$\vdots$	$\vdots$	$\vdots$

The overall result of this final algebraic classification is indicated in Table 2.3. For comparison purposes the PN and NPN statistics are also included. However, it must be concluded that the actual expressions for the canonic SD classification entries become formidable Boolean equations for n greater than, say, 3. As a result numerical classification methods as distinct from these algebraic methods show considerable advantages, as we shall explore in the sections following.

Table 2.3. The PN, NPN, and SD classification statistics for binary functions of $n \leq 6$

Number of input variables n	1	2	3	4	5	6
Total number of functions of $\leq n$ variables	4	16	256	65536	$\approx 4.3 \times 10^9$	$\approx 1.8 \times 10^{19}$
PN classification	3	6	22	402	1228158	$\approx 4 \times 10^{14}$
NPN classification	2	4	14	222	616126	$\approx 2 \times 10^{14}$
SD classification	1	3	7	83	109958	*
Total number of functions of exactly n variables	2	10	218	64594	$\approx 4.3 \times 10^9$	$\approx 1.8 \times 10^{19}$
PN classification	1	3	16	380	1227756	$\approx 4 \times 10^{14}$
NPN classification	1	2	10	208	615904	$\approx 2 \times 10^{14}$
SD classification	0	2	4	76	109875	*

Note: (1) The SD classification includes the hyperfunctions of $n+1$ variables.
(2) * = not computed.

2.3. A numerical classification for linearly separable functions

Numerical classification methods have been extensively investigated for linearly separable functions (see Fig. 1.18) in an attempt to list all such functions in a compact numerical form together with the representative threshold-logic specification which may be used to realize each classified entry. The main objective of this exercise was to enable any given binary function to be processed and its appropriate numerical characterizing parameters determined, which values could then be compared with the standard listing of all possible linearly separable functions. If the determined parameter values were listed, the given function was a linearly separable function which could then be realized by a threshold-logic gate of appropriate input weightings and output threshold which the tables would also provide. If the determined values were not listed, then the given function was not linearly separable and hence could not be realized by the use of merely one threshold-logic gate. This technique is generally illustrated in Table 2.4, and will be extensively referred to again in Chapter 4.

Table 2.4. The main area of application for the numerical classification ("look-up") tables for linearly separable functions

---
Given binary function $f(x)$
↓
Determination of its numerical characterizing parameters
↓
Check whether these parameter values are listed in the standard look-up tables for linearly separable functions

| If yes, use the look-up tables to give the optimum threshold-logic realization for $f(x)$, i.e., $f(x) = \langle a_1 x_1 + \ldots + a_n x_n \rangle_t$ | If no, then there is no single threshold-logic-gate realization and a more complex network is necessary to realize $f(x)$ |

This analysis of linearly separable functions and the publication of tabulations for all such functions of $n \leq 7$ has been successfully completed. Beyond $n = 7$ the listing becomes too lengthy to reproduce in detail, although the statistics for $n = 8$ have been fully investigated and documented. These numerical parameters which are used to classify the linearly separable functions are now normally referred to as the Chow parameters.

The Classification of Logic Functions

Although this particular classification was developed specifically for linearly separable functions, which form a decreasing percentage of all possible binary functions as the number of input variables increases (see Table 2.7 later), it is significant to consider this classification technique in detail as it forms an important link with a more powerful classification procedure for all possible binary functions, whether linearly separable or not. Hence in the following sections we shall examine in some detail the Chow-parameter classification procedure, before going on in subsequent sections to an all-embracing general case.

2.3.1. Some Preliminary Considerations

Before formally considering the Chow-parameter classification, let us look at two related aspects which arise when considering the possible use of threshold-logic gates. One immediate difference which is present between a threshold-logic gate and a normal vertex gate, say a NAND gate, is that the inputs to a threshold-logic gate may not all have the same "importance" in determining the gate output state, as was shown in Chapter 1. Hence connecting together or otherwise energizing the gate inputs can give rise to many more different output functions from such a gate than is possible from a simple vertex gate.

However, let us start by considering a simple threshold-logic gate, say a four-input one with input weightings of 3, 2, 2, and 1 as shown in Fig. 2.1a. With the particular input connections shown, the output function is

$$f(x) = \langle 3x_1 + 2x_2 + 2x_3 + x_4 \rangle_4 \quad \text{(threshold)}$$
$$= [x_1(x_2 + x_3 + x_4) + x_2 x_3] \quad \text{(Boolean)}.$$

Reconnecting the four inputs (Permutation of the input variables) yields similar output functions, which clearly fall into the same classification entry under such input-variable permutation operations. Negation of the input variables also yields similar functions, with $\bar{x}_1$ instead of x_1 etc., appearing in the expression for $f(x)$, these two operations together therefore constituting a PN invariance classification for all the linearly separable functions of this form. Figure 2.1b, c illustrates two functions which result from two simple PN invariance operations.[3]

---

[3] The term "invariance," first used by Dertouzos[6], covers the operations which relate all the different binary functions within one classification entry, that is permutation, negation, etc.

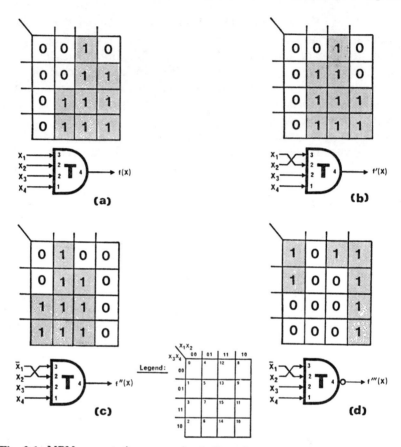

Fig. 2.1. NPN permutations on a simple linearly separable function.

(a) Given function $f(x) = \langle 3x_1 + 2x_2 + 2x_3 + x_4 \rangle_4$
$= [x_1(x_2 + x_3 + x_4) + x_2x_3]$

(b) Permutation of x_1 with x_2, giving
$f'(x) = \langle 2x_1 + 3x_2 + 2x_3 + x_4 \rangle_4$
$= [x_2(x_1 + x_3 + x_4) + x_1x_3]$

(c) Negation of x_1 in $f'(x)$, giving
$f''(x) = \langle 2\bar{x}_1 + 3x_2 + 2x_3 + x_4 \rangle_4$
$= [x_2(\bar{x}_1 + x_3 + x_4) + \bar{x}_1x_3]$

(d) Negation of the output of $f''(x)$, giving
$f'''(x) = \langle 2x_1 + 3\bar{x}_2 + 2\bar{x}_3 + \bar{x}_4 \rangle_5$
$= [\bar{x}_2(x_1 + \bar{x}_3) + x_1\bar{x}_3\bar{x}_4]$

The Classification of Logic Functions

Similarly if the output of the gate is complemented, this corresponds to the second *N*egation operation in the NPN classification procedure. Nothing new has been introduced so far in comparison with the NPN classification procedure for any binary function, whether linearly separable or not. Figure 2.1*d* illustrates the overall negation operation to complete this initial illustration.

The output-detection threshold value of the threshold-logic gate is another parameter of the gate which we have not considered so far. Can it somehow be brought into the permutation operations which allow as many different possible functions $f(x)$ to be categorized under one common classification entry? By way of a non-mathematical introduction, consider an early and now completely obsolete type of threshold-logic gate, that of the magnetic-core circuit element which was introduced in the middle 1950's. It has of course now been utterly superseded by solid-state circuits, but nevertheless serves to illustrate some interesting and revealing features of threshold-logic working[8,9].

Figure 2.2 shows the essentials of a simple magnetic-core element. It consists of n input windings N_1 to N_n, an output winding, and a "bias" or "threshold" winding N_t. The basic action of this gate is that

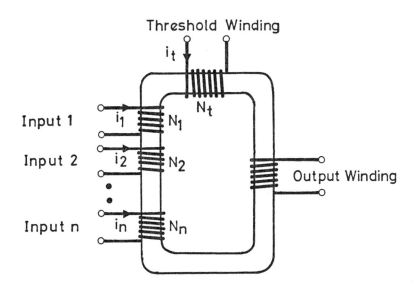

Fig. 2.2. Schematic arrangement of a magnetic-core threshold element of c. 1955, using a square-hysteresis-loop magnetic core.

a certain minimum total input ampere-turns $\sum_{i=1}^{n} i_i N_i$ is necessary in order to saturate the core in a positive direction, against the action of the opposite polarity biasing or threshold ampere-turns $i_t N_t$. Effectively 0, 1 binary input signals are applied to each of the input windings, the saturation of the core in a positive direction giving the logic output change from 0 to 1 on the output winding. Notice that there are $n+1$ control windings for n input signals.

Now any interchange of input signals applied to the various input windings N_1 to N_n corresponds to the *Permutation* operation of the normal classification method. Similarly, reversing the connections of the input signal to any input winding N_i may be regarded as a *Negation* operation on this input, whilst reversing the connections from the output winding may be considered as an output *Negation* operation. Thus the three NPN classification operations may be illustrated very neatly by an ideal magnetic-core device. However, we still have not brought into these invariance considerations the $(n+1)^{\text{th}}$ control winding, namely the threshold winding N_t.

Should we now consider the interchange of any input winding N_i with the threshold winding N_t, then, assuming $N_i \neq N_t$, a new family of threshold-logic functions becomes possible from the same magnetic-core element. We have effectively performed an invariance operation between a gate input and the gate threshold, and the single threshold logic element with its $n+1$ windings (gate parameters) becomes the common factor between all these possibilities. Clearly we have now produced for the class of linearly separable functions an even more compact classification than that which the NPN classification provides. Notice also that should all n input windings N_1 to N_n be dissimilar to each other and to N_t, then n possible interchanges and n possible new families of linearly separable functions may be produced, each of which is freely subject to the full NPN invariance possibilities. Figure 2.3*a, b* illustrates one such possibility.

In practice such a development would only be possible with perfect magnetic devices. Nevertheless this theoretical exercise does illustrate some extra possibility in the classificiation of linearly separable functions, involving the consideration of *n+1 gate parameters for an n-input gate*, which does not arise in the NPN classification procedure. The Chow parameters which we shall formally consider in a moment do in fact embrace $n+1$ parameters for n-variable linearly separable functions.

Let us, however, look at yet another aspect of threshold-logic. If we take an n-input threshold-logic gate and use it with less than n signal

The Classification of Logic Functions

inputs, say p inputs, where $p < n$, then the unused $(n - p)$ gate inputs may be used in a variety of ways. With normal vertex gates little logic flexibility results in using a gate with a higher fan-in capacity than necessary; an n-input NAND gate, for example, always remains a NAND gate to p inputs (assuming the unusued inputs are not clamped at logic 0) and similarly for all other vertex gates.

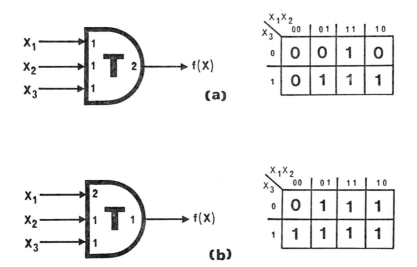

Fig. 2.3. An example invariance operation directly between an input weighting value and the gate output threshold value t:

(a) gate $\langle 1, 1, 1 \rangle_2$
(b) gate $\langle 2, 1, 1 \rangle_1$

(Note: This simple invariance operation of directly interchanging the value of an input weighting with the gate threshold value is not precisely the invariance operation which the subsequent Chow-parameter classification procedure will provide. The Chow invariance procedures always produce minimum-integer functions which this simple procedure does not.)

However, consider the four-input threshold-logic gate originally illustrated in Fig. 2.1a. Suppose we have only three input signals present. Then the possibilities which arise with this gate are (i) apply a steady logic 0 to the unwanted gate input, (ii) apply a steady logic 1 to the unwanted gate input, or (iii) connect the unwanted input to one of the other used inputs. These three possibilities mean that (i) the used inputs are equated against an unchanged gate threshold value, (ii) the used inputs are equated against effectively a reduced gate threshold, and (iii) one gate input is given an increased weighting value. Hence if, for example, input x_4 is unused in the gate of Fig. 2.1a, the following possibilities arise:

original function $f(x) = \langle 3x_1 + 2x_2 + 2x_3 + x_4 \rangle_4$

modification (i): $f(x) = \langle 3x_1 + 2x_2 + 2x_3 \rangle_4$
$= [x_1(x_2 + x_3) + x_2 x_3]$

modification (ii): $f(x) = \langle 3x_1 + 2x_2 + 2x_3 \rangle_3$
$= [x_1 + x_2 x_3]$

modification (iii): $f(x) = \langle 3x_1 + 2x_2 + 3x_3 \rangle_4$ assuming gate inputs x_3 and x_4 are connected together
$= [x_1 x_2 + x_2 x_3 + x_1 x_3]$.

We must be very careful to distinguish between this latter area and the previous linearly separable function classification. In the *function classification* we are concerned with appropriately listing all possible linearly separable functions of n variables in some form which may not have a single realization. However, in the latter threshold-logic *gate usage* considerations, we are specifically considering how a particular n-input threshold-logic gate may be used if not all of its n inputs are employed for independent binary input signals. Thus a classification entry specifically covering, say, a six-input function does not directly show how the six-input threshold-logic gate required to realize this function may be employed for situations requiring less than six input variables. Nevertheless the function classification is the key area, as from this all the other considerations, including the choice of the most versatile n-input threshold-logic gate, may be developed. Table 2.5 illustrates this distinction, which will be amplified when we have covered the Chow-parameter method of classification and its subsequent use in following chapters.

The Classification of Logic Functions

Table 2.5. The distinction between the Chow-parameter classification for linearly separable functions and the linearly separable functions which may be realized by one threshold-logic gate

(a)

Chow-parameter function classification, each classification entry assuming some free mathematical invariance operation amongst the gate input weightings and the gate threshold (not directly practical)

Actual threshold-logic gate with one specific output threshold value t.	Further threshold-logic gate(s), each with their own specific output threshold value t
All linearly separable functions which are NPN variants on these n inputs and one output	All other linearly separable functions which are NPN variants on each n-input–one-output gate.

(b)

Given threshold-logic gate with n inputs and one output threshold value t

All linearly separable functions of n variables = NPN variants on the full n inputs and one output	All other linearly separable functions of $<n$ inputs = NPN variants on every configuration of $<n$ signal inputs.

(a) The function classification meaning.

(b) Functions which may be realized by one gate, not necessarily using all its inputs for separate input signals.

2.3.2. The Chow Parameters

The numbers which may be used to define and classify any linearly separable function $f(x)$, nowadays usually referred to as the Chow parameters, follow from the work of C.K. Chow, who first formally proved that only $n + 1$ numbers were necessary to define fully any given linearly separable function of n variables[10]. This development furthered earlier work of Golomb[11] and Ninomiya[12], both of whom were concerned with a larger set of 2^n numbers. Work in parallel with and subsequent to that of Chow, particularly by Dertouzos[6], Winder[13,14], and Muroga[7,15], has refined and finalized this $n + 1$ numerical classification procedure. The present result is that any given linearly separable function may be defined by a series of $n + 1$ integer numbers. If these numbers are subsequently rearranged in a given ordering, conventionally one of decreasing magnitude, then this *positive canonic weight-threshold* information is the Chow-parameter classification for the given function. Tables of all such function classifications for $n \leq 6$ were first published by Dertouzos[6] and for $n \leq 7$ by Winder[14] and enumerated and analysed for $n \leq 8$ by Muroga et al.[15].

To give the full meaning of the Chow parameters, from the point of view both of classifying linearly separable functions and of attempting to derive the optimum (or *minimum integer*[4]) threshold realizations for such functions, let us review in the following pages the principal algebraic developments in this area.

Firstly, instead of the more usual binary values of 0 and 1 which we normally use to represent false and true, it is more convenient in this development to use the two values -1 and $+1$ instead of 0 and 1, respectively. We shall also find this change helpful when we subsequently consider the classification of non-linearly separable functions and spectral techniques.

Now as by definition a threshold-logic gate is concerned with the arithmetic summation $\sum_{i=1}^{n} a_i x_i$ of the weighted inputs $x_i, i = 1 \ldots n$, the choice of $-1, +1$ instead of $0, 1$ will mean that the summation can take a positive or negative value. The sign of the summation can in fact now be indicative of the output state. Let us use the symbols $y_1, \ldots, y_n$ instead of $x_1, \ldots, x_n$ to identify the binary input variables when we are using the binary values $-1, +1$, and similarly $f(y)$ instead of $f(x)$ when we consider the function output to take these two values. Numerically it therefore follows that

$$x_i = \tfrac{1}{2}(y_i + 1) \quad i = 1 \ldots n.$$

Hence redefining the basic threshold-logic output equation we have

---

[4] A "minimum integer" realization is when the input weightings and the gate threshold are integer values of the smallest value necessary to realize correctly the given linearly separable function. All the examples in Chapter 1 were minimum-integer threshold realizations.

The Classification of Logic Functions

$$f(y) = +1 \text{ when } \sum_{i=1}^{n} a_i \tfrac{1}{2}(y_i + 1) \geq t$$

$$= \left(\sum_{i=1}^{n} a_i y_i\right) + \left(\sum_{i=1}^{n} a_i\right) - 2t \geq 0.$$

Similarly

$$f(y) = -1 \text{ when } \left(\sum_{i=1}^{n} a_i y_i\right) + \left(\sum_{i=1}^{n} a_i\right) - 2t < 0.$$

$\hspace{20em}$ (2.1)

Let us now define a further "weight" a_0 as follows:

$$a_0 = \left(\sum_{i=1}^{n} a_i\right) - 2t + 1.$$

Rearranging,

$$t = \tfrac{1}{2}\left\{\left(\sum_{i=1}^{n} a_i\right) - a_0 + 1\right\}. \hspace{3em} (2.2)$$

Substituting (2.2) back into equation (2.1) now gives

$$f(y) = +1 \text{ when } \left(\sum_{i=1}^{n} a_i y_i\right) + a_0 - 1 \geq 0$$

and

$$f(y) = -1 \text{ when } \left(\sum_{i=1}^{n} a_i y_i\right) + a_0 - 1 < 0.$$

Introducing a further input "variable" y_0, which by definition always takes the value $+1$, we may bring the above a_0 term within the summation bracket, yielding the two final equations

$$f(y) = +1 \text{ when } \left(\sum_{i=0}^{n} a_i y_i\right) \geq +1.0$$

and

$$f(y) = -1 \text{ when } \left(\sum_{i=0}^{n} a_i y_i\right) < +1.0$$

$\hspace{20em}$ (2.3)

where a_0 is as defined above and $y_0 = +1.0$ in all circumstances. Notice the two particular features of this development as follows.

(i) The a_i's, $i = 1 \ldots n$, have not been changed at all in the above conversions from the 0, 1 to the $-1, +1$ configuration. Thus if all the a_i's, $i = 0 \ldots n$, of the latter configuration are known, then the conventional 0, 1 threshold realization $f(x) = \langle a_1 x_1 + a_2 x_2 + \ldots a_n x_n \rangle_t$ is immediately available, because (a) $a_1, a_2, \ldots, a_n$ remain the same and (b) t is the evaluation of equation (2.2) above.

(ii) The above development has introduced a $(n + 1)^{\text{th}}$ parameter, or "dimension," over and above the normal n-input variables a_i, x_i, y_i, $i = 1 \ldots n$. This feature of n-input threshold-logic functions being characterized by a set of $(n + 1)$ parameters is a feature which has already been indicated.

However, as in practice we are always interested in a minimum-integer solution for the a_i's and the gate threshold t, we may qualify equation (2.3) above for such conditions. With optimum minimum-integer conditions the lowest input summation $\sum_{i=1}^{n} a_i x_i$ to give $f(x) = 1$ will be exactly t, whilst the highest input summation to give $f(x) = 0$ should be $t - 1.0$. This may be checked in any of the examples of threshold-logic gates given in Chapter 1. There is therefore a *threshold-gap* of unity between the $f(x) = 0$ and the $f(x) = 1$ conditions in the 0, 1 domain.

However, in the $-1, +1$ domain the steps in the input summation are twice those in the 0, 1 domain, as where any input, say y_1, changes from -1 to $+1$ it varies the total gate input summation by $2a_1$ instead of merely a_1 as in the 0, 1 case. Hence the threshold gap in the minimum-integer $-1, +1$ case becomes 2.0 rather than 1.0. Therefore the highest input summation for $f(y) = -1$ that can arise is 2.0 less than the minimum $f(y) = +1$ value. Hence for the minimum-integer case, equations (2.3) may be rewritten

$$\left. \begin{array}{l} f(y) = +1 \text{ when } \left(\sum_{i=0}^{n} a_i y_i \right) \geq +1.0 \\ \text{and} \\ f(y) = -1 \text{ when } \left(\sum_{i=0}^{n} a_i y_i \right) \leq -1.0. \end{array} \right\} \quad (2.3a)$$

Notice that the $\geq +1.0, < +1.0$ limits of equations (2.3) only arose owing to our including a $+1$ in the (arbitrary) definition for a_0. If we had omitted this value (made it zero valued), then equations (2.3) would have become $\geq 0, < 0$, respectively, but the value which t would have taken from equation (2.2) would in practice be found to be fractional, i.e. $\frac{1}{2}$, $1\frac{1}{2}$, $2\frac{1}{2}$, etc. Therefore the $+1$ included during the initial definition of a_0 provides an integer value for t, assuming the a_i's, $i = 1 \ldots n$, are also minimum-integer values. This additional $+1$ is not always included by all authors in the definition for a_0, for example see Dertouzos[6], in which case threshold values such as 1.5 occur in their developments.

Continuing our development, in an attempt to synthesize given linearly separable functions in a minimum-integer form, we may proceed as follows, continuing to work in the $-1, +1$ domain.

Let a_1, $i = 0 \ldots n$, be the set of minimum-integer values which successfully realize the given function $f(y)$, recalling that the gate-threshold value t is obtainable from equation (2.2). Then

$$\sum_{i=0}^{n} a_i y_i = \text{positive at all minterms } p \text{ where } f(y) = +1$$
$$= \text{negative at all minterms } p \text{ where } f(y) = -1$$

with $y_0 = +1$ at all minterms as before.

Therefore as the sign of $f(y)$ is always the same as the sign of $\sum_{i=0}^{n} a_i y_i$ at all minterms when the a_i correctly realize $f(y)$, i.e. both negative or both positive, then

$$\left\{ f(y) \cdot \sum_{i=0}^{n} a_i y_i \right\} = + \left| \sum_{i=0}^{n} a_i y_i \right| \tag{2.4}$$

at every minterm p. Therefore summation of both sides of equation (2.4) over all 2^n minterms 0 to $2^n - 1$ gives

$$\sum_{p=0}^{2^n-1} \left\{ f(y) \cdot \sum_{i=0}^{n} a_i y_i \right\} = \sum_{p=0}^{2^n-1} \left\{ + \left| \sum_{i=0}^{n} a_i y_i \right| \right\}. \tag{2.5}$$

This relationship may be shown to be a necessary and a sufficient condition for linear separability. Necessity is shown by the above; sufficiency may be proved by contradiction.

Rearranging the left-hand summations of equation (2.5) we have

$$\sum_{i=0}^{n} a_i \left(\sum_{p=0}^{2^n-1} \{ f(y) \cdot y_i \} \right) = \sum_{p=0}^{2^n-1} \left\{ + \left| \sum_{i=0}^{n} a_i y_i \right| \right\}. \tag{2.6}$$

Let us now define a further parameter, one which we shall subsequently have many occasions to refer to in several guises, as follows:

$$b_i, \ i = 0 \ldots n, \triangleq \sum_{p=0}^{2^n-1} \{ f(y) \cdot y_i \} \tag{2.7}$$

where y_0 is, as before, always $+1.0$. From equation (2.6) we obtain

$$\sum_{i=0}^{n} a_i b_i = \sum_{p=0}^{2^n-1} \left\{ + \left| \sum_{i=0}^{n} a_i y_i \right| \right\}. \tag{2.8}$$

In this final equation all factors except the a_i's are available from the given Boolean function truth table, converting the usual 0, 1 input–output truth values to $-1, +1$ respectively.

It may now be shown that if a set of a_i's are chosen which do not realize the given function, then

$$\sum_{i=0}^{n} a_i b_i < \sum_{p=0}^{2^n-1} \left\{ + \left| \sum_{i=0}^{n} a_i y_i \right| \right\}.$$

Thus if equation (2.8) is rearranged as

$$I = \sum_{p=0}^{2^n-1} \left\{ + \left| \sum_{i=0}^{n} a_i y_i \right| \right\} - \sum_{i=0}^{n} a_i b_i \tag{2.8a}$$

the problem of producing a minimum-integer threshold-logic realization for the given Boolean function has been reduced to that of minimizing the values in equation (2.8a) by adjustment of the chosen values for the a_i's, with $I = 0$ as the necessary overriding condition for correct realization of the given function.

In Chapter 4 we shall continue this development to show the problems of determining the minimum-integer a_i values. Equation (2.8a) does not provide a direct means of determination; should the given function $f(x)$ not be linearly separable then by definition no set of a_i's can be found to realize $f(x)$, but if $f(x)$ is linearly separable then there are (theoretically) an infinite set of a_i's which will satisfy the equation $I = 0$. The minimum-integer set is unfortunately not directly obtainable from this or any other algebraic expression as we are dealing with a difficult discontinuous form of arithmetic summation.

However, to return to our more immediate interests, this development has now introduced the additional parameters a_0 and y_0 and the $n + 1$ parameters, b_i, $i = 0 \ldots n$. In particular let us look at the b_i's. Recall that by definition they are

$$b_i, i = 0 \ldots n, \triangleq \sum_{p=0}^{2^n-1} \{f(y) \cdot y_i\}$$

that is, each b_i is the arithmetic sum of the product of the value of the function output $f(y)$ and the input variable y_i at each minterm p, summed over all 2^n minterms. Thus when the function output $f(y)$ and the input variable y_i agree in value, i.e. both are -1 or both are $+1$, then the product value at such a minterm is $+1$, whilst when they disagree the product value must be -1. Hence each b_i is a measure of the *agreement* (or *correlation*) between the function output value and an input variable, and may be redefined as follows:

$b_i, i = 0 \ldots n \triangleq$ {(number of agreements between the value of the input variable y_i and the output function value $f(y)$) − (number of disagreements between the value of the input variable y_i and the output value $f(y)$)}.

Note that in the case of the b_0 term y_0 is by definition always $+1$, and therefore b_0 reduces to
{(number of minterms where $f(y)$ is $+1$) − (number of minterms where $f(y)$ is -1)}
= {(number of true minterms) − (number of false minterms)}.

Notice now that the latter definitions are equally and directly applicable to the more usual 0, 1 Boolean domain, as Table 2.6 illustrates. The equivalent of y_0 in the 0, 1 domain is x_0, where $x_0 \triangleq$ always 1.

The Classification of Logic Functions

Table 2.6 The evaluation of the b_i parameters for the Boolean function $f(x) = [\bar{x}_1 x_2 + x_3]$, parameter values being $b_0 = +2$, $b_1 = -2$, $b_2 = +2$, and $b_3 = +6$

(a) Evaluation from the $-1, +1$ $y_i, f(y)$ data.

Minterm p	Inputs				Output	Product terms			
	y_0	y_1	y_2	y_3	$f(y)$	$f(y) \cdot y_0$	$f(y) \cdot y_1$	$f(y) \cdot y_2$	$f(y) \cdot y_3$
0	+1	−1	−1	−1	−1	−1	+1	+1	+1
1	+1	−1	−1	+1	+1	+1	−1	−1	+1
2	+1	−1	+1	−1	+1	+1	−1	+1	−1
3	+1	−1	+1	+1	+1	+1	−1	+1	+1
4	+1	+1	−1	−1	−1	−1	−1	+1	+1
5	+1	+1	−1	+1	+1	+1	+1	−1	+1
6	+1	+1	+1	−1	−1	−1	−1	−1	+1
7	+1	+1	+1	+1	+1	+1	+1	+1	+1
				$\sum_{p=0}^{2^n-1} \{f(y) \cdot y_i\}$:		+2	−2	+2	+6

(b) Evaluation from the 0, 1 $x_i, f(x)$ data.

Minterm p	Inputs				Output	Agreements/disagreements			
	x_0	x_1	x_2	x_3	$f(x)$	With x_0	With x_1	With x_2	With x_3
0	1	0	0	0	0	d	a	a	a
1	1	0	0	1	1	a	d	d	a
2	1	0	1	0	1	a	d	a	d
3	1	0	1	1	1	a	d	a	a
4	1	1	0	0	0	d	d	a	a
5	1	1	0	1	1	a	a	d	a
6	1	1	1	0	0	d	d	d	a
7	1	1	1	1	1	a	a	a	a
{agreements − disagreements}:						+2	−2	+2	+6

It will be noticed from this simple example that the b_i values will always be *even-value* positive or negative numbers. It would be possible to divide them all by 2 to reduce their magnitude, but this has little merit and does not add or subtract anything to their basic information content. Hence in this book we shall continue to use the numbers as here defined; this may be dissimilar to certain published papers, but currently represents the most common practice. Another basic point worthy of notice at this stage is that if the output function $f(x)$ (or $f(y)$) is exactly the same as any single-input variable, say $f(x) = x_3$, then the value for b_3 will take the maximum possible value, this value being $+2^n$, i.e. the maximum possible correlation between $f(x)$ and x_3. All other b_i values will be zero. Should $f(x)$ be merely $\bar{x}_3$, then b_3 will be -2^n, all other b_i values still zero. These features will be referred to again, particularly in Chapter 5.

Now for any given linearly separable function $f(x)$ there is a unique set of b_i values. Hence the numerical b_i values uniquely describe the given function, provided it is linearly separable and hence realizable by one threshold-logic gate of appropriate input weightings a_i and threshold t. These b_i values are usually referred to as the *Chow parameters* for $f(x)$ following the original work of Chow in this area, although the definition we have taken for the b_i's in the preceding development is not precisely Chow's original but that used by Dertouzos and others in subsequent development of this work. The following three very simple examples, all of which are in the same PN classification as the function used in Table 2.6, illustrate this feature of the b_i values uniquely defining $f(x)$.

Example 1 $f(x) = [x_1 + x_2\bar{x}_3]$
$= \langle 2x_1 + x_2 + \bar{x}_3 \rangle_2$

Chow parameters $b_0 = +2$, $b_1 = +6$, $b_2 = +2$, $b_3 = -2$, which would usually be written in the ordered form 2, 6, 2, −2.

Example 2 $f(x) = [\bar{x}_1\bar{x}_2 + \bar{x}_3]$
$= \langle \bar{x}_1 + \bar{x}_2 + 2\bar{x}_3 \rangle_2$

Chow parameters 2, −2, −2, −6.

Example 3 $f(x) = [x_2 + x_1\bar{x}_3]$
$= \langle x_1 + 2x_2 + \bar{x}_3 \rangle_2$

Chow parameters 2, 2, 6, −2.

Consider now taking the complement of any function, say the complement of Example 1 above. This now gives the NPN variant

$f(x) = \overline{[x_1 + x_2\bar{x}_3]}$
$= [\bar{x}_1(\bar{x}_2 + x_3)]$
$= [\bar{x}_1\bar{x}_2 + \bar{x}_1 x_3]$
$= \langle 2\bar{x}_1 + \bar{x}_2 + x_3 \rangle_3$

The Classification of Logic Functions

Chow parameters $-2, -6, -2, 2$.

Notice that the *N*egation of the function, the third of the NPN operations, reverses the sign of *all* the b_i's of $\overline{f(x)}$ compared with those of $f(x)$.

All the NPN variants we have examined in the above example have Chow parameter values which involve the numbers 2, 2, 2, and 6 in some sequence or other, each number being positive or negative. We have in fact illustrated the numerical classification method which is available to classify all possible linearly separable functions. However, let us define this classification more formally, and point out one very significant further feature not illustrated so far.

The full classification method for linearly separable functions consists of the *magnitude of the* $n + 1$ *Chow parameters, arranged (conventionally) in decreasing magnitude order*. For example the classification entry to which all the above example functions belong is the entry

6, 2, 2, 2.

This classification covers all the possible three-variable linearly separable functions which possess these numbers as their Chow parameters, in any order and each value a positive or a negative value.

Now in the examples so far considered in this classification entry, the b_0 value has always been ± 2, with the remaining 6, 2, and 2 values distributed among the b_1, b_2, and b_3 terms. However, let us take the value 6 as the b_0 value. This now gives a further possible range of functions whose Chow parameters are any sign combination of

$$b_0 = \pm 6, \quad b_1 = \pm 2, \quad b_2 = \pm 2, \quad b_3 = \pm 2.$$

One such function, parameter values $-6, 2, -2, -2$ is the function

$$f(x) = x_1 \bar{x}_2 \bar{x}_3 \text{ (Boolean)} = \langle x_1 + \bar{x}_2 + \bar{x}_3 \rangle_3 \text{ (threshold)}.$$

Notice that this simple example illustrates an interesting feature which we shall pursue in Chapter 4, namely that it is straightforward to determine the particular Chow parameters for any given function $f(x)$ from the definition of the b_i terms, but to determine $f(x)$ from given Chow-parameter values is not a direct procedure. It is not a linear mathematical process to go from the b_i's back to the two-value Boolean domain of $f(x)$.

Summarizing this example, the classification entry 6, 2, 2, 2 allows full invariance operations of these numbers between all the b_i terms, including b_0, with positive or negative values of each. The b_0 term effectively brings the gate threshold into this numerical classification procedure by way of equations (2.7) and (2.2), and provides the $(n + 1)^{th}$ term in the classification entry as was suggested may be possible by the preliminary discussions centered around the magnetic-core element of Fig. 2.2. For any particular value of b_0 chosen from the classification entry we then have the following.

(i) Change of sign of all the b_i's corresponds to Negation of the function $f(x)$. Notice that changing the sign of b_0 alone (assuming $b_0 \neq$ zero) does not give $\overline{f(x)}$, but instead gives another value of t from equation (2.2) and hence a further family of binary functions.

(ii) Permutation of the b_i's, $i \neq 0$, corresponds to corresponding permutations of the input variables x_i.

(iii) Change of sign of any b_i, $i \neq 0$, corresponds to Negation of the particular input variable x_i.

Thus the full NPN algebraic classification procedure is incorporated in the invariance operations on these numerical parameters. In addition the choice of the b_0 value corresponds to the additional SD algebraic classification of Goto and Takahasi. For a more detailed treatment of the subject, including the reasons why hyperfunctions and hence the SD classification are equivalent to the $n + 1$ Chow parameters, references may be made to the many publications in this area[6,7,10–16], in particular Dertouzos[6], Muroga[7], and Winder[16].

The compactness of the Chow-parameter classification for linearly separable functions is unsurpassed. For $n \leq 3$ there are merely three entries, namely

and
$$8, 0, 0, 0$$
$$6, 2, 2, 2$$
$$4, 4, 4, 0$$

These three entries cover a total of 104 different linearly separable functions. The full Chow-parameter classification tables for $n \leq 6$, which we shall have numerous occasions to refer to in later chapters of this book, will be found in Appendix A.

Finally it must be re-emphasized that this classification procedure is applicable only to linearly separable Boolean functions, such functions constituting only a small majority of all the 2^{2^n} possible functions when n is large. Table 2.7 gives the statistics of the classification procedures which we have now considered.

Table 2.7. The total number of binary functions and the total number in the subclass of linearly separable (threshold) functions for $n = 3$ through $n = 7$ together with the number of listings in the various invariance classifications

	n	3	4	5	6	7
Unclassified	Boolean functions					
	total	256	65536	$\simeq 4.3 \times 10^9$	$\simeq 1.8 \times 10^{19}$	$\simeq 3.4 \times 10^{38}$
	non-degenerate	218	64594	$\simeq 4.3 \times 10^9$	$\simeq 1.8 \times 10^{19}$	$\simeq 3.4 \times 10^{38}$
	degenerate	38	942	*	*	*
	Threshold functions					
	total	104	1882	94572	15028134	8378070864
	non-degenerate	72	1536	86080	14487040	8274797440
	degenerate	32	346	8492	541094	103273424
Classified	PN Boolean types					
	total	22	402	1228158	$\simeq 4 \times 10^{14}$	*
	PN threshold types					
	total	10	27	119	1113	29375
	non-degenerate	5	17	92	994	28262
	degenerate	5	10	27	119	1113
	NPN Boolean types					
	total	14	222	616126	$\simeq 2 \times 10^{14}$	*
	NPN threshold types					
	total	6	15	63	567	14755
	non-degenerate	3	9	48	504	14188
	degenerate	3	6	15	63	567
	Self-dual Boolean types					
	total	7	83	109958	*	*
	SD threshold types (Chow-parameter listings)					
	total	3	7	21	135	2470
	non-degenerate	1	4	14	114	2335
	degenerate	2	3	7	21	135

Note: (i) The SD binary function classification for any n includes the hyperfunctions of $n + 1$ variables.
(ii) * Not computed.
(iii) For more detailed statistics, including $n = 8$, see Muroga et al.[15]

2.3.3. Two Further Comments

Before we proceed to consider the possibility of establishing some numerical classification technique suitable for all binary functions, and not just the linearly separable class covered by the Chow classification, let us just look at two points which follow from the preceding section.

Firstly, if the literature on the Chow parameters is consulted, several definitions of the Chow parameters may be encountered. At first sight some may look different from our definition in 2.3.2, but on closer examination turn out to be equivalent. The definition chosen in equation (2.7) was

$$b_i, i = 0 \ldots n, \triangleq \sum_{p=0}^{2^n-1} \{f(y) \cdot y_i\}$$

which was subsequently shown to be

$$\{ \text{(the number of agreements} \ldots) - \text{(the number of disagreements} \ldots) \}$$

taken over all 2^n minterms.

An alternative way in which the Chow parameters may be defined is

$$b_0 \triangleq m_f - 2^{n-1}$$
$$b_i, i = 1 \ldots n, \triangleq m_{x_i} - m_{\bar{x}_i}$$

where m_f is the number of true minterms in the given function, m_{x_i} is the number of true minterms where x_i is also true, and $m_{\bar{x}_i}$ is the number of true minterms where x_i is false.

These two equations may therefore be expressed verbally as

$b_0 = \{$the number of true minterms $- 2^{n-1}\}$
$b_i, i = 1 \ldots n = \{$(the number of minterms where both $f(x)$ is 1 and x_i is 1) $-$ (the number of minterms where $f(x)$ is 1 and x_i is 0)$\}$.

The numerical relationship between these two sets of definitions may be shown as follows.

Let a be the number of minterms where $f(x) = 1$ and $x_i = 1$, $(f(y) = +1$ and $y_i = +1)$; b be the number of minterms where $f(x) = 0$ and $x_i = 1$, $(f(y) = -1$ and $y_i = +1)$; c be the number of minterms where $f(x) = 1$ and $x_i = 0$, $(f(y) = +1$ and $y_i = -1)$; and d be the

The Classification of Logic Functions

number of minterms where $f(x) = 0$ and $x_i = 0$ ($f(y) = -1$ and $y_i = -1$). Then

$$a + b + c + d = 2^n.$$

For the b_0 term $a + b = 2^n$, as the c and d terms are both zero, whilst for all remaining b_i terms, $i \neq 0$,

$$(a + b) = (c + d) = \tfrac{1}{2}2^n, = 2^{n-1}.$$

Considering the definitions for b_0, we have the following:

b_0 as used in this text	b_0 as alternatively defined above
$b_0 = a - b$ $= a - (2^n - a)$ $= 2a - 2^n$ $= 2(a - 2^{n-1})$	$a - 2^{n-1}$

Considering the definitions for the b_i terms, $i \neq 0$, we have the following:

b_i as used in this text	b_i as alternatively defined above
$b_i = (a + d) - (b + c)$ $= (a - c) - (b - d)$ $= (a - c) - \{(2^{n-1} - a) - (2^{n-1} - c)\}$ $= (a - c) + (a - c)$ $= 2(a - c)$	$a - c$

Therefore it is seen that these two definitions are numerically similar, except for a constant factor 2. It is of no fundamental significance which is used; in fact the alternative definitions may be found multiplied by a factor of 2 in some texts to make them numerically the same as in this book, or alternatively our definitions may be multiplied by a factor of $\tfrac{1}{2}$ to achieve numerical conformity with these alternative definitions.

The final point of interest concerns the invariance operations over all the $n + 1$ Chow parameters and their practical meaning. Consider some representative four-variable functions, all of which are covered by the single Chow-parameter classification of say 8 8 4 4 4. With any chosen value for the b_0 term (i.e. either 8 or 4) the effect of invariance operations among the remaining four parameters is readily demonstrated and appreciated. Such NPN operations show up graphically on the Karnaugh map plots of Fig. 2.4 a, b, c. Notice that the shape of the

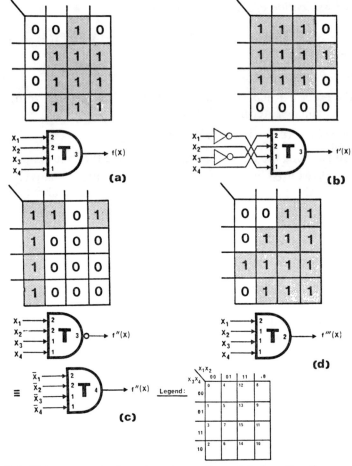

Fig. 2.4. Chow-parameter invariance operations: (*a*) given function
$$f(x) = \langle 2x_1 + 2x_2 + x_3 + x_4 \rangle_3$$
$$= [x_1(x_2 + x_3 + x_4) + x_2(x_3 + x_4)]$$
(*b*) negation of x_1 and x_3 and interchange of $\bar{x}_1, \bar{x}_3$; x_2, x_4, giving
$$f'(x) = \langle \bar{x}_1 + x_2 + 2\bar{x}_3 + 2x_4 \rangle_3$$
$$= [\bar{x}_3(x_4 + \bar{x}_1 + x_2) + x_4(\bar{x}_1 + x_2)]$$
(*c*) negation of original function $f(x)$, giving
$$f''(x) = \langle 2\bar{x}_1 + 2\bar{x}_2 + \bar{x}_3 + \bar{x}_4 \rangle_4$$
$$= [\bar{x}_1(\bar{x}_2 + \bar{x}_3\bar{x}_4) + \bar{x}_2\bar{x}_3\bar{x}_4]$$
(*d*) invariance operation on $f(x)$ between b_0 and b_2, giving
$$f'''(x) = \langle 2x_1 + x_2 + x_3 + x_4 \rangle_2$$
$$= [x_1 + x_2(x_3 + x_4) + x_3x_4].$$

The Classification of Logic Functions

true/false minterm pattern on the Karnaugh map layout remains unchanged in all these permutations, its position and orientation on the map, however, conforming to the NPN permutations of the Chow parameters.

However, when we choose another value for b_0, and hence produce a further family of functions by the NPN invariance operations, the characteristic shape of this further family of functions is quite dissimilar from that of the first family. Fig. 2.4d illustrates this feature. This is a very interesting topological feature of the Chow-parameter classification as no obvious means of logically relating such dissimilar covering patterns appears possible from the binary information. They are in fact linked by appropriate exclusive-OR and negation operations as will be illustrated in Appendix C, this relationship being based upon developments which are covered in the following section.

2.4. A Numerical Classification for all Binary Functions

2.4.1. Limitations of the $n + 1$ Chow Parameters

Looking at the wider case of all possible binary functions, and not merely the linearly separable class, the $n + 1$ Chow parameters do not provide sufficient information content to classify all functions. This may be readily demonstrated.

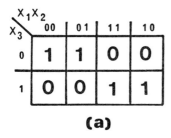

Fig. 2.5. Two simple non-linearly separable functions with identical Chow-parameter values:

(a) $f(x) = [\bar{x}_1\bar{x}_3 + x_1x_3]$
(b) $f(x) = [\bar{x}_1(x_2\bar{x}_3 + \bar{x}_2x_3) + x_1(x_2x_3 + \bar{x}_2\bar{x}_3)]$.

For example, consider the two functions illustrated in Fig. 2.5, neither of which is linearly separable and hence realizable with one threshold-logic gate. Evaluation of the b_0, b_1, b_2, and b_3 parameters for these two functions will show that the values for each are 0, 0, 0, 0. There is in fact a balanced spread of true/false minterms in these two deliberately chosen examples, which results in all the Chow parameters being zero valued. So clearly, as these two simple examples have identical b_i parameters, these parameters are inadequate to define the functions. Many other similar examples may be found in which the Chow-parameter values for two dissimilar non-linearly separable functions are identical, although not necessarily zero valued. Therefore to cover all classes of binary functions we must either augment these $n + 1$ Chow parameters with some additional information or derive some more powerful classification procedure.

The method which has been proposed, and on which research is still being pursued, is to employ a larger set of characteristic numbers, 2^n in number for any function $f(x)$ of n input variables, compared with the $n + 1$ which were sufficient for the linearly separable case. By appropriate choice of scaling factors, the first $n + 1$ of these 2^n numbers can be made numerically equal to the Chow parameters, with the remaining $2^n - (n + 1)$ as additional information covering the non-linearly separable cases. There is, however, a degree of redundancy in the full 2^n set of numbers, as they are not all independent, but to date research to define a best minimum-necessary set from amongst full 2^n set has not reached general agreement.

The 2^n numbers which may be used to specify any binary function $f(x)$ are usually termed the *"spectrum"* of $f(x)$, with the methods of deriving and applying these numbers being referred to as "spectral techniques." We shall be closely examining the possiblities of using spectral techniques for network design in a subsequent chapter of this book, but first we should now look at some of the mathematical background of this area.

2.4.2. Orthogonal Transforms

What we are now beginning to consider is the transformation of our conventional binary data, normally expressed in Boolean algebraic form or by truth tables involving just the two binary values, conventionally 0 and 1, into an entirely different mathematical domain, a domain in which the numbers which define our data will not be confined to two binary values only. Table 2.8 illustrates our considerations.

The Classification of Logic Functions

Table 2.8. The spectral transform requirements for binary data.

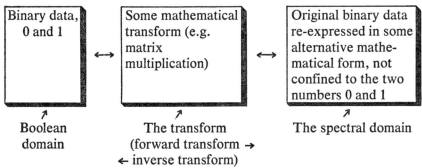

A great deal of research in this transform area has been involved with pattern recognition and image coding, etc., whereby input information is digitized and transformed into a set of coefficients of orthogonal functions, this set being used in some appropriate way for pattern-recognition, pattern-transmission, or pattern-enhancement purposes, or whatever particular requirement is being sought. We are not, however, concerned in this particular field, except that the mathematical transforms used may be related to those we shall consider for our function classification or function synthesis procedures.

The transforms which are most advantageous in these areas are special *orthogonal matrix transforms*. We shall define and illustrate the terminology "orthogonal" in a moment. The particular advantage of these transforms is that the matrix multiplications necessary may be performed with $N \log_2 N$ or fewer arithmetic operations for an $N \times N$ matrix, compared with N^2 generally necessary for this size matrix. Further, if the matrix can be constructed from $+1$'s and -1's then maximum computational simplicity may be achieved. Finally, if the matrix and its inverse can be made identical, except possibly for some constant normalizing factor, then the same computational procedure may be used in each direction of transform of the system illustrated in Table 2.8.

Let us define our first term "orthogonal." Two functions, say i, θ and j, θ, are said to be orthogonal over some complete range or interval, say 0 to θ, if they possess the property

$$\int_0^\theta (i,\theta)(j,\theta)\, d\theta = 0 \text{ for } i \neq j$$
$$= K \text{ for } i = j \tag{2.9}$$

where K is some constant. This may also be written as

$$\int_0^\theta (i,\theta)(j,\theta)\, d\theta = K\delta_{i,j} \tag{2.9a}$$

where $\delta_{i,j}$ is the Kronecker delta function

$$\triangleq 0 \text{ when } i \neq j,$$
$$= 1 \text{ when } i = j.$$

If $K = 1$, then the two functions are said to be "orthonormal." Sine and cosine functions taken over the fundamental interval 0° to 360° are simple examples of orthogonal functions.

If discrete-valued functions or vectors are considered instead of continuously varying functions, then the above integration may be replaced by an appropriate summation. For example consider, say, three vectors $\vec{A}_1, \vec{A}_2, \vec{A}_3$, each of length L, whose co-ordinates along three mutually-perpendicular axes are

$$\vec{A}_1 = a_{1(1)} + a_{1(2)} + a_{1(3)}$$
$$\vec{A}_2 = a_{2(1)} + a_{2(2)} + a_{2(3)}$$
and
$$\vec{A}_3 = a_{3(1)} + a_{3(2)} + a_{2(3)}$$

then the condition for orthogonality is

$$\sum_{k=1}^{3} a_{i(k)} a_{j(k)} = L^2 \delta_{i,j}. \tag{2.10}$$

Vectors which are mutually at right-angles to each other in n-space are therefore orthogonal vectors. Notice that orthogonality requires zero correlation between each pair of functions, that is, no information about one of them can be constructed from one other.

Continuing our vector considerations a little further, if the vector co-ordinates of the above are written in the form of a 3 × 3 matrix, we have, dropping the parentheses,

$$[M] = \begin{bmatrix} a_{11} & a_{12} & a_{13} \\ a_{21} & a_{22} & a_{23} \\ a_{31} & a_{32} & a_{33} \end{bmatrix}.$$

If orthogonality holds, then the matrix multiplied by its transpose is

$$M \cdot M^t = \begin{bmatrix} a_{11} & a_{12} & a_{13} \\ a_{21} & a_{22} & a_{23} \\ a_{31} & a_{32} & a_{33} \end{bmatrix} \begin{bmatrix} a_{11} & a_{21} & a_{31} \\ a_{12} & a_{22} & a_{32} \\ a_{13} & a_{23} & a_{33} \end{bmatrix}$$

$$= \begin{bmatrix} |A_1|^2 & 0 & 0 \\ 0 & |A_2|^2 & 0 \\ 0 & 0 & |A_3|^2 \end{bmatrix} \qquad (2.11)$$

which, if all vectors have equal length L, becomes

$$\mathbf{M} \cdot \mathbf{M}^t = L^2 \mathbf{I} \qquad (2.12)$$

where $\mathbf{I}$ is the unit matrix. From (2.12) we have

$$\mathbf{M}^t = \frac{L^2 \mathbf{I}}{\mathbf{M}} = L^2 \times \mathbf{M}^{-1}$$

where $\mathbf{M}^{-1}$ is the inverse matrix. Therefore a property of orthogonal matrices is that their transpose is equal to their inverse multiplied by some normalizing factor if necessary. If the normalizing factor is unity ($L = 1.0$), then $\mathbf{M}^t = \mathbf{M}^{-1}$, and the orthogonal matrix $\mathbf{M}$ has now the additional property of being orthonormal. Finally, if we can make our orthogonal matrix symmetric, that is, going back to (2.11), if $a_{12} = a_{21}$, $a_{13} = a_{31}$, etc., then the matrix $\mathbf{M}$ equals its transpose $\mathbf{M}^t$ equals its inverse $\mathbf{M}^{-1}$, and the transform computation in each direction of Table 2.6 becomes an identical computational procedure.

Now the class of orthogonal matrices which satisfy the reduced number of arithmetic operations criteria are the $N \times N$ matrices which may be factored into a Kronecker product of a reduced set of matrices[17], where the Kronecker product of two matrices $\mathbf{M}$ is defined as

$$[\mathbf{M}] \times_k [\mathbf{M}] \triangleq \begin{bmatrix} \mathbf{M} & \mathbf{M} \\ \mathbf{M} & -\mathbf{M} \end{bmatrix}$$

and where $\times_k$ represents the Kronecker product operation.

Now the smallest ("core") Kronecker matrix, of order 2×2, will take the form

$$\begin{bmatrix} A & B \\ C & D \end{bmatrix}.$$

The Kronecker orthogonality constraints may now be shown to require

$$\left. \begin{array}{l} A^2 + B^2 = 1 \\ C^2 + D^2 = 1 \\ AC + BD = 0 \end{array} \right\} \qquad (2.13)$$

which, if an additional constraint is applied to achieve symmetry as well as orthogonality, requires

$$\left. \begin{array}{l} B = C \\ A = -D. \end{array} \right\} \qquad (2.13a)$$

One solution to the requirements is

$$\begin{bmatrix} \cos\theta & \sin\theta \\ \sin\theta & -\cos\theta \end{bmatrix}$$

Taking the particular case of $\theta = 45°$ yields

$$\frac{1}{\sqrt{2}} \begin{bmatrix} 1 & 1 \\ 1 & -1 \end{bmatrix}. \tag{2.14}$$

Omitting the $1/\sqrt{2}$ normalizing factor, this matrix now becomes the lowest-order Hadamard matrix

$$\mathbf{H}_2 = \begin{bmatrix} 1 & 1 \\ 1 & -1 \end{bmatrix} \tag{2.15}$$

The Hadamard matrix of order $N = 4$ is

$$\mathbf{H}_4 = \begin{bmatrix} \mathbf{H}_2 & \mathbf{H}_2 \\ \mathbf{H}_2 & -\mathbf{H}_2 \end{bmatrix}$$

$$= \begin{bmatrix} 1 & 1 & 1 & 1 \\ 1 & -1 & 1 & -1 \\ 1 & 1 & -1 & -1 \\ 1 & -1 & -1 & 1 \end{bmatrix} \begin{matrix} \text{Sequency} \\ 0 \\ 3 \\ 1 \\ 2 \end{matrix} \tag{2.16}$$

Similarly for $N = 8$ we have

$$\mathbf{H}_8 = \begin{bmatrix} 1 & 1 & 1 & 1 & 1 & 1 & 1 & 1 \\ 1 & -1 & 1 & -1 & 1 & -1 & 1 & -1 \\ 1 & 1 & -1 & -1 & 1 & 1 & -1 & -1 \\ 1 & -1 & -1 & 1 & 1 & -1 & -1 & 1 \\ 1 & 1 & 1 & 1 & -1 & -1 & -1 & -1 \\ 1 & -1 & 1 & -1 & -1 & 1 & -1 & 1 \\ 1 & 1 & -1 & -1 & -1 & -1 & 1 & 1 \\ 1 & -1 & -1 & 1 & -1 & 1 & 1 & -1 \end{bmatrix} \begin{matrix} \text{Sequency} \\ 0 \\ 7 \\ 3 \\ 4 \\ 1 \\ 6 \\ 2 \\ 5 \end{matrix} \tag{2.17}$$

Looking at these final +1, −1 matrices, the number of changes from +1 to −1 and *vice versa* along each row of the matrix will be seen to be dissimilar. Thus a "frequency" interpretation may be given to each row, depending upon the total number of sign changes. The term *sequency* has been coined by Harmuth for this property[18], the sequency values being written against the rows in the above $N = 4$ and $N = 8$ examples.

The Classification of Logic Functions

Reverting back for a moment to Table 2.8, if the transforms used in any transformation process are to be of practical use, all the information content contained in one domain must be transferred into the second domain. We must not lose any information in applying the forward or the inverse transform. The information content may be contained in a completely dissimilar set of numbers or coefficients and distributed differently in the two domains, but it must be possible to reconstruct either set of data from the other set. Now *complete orthogonal matrices* possess this essential property for our purpose. The +1, −1 Hadamard matrix in particular is a complete orthogonal matrix, and a $2^n \times 2^n$ Hadamard matrix is therefore suitable as the transform operator in Table 2.8 for any n-variable binary function. Let us therefore continue our discussions around this Hadamard matrix, which will mean that we shall not be considering at all any other transforms which may possess the property of orthogonality, such as the Harr transform and others. Details of other transforms and further elaboration of the principles introduced here may be found in the published literature[18-24]. A particularly useful bibliography will be found in Barrett *et al*.[24].

2.4.3. Rademacher–Walsh Functions

The previous purely mathematical development leading to the +1, −1 Hadamard matrix has so far been given no specific meaning or significance—it has merely been a mathematically very satisfying orthogonal matrix. However, as it is composed of two-valued entries, then intuitively it may have some application or relevance to two-valued digital work.

In fact the orthogonal Hadamard matrix may be directly linked with Walsh functions, these being a set of orthogonal functions first proposed by the American mathematician J. L. Walsh in 1923[25]. However, as we shall shortly see, the row ordering of the original Walsh set is dissimilar to that of the Hadamard matrix row ordering, but, although symmetry or other features may be lost in any re-ordering of the rows (or columns) of an orthogonal matrix, orthogonality is not lost and hence full information content is retained when using any re-ordered orthogonal set as a transform.

The Walsh functions form a complete or "closed" set of orthogonal functions. As proposed by Walsh they are the N-length, N in number (where N is any integer power of 2) two-valued functions given by

$$\text{Wal}(j,k) = \prod_{r=0}^{r=n-1}\left\{(-1)^{\{k_{n-r}+k_{n-r-1}\}j_r}\right\} \tag{2.18}$$

where $j = 0, 1, 2, \ldots, N-1$, $k = 0, 1, 2, \ldots, N-1$, $N = 2^n$,
and where j_r, $r = 0$, = least significant binary digit of j when j is expressed in a natural binary number
$\quad j_r, r = 1,$ = next significant binary digit of j

$$\vdots$$

$\quad j_r, r = n - 1,$ = most significant binary digit of j
$\quad j_r, r = n, \triangleq 0$
and similarly for
$\quad k_r, r = 0, 1, \ldots, n.$

For $N = 8$ ($n = 3$) this gives the following expression:

$$\begin{aligned}\text{Wal}(j,k), j &= j_2 j_1 j_0 = \text{000 through to 111} \\ k &= k_2 k_1 k_0 = \text{000 through to 111} \\ &= (-1)^{\{k_2\}j_0} \cdot (-1)^{\{k_2+k_1\}j_1} \cdot (-1)^{\{k_1+k_0\}j_2} \\ &\equiv (-1)^{\{j_2\}k_0} \cdot (-1)^{\{j_2+j_1\}k_1} \cdot (-1)^{\{j_1+j_0\}k_2}\end{aligned}$$

which gives the $+1$, -1 matrix as follows:

$$\text{Wal}(j,k) = \begin{array}{c} j=0 \\ \\ \\ \\ \\ \\ \\ \\ \\ \end{array}\begin{bmatrix} 1 & 1 & 1 & 1 & 1 & 1 & 1 & 1 \\ 1 & 1 & 1 & 1 & -1 & -1 & -1 & -1 \\ 1 & 1 & -1 & -1 & -1 & -1 & 1 & 1 \\ 1 & 1 & -1 & -1 & 1 & 1 & -1 & -1 \\ 1 & -1 & -1 & 1 & 1 & -1 & -1 & 1 \\ 1 & -1 & -1 & 1 & -1 & 1 & 1 & -1 \\ 1 & -1 & 1 & -1 & -1 & 1 & -1 & 1 \\ 1 & -1 & 1 & -1 & 1 & -1 & 1 & -1 \end{bmatrix} \begin{array}{c} k= \\ 0 \\ 1 \\ 2 \\ 3 \\ 4 \\ 5 \\ 6 \\ 7 \end{array} \begin{array}{c} \text{Sequency} \\ 0 \\ 1 \\ 2 \\ 3 \\ 4 \\ 5 \\ 6 \\ 7 \end{array} \quad (2.19)$$

The square-wave functions that this matrix represents are illustrated in Fig. 2.6.

It will be noticed that these Walsh functions take the same values as the rows in the Hadamard matrix of the same size, but in a different order. Here they are sequency ordered, and therefore do not obey the recursive Kronecker product construction which was used to build up the matrices (2.15) to (2.17).

However, the Walsh functions may be redefined in many different ways which results in several dissimilar orderings of the rows[24,26-30].

The Classification of Logic Functions

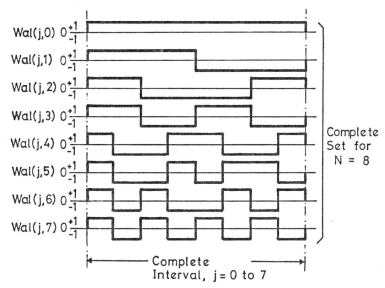

Fig. 2.6. The Walsh functions for $N = 8$ ($n = 3$) in sequency order.

From the point of view of our subsequent usage of them, it is most convenient and instructive to redefine them in terms of a reduced set of functions, the Rademacher functions[31]. The Rademacher functions are orthogonal, but do not constitute a complete set.

The Rademacher functions themselves may be defined in several dissimilar but functionally identical ways. All definitions define a series of square waves with two discrete values, normally ± 1, over a complete interval which is usually taken as from either $-\frac{1}{2}$ to $+\frac{1}{2}$, or from 0 to 1.

One definition for the Rademacher functions $\vec{R_n}(\theta)$, $n = 0, 1, 2, \ldots$, over the interval from 0 to 1 is

$$\vec{R_n}(\theta) \triangleq \text{sign}\left\{\sin(2^n \pi \theta)\right\} 1.0 \tag{2.20}$$

where $0 \leq \theta \leq 1$.

An alternative definition over the same interval is

$$\vec{R_n}(\theta) \triangleq +1 \text{ if } \frac{m}{2^n} \leq \theta \leq \frac{m+1}{2^n}, \ m \text{ an even integer}$$

$$= -1 \text{ if } \frac{m}{2^n} \leq \theta \leq \frac{m+1}{2^n}, \ m \text{ an odd integer} \tag{2.20a}$$

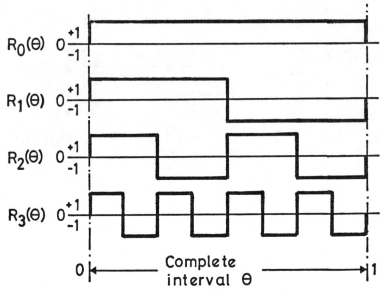

Fig. 2.7. The orthogonal set of basic Rademacher functions for $N = 8$ ($n = 3$).

Both these definitions yield the functions illustrated in Fig. 2.7. Alternative definitions based on the interval $-\frac{1}{2} \leq \theta \leq +\frac{1}{2}$ yield identical functions, but displaced so as to be central about $\theta = 0$. It will be apparent from Fig. 2.7 that the Rademacher functions $\vec{R}_1$, $\vec{R}_2$, etc. correspond to the input variables x_1, x_2, etc. of a binary system, where the full 2^n minterms of the binary system are contained in the interval $\theta = 0$ to 1. Indeed if we use a $+1, -1$ binary notation instead of the more usual 0, 1 notation, then the Rademacher functions may be redefined as[5]

$$\vec{R}_0 \triangleq +1, = x_0$$
$$\vec{R}_i, i = 1\ldots n, \triangleq z_i \tag{2.20b}$$

where

$$z_i = -(2x_i - 1), \quad x_i \in \{0, 1\}.$$

The complete set of Walsh functions may now be defined in terms of this reduced set of Rademacher functions. The complete set is given

---

[5] The superscript → above the functions R_0, R_1, etc. indicates that they are vector quantities. However, we shall follow general practice and omit this superscript from here on.

The Classification of Logic Functions

by all different possible inner vector products of the Rademacher functions, taken two at a time, three at a time, up to n at a time. For $n = 3$ this yields the following (the juxtaposition R_1R_2 indicates the appropriate operation between R_1 and R_2, and so on).

$$
\begin{array}{l}
R_0 \\
R_1 \\
R_2 \\
R_3 \\
R_1R_2 \\
R_1R_3 \\
R_2R_3 \\
R_1R_2R_3
\end{array}
\begin{bmatrix}
1 & 1 & 1 & 1 & 1 & 1 & 1 & 1 \\
1 & 1 & 1 & 1 & -1 & -1 & -1 & -1 \\
1 & 1 & -1 & -1 & 1 & 1 & -1 & -1 \\
1 & -1 & 1 & -1 & 1 & -1 & 1 & -1 \\
1 & 1 & -1 & -1 & -1 & -1 & 1 & 1 \\
1 & -1 & 1 & -1 & -1 & 1 & -1 & 1 \\
1 & -1 & -1 & 1 & 1 & -1 & -1 & 1 \\
1 & -1 & -1 & 1 & -1 & 1 & 1 & -1
\end{bmatrix}
\begin{array}{l}
\text{Sequency} \\
0 \\
1 \\
3 \\
7 \\
2 \\
6 \\
4 \\
5
\end{array}
\quad (2.21)
$$

Notice that, owing to the orthogonality of the basic set of Rademacher functions, we have

$R_0 \times R_i, \quad i \neq 0, \ = R_i$

$R_i \times R_i, \quad i = 0 \ldots n, \ = R_0$

$R_0 \times R_i \times R_j, \quad i, j \neq 0, \ = R_i \times R_j$

Therefore n Rademacher functions R_0 to R_n will always produce exactly 2^n dissimilar final functions. However, unlike the previous sequency-ordered matrix of (2.19), the resultant $2^n \times 2^n$ non-sequency-ordered matrix does not retain the property that the matrix is equal to its transpose, and therefore equal to its inverse with an appropriate scaling factor.

Thus the Walsh functions generated from the Rademacher set form a complete orthogonal matrix, but in a different sequency order from the tabulation of (2.19). Because of the ready generation of this complete set from the fundamental Rademacher set, the row entries of this ordered set are normally referred to in engineering circles as the Rademacher–Walsh functions. The first $n + 1$ (Rademacher) functions may be termed the "primary" input set, with the remaining $2^n - (n + 1)$ functions as the "secondary" input set[6].

Let us now apply this complete function matrix to our problem of transforming from the binary domain to the spectral domain. Suppose, for example, we have the simple three-variable binary function

$$f(x) = [\bar{x}_1 x_2 + x_1 \bar{x}_2 + x_2 x_3].$$

Its output truth table is shown in Table 2.9, together with the conversion of the output from the $f(x)$ 0, 1 values to the $f(z)$ +1, −1 values. This is the same numerical conversion as previously noted between the x_i inputs and the Rademacher R_i inputs. The column vector $f(z)$ will now represent our binary function data.

Table 2.9. The input–output truth-table for the binary function $f(x) = [\bar{x}_1 x_2 + x_1 \bar{x}_2 + x_2 x_3]$

Minterm p	Inputs			Output	
	x_1	x_2	x_3	$f(x)$ $f(x) \in \{0,1\}$	$f(z)$ $f(z) \in \{+1,-1\}$
0	0	0	0	0	1
1	0	0	1	0	1
2	0	1	0	1	−1
3	0	1	1	1	−1
4	1	0	0	1	−1
5	1	0	1	1	−1
6	1	1	0	0	1
7	1	1	1	1	−1

The transform procedure is now to take the 2^n column vector $f(z)$, = **F**], and multiply it by the complete orthogonal Rademacher–Walsh matrix, = [**T**]. This matrix multiplication will yield the column vector **S**], the numerical values in which will uniquely constitute the spectrum of the original function $f(x)$. Each value in **S**] represents a spectral coefficient of $f(x)$.

So, applied to our example function, we have

$$[\mathbf{T}] \times \mathbf{F}] = \mathbf{S}]$$

$$\begin{bmatrix} 1 & 1 & 1 & 1 & 1 & 1 & 1 & 1 \\ 1 & 1 & 1 & 1 & -1 & -1 & -1 & -1 \\ 1 & 1 & -1 & -1 & 1 & 1 & -1 & -1 \\ 1 & -1 & 1 & -1 & 1 & -1 & 1 & -1 \\ 1 & 1 & -1 & -1 & -1 & -1 & 1 & 1 \\ 1 & -1 & 1 & -1 & -1 & 1 & -1 & 1 \\ 1 & -1 & -1 & 1 & 1 & -1 & -1 & 1 \\ 1 & -1 & -1 & 1 & -1 & 1 & 1 & -1 \end{bmatrix} \begin{bmatrix} 1 \\ 1 \\ -1 \\ -1 \\ -1 \\ -1 \\ 1 \\ -1 \end{bmatrix} = \begin{bmatrix} -2 \\ +2 \\ +2 \\ +2 \\ +6 \\ -2 \\ -2 \\ +2 \end{bmatrix}$$

The spectrum of the given function $f(x)$ is therefore the spectral coefficients

$$-2, 2, 2, 2, 6, -2, -2, 2$$

written in descending order of the vector $\mathbf{S}]$. We would conventionally label these coefficients r_0, r_1, etc., thus giving for our example the detailed resultant coefficient values

r_0	r_1	r_2	r_3	r_{12}	r_{13}	r_{23}	r_{123}
-2	2	2	2	6	-2	-2	2

This simple example represents an optimum means whereby binary data may be transformed into the spectral domain. We shall not pursue here any further the mechanics of the matrix multiplications and summations; it suffices to note that because of the build-up of these matrices extremely fast computer transform algorithms can be proposed. Details of these will be found published.[24,30,32,33]

Before we finally consider our present goal of function classification, however, let us show a little more clearly in logic terms what these spectral coefficient values mean, and also provide a link back to our previous Chow parameters. If we convert all the $+1$, -1 values of the Rademacher–Walsh matrix back into our more familiar binary numbers of 0 and 1, by direct substitution of 0 for $+1$ and 1 for -1, we obtain the following. The example below is of course for the $n = 3$ matrix detailed in (2.21).

Rademacher–Walsh functions	$\{+1, -1\}$ notation							
R_0	1	1	1	1	1	1	1	1
R_1	1	1	1	1	-1	-1	-1	-1
R_2	1	1	-1	-1	1	1	-1	-1
R_3	1	-1	1	-1	1	-1	1	-1
R_{12}	1	1	-1	-1	-1	-1	1	1
R_{13}	1	-1	1	-1	-1	1	-1	1
R_{23}	1	-1	-1	1	1	-1	-1	1
R_{123}	1	-1	-1	1	-1	1	1	-1

118   The Logical Processing of Digital Signals

Converted Rademacher–Walsh functions	$\{0, 1\}$ notation							
R'_0	0	0	0	0	0	0	0	0
R'_1	0	0	0	0	1	1	1	1
R'_2	0	0	1	1	0	0	1	1
R'_3	0	1	0	1	0	1	0	1
R'_{12}	0	0	1	1	1	1	0	0
R'_{13}	0	1	0	1	1	0	1	0
R'_{23}	0	1	1	0	0	1	1	0
R'_{123}	0	1	1	0	1	0	0	1

Notice that this 0, 1 matrix is *no longer an orthogonal matrix* and therefore cannot be used as it stands as a mathematical transform in the same way as the orthogonal +1, −1 matrix.

However, if we examine the definitions of the R_i functions in the more familiar context of 0's and 1's, we can see that they are related as follows.

$\{+1, -1\}$ function	Equivalent $\{0, 1\}$ logical relationship
R_1	x_1
R_2	x_2
R_3	x_3
R_{12}	$x_1 \oplus x_2$
R_{13}	$x_1 \oplus x_3$
R_{23}	$x_2 \oplus x_3$
R_{123}	$x_1 \oplus x_2 \oplus x_3$

Hence our 2^n Rademacher–Walsh functions correspond to the following:

$R_0 \triangleq 1.0$

$R_i = i = 1 \cdots n$: the respective binary inputs x_1 to x_n respectively

$R_i, i = 12, 13, \cdots, 12..n$: all possible different exclusive-OR combinations of the binary inputs, taken two at a time, three at a time, up to n at a time.

Now if we look back at the mathematics of our +1, −1 transform and consider what happens in generating the final spectral coefficient values, we shall see that the matrix multiplication of the +1's and −1's is effectively computing the *number of agreements minus the number*

The Classification of Logic Functions

of disagreements of the values in the column vector F], which represents the given function $f(x)$, with the rows of the transform [T]. Hence we may define the resulting values of the spectral coefficients in S] as follows:

$$r_0 = \{(\text{number of } +1\text{'s in F]}) - (\text{number of } -1\text{'s in F]})\}$$
$$= \{(\text{number of false minterms in } f(x)) - (\text{number of true minterms in } f(x))\}$$

$r_i, i \neq 0 = \{$(number of agreements between F] and the appropriate row of [T]) $-$ (number of disagreements between F] and the appropriate row of [T])$\}$.

Breaking down the latter into the primary r_1 to r_n coefficients and the remaining higher-order coefficients, we have

$r_i, i = 1 \cdots n = \{$(number of agreements between $f(x)$ and the input variable x_i) $-$ (number of disagreements between $f(x)$ and x_i)$\}$

and

$r_i, i = 12, 13, \cdots, 12..n = \{$(number of agreements between $f(x)$ and the appropriate exclusive-OR of the x_i input variables) $-$ (number of disagreements between $f(x)$ and the appropriate exclusive-OR of the x_i inputs)$\}$

Notice now the relationship which exists between these spectral coefficients and the previous Chow classification b_i parameters for linearly separable functions:

$$r_0 = -b_0$$
$$r_i, i = 1 \cdots n, = b_i, i = 1 \ldots n$$
$r_i, i = 12, 13, \cdots, 12..n$: not present in the Chow-parameter scheme.

The difference in sign of the r_0 coefficient compared with b_0 arises from the sign of R_0 in the Rademacher–Walsh matrix of (2.21). It would be possible to redefine $R_0 \triangleq -1$ instead of $+1$, or indeed change the signs of any rows or columns of this matrix, without losing orthogonality, but here we shall continue to use the normal sign convention as previously detailed. Hence the sign of our coefficient r_0 in the Rademacher–Walsh spectral domain will remain opposite to that of our Chow parameter b_0 coefficient. Care must, however, be taken when

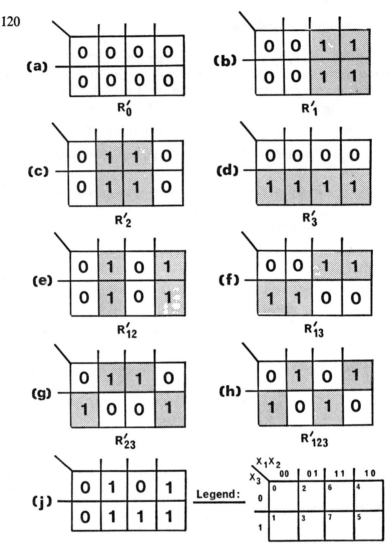

Fig. 2.8 Maps of the seven Rademacher–Walsh functions of $n = 3$, together with the example function $f(x) = [\bar{x}_1 x_2 + x_1 \bar{x}_2 + x_2 x_3]$. The {agreement − disagreement} counts may be made by effectively overlaying the function map on each of the Rademacher–Walsh maps in turn: (a)–(h) the R'_0 to R'_{123} maps; (j) the example function $f(x)$.

reading other published literature as other sign conventions and resulting coefficient signs may be encountered.

Just to illustrate the equivalence of the {agreement − disagreement} definitions for the r_i's with the mathematical transform, consider again the simple function $f(x) = [\bar{x}_1 x_2 + x_1 \bar{x}_2 + x_2 x_3]$ which had the

The Classification of Logic Functions

spectral coefficient values −2, 2, 2, 2, 6, −2, −2, 2. The detailed {agreement − disagreement} tally is illustrated in Table 2.10, which may be compared with similar tabulation in Table 2.6 for the b_i parameter computation.

Table 2.10. The computation of the spectral coefficient values for
$$f(x) = [\bar{x}_1 x_2 + x_1 \bar{x}_2 + x_2 x_3]$$
using {agreement − disagreement} counting.

(a)

Minterm p	Primary input set				Secondary input set				Function output $f(x)$
	R'_0	R'_1 $=x_1$	R'_2 $=x_2$	R'_3 $=x_3$	R'_{12} $=x_1\oplus x_2$	R'_{13} $=x_1\oplus x_3$	R'_{23} $=x_2\oplus x_3$	R'_{123} $=x_1\oplus x_2\oplus x_3$	
0	0	0	0	0	0	0	0	0	0
1	0	0	0	1	0	1	1	1	0
2	0	0	1	0	1	0	1	1	1
3	0	0	1	1	1	1	0	0	1
4	0	1	0	0	1	1	0	1	1
5	0	1	0	1	1	0	1	0	1
6	0	1	1	0	0	1	1	0	0
7	0	1	1	1	0	0	0	1	1

(b)

Minterm p	Agreements/disagreements between $f(x)$ and R'_i							
	with R'_0	with R'_1	with R'_2	with R'_3	with R'_{12}	with R'_{13}	with R'_{23}	with R'_{123}
0	a	a	a	a	a	a	a	a
1	a	a	a	d	a	d	d	d
2	d	d	a	d	a	d	a	a
3	d	d	a	a	a	a	d	d
4	d	a	d	d	a	a	d	a
5	d	a	d	a	a	d	a	d
6	a	d	d	a	a	d	d	a
7	d	a	a	a	d	d	d	a
$\sum (a-d)$	−2	+2	+2	+2	+6	−2	−2	+2

(a) The function truth table in terms of the complete set of Rademacher–Walsh functions R'_i (0, 1 notation).
(b) The {agreement − disagreement} count.

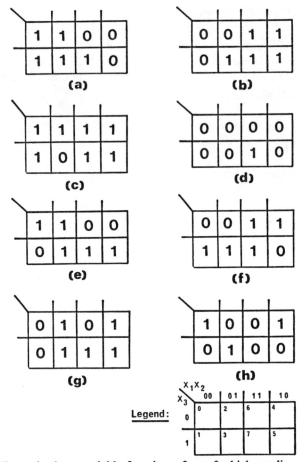

Fig. 2.9. Example three-variable functions, four of which are linearly separable and four of which are non-linearly separable.

(a) $f(x) = [\bar{x}_1 + x_2 x_3]$
(b) $f(x) = [x_1 + x_2 x_3]$ ⎫
(c) $f(x) = [x_1 + \bar{x}_2 + \bar{x}_3]$ ⎬ all linearly separable
(d) $f(x) = [x_1 x_2 x_3]$ ⎭

(e) $f(x) = [\bar{x}_1 \bar{x}_3 + \bar{x}_1 x_2 + x_1 x_3]$ ⎫
(f) $f(x) = [x_1 \bar{x}_3 + x_1 x_2 + \bar{x}_1 x_3]$ ⎬ non-linearly separable
(g) $f(x) = [x_1 \bar{x}_2 + \bar{x}_1 x_2 + x_2 x_3]$ ⎭
(h) $f(x) = [\bar{x}_2 \bar{x}_3 + \bar{x}_1 x_2 x_3]$

This {agreement − disagreement} counting procedure may also be conveniently performed for small n using Karnaugh maps; Fig. 2.8 illustrates this same example function.

The spectral coefficients may be found defined in several different

The Classification of Logic Functions

but basically identical ways, as we saw was possible with the b_i Chow parameters. Details may be found published elsewhere[24,30,34,35], but here we shall continue to use the latter definitions unless otherwise noted.

2.4.4. Function Classification Using Rademacher–Walsh Spectra

Having covered the basic mathematical background of the Rademacher–Walsh spectral coefficients, we may now continue with the main purpose of this section, that of function classification. We shall, however, have extensive involvement with the Rademacher–Walsh coefficients for other purposes in subsequent chapters of this book.

Like the Chow parameters the Rademacher–Walsh spectral coefficients are seen to be a form of correlation between function inputs and output. Unlike the reduced set of Chow parameters, however, they provide a unique set of coefficient values for any given function $f(x)$ owing to the completeness of the underlying transform. Can we, therefore, take some or all of the r_i values for different functions $f(x)$ and rearrange them in some preferred ordering as we did with the Chow b_i parameters, this preferred ordering then representing a classification entry for all related binary functions?

Firstly let us look at some very simple examples and their r_i spectral coefficient values.

Figure 2.9 illustrates four linearly separable functions, which possess the same Chow parameter classification, and four further functions which are not linearly separable. Using the definitions for the Rademacher–Walsh spectral coefficients which give the same values for the r_i's as the Chow parameter b_i values except for the sign r_0 these eight functions possess the following coefficient values.

Function	Chow coefficients				Rademacher–Walsh coefficients							
	b_0	b_1	b_2	b_3	r_0	r_1	r_2	r_3	r_{12}	r_{13}	r_{23}	r_{123}
(1) $f(x) = [\bar{x}_1 + x_2 x_3]$	2	−6	2	2	−2	−6	2	2	−2	−2	−2	2
(2) $f(x) = [x_1 + x_2 x_3]$	2	6	2	2	−2	6	2	2	2	2	−2	−2
(3) $f(x) = [x_1 + \bar{x}_2 + \bar{x}_3]$	6	2	−2	−2	−6	2	−2	−2	−2	−2	2	2
(4) $f(x) = [x_1 x_2 x_3]$	−6	2	2	2	6	2	2	2	−2	−2	−2	2
(5) $f(x) = [\bar{x}_1 \bar{x}_3 + \bar{x}_1 x_2 + x_1 x_3]$	2	−2	2	2	−2	−2	2	2	2	−6	−2	−2
(6) $f(x) = [x_1 \bar{x}_3 + x_1 x_2 + \bar{x}_1 x_3]$	2	2	2	2	−2	2	2	2	−2	6	−2	2
(7) $f(x) = [x_1 \bar{x}_2 + \bar{x}_1 x_2 + x_2 x_3]$	2	2	2	2	−2	2	2	2	6	−2	−2	2
(8) $f(x) = [\bar{x}_2 \bar{x}_3 + \bar{x}_1 x_2 x_3]$	−2	−2	−2	−2	2	−2	−2	−2	2	2	−6	−2

Rearranging the b_i values in descending magnitude order indicates that the first four functions belong to the Chow classification entry

6, 2, 2, 2

whilst the latter four functions belong to a classification entry of

2, 2, 2, 2

Only the former is, however, listed as a valid classification for linearly separable functions. The 2, 2, 2, 2 classification is not a listed entry (see Appendix A), and therefore these latter functions are (i) not in the class of linearly separable functions and (ii) not therefore uniquely defined and classified by these $n + 1$ parameters only. Indeed this last point is deliberately illustrated by the sixth and seventh example functions, both of which have absolutely identical first $n + 1$ spectral coefficient values.

However, the full set of 2^n numbers 6, 2, 2, 2, 2, 2, 2, 2 is somehow characteristic of all these functions, and may represent a classification entry. Indeed this is the basis of our numerical classification procedure for all binary functions, and is not confined to just the linearly separable subclass.

However, may we take the full set of 2^n coefficient values for any given function $f(x)$ and freely rearrange them in, say, decreasing magnitude as we did with the Chow b_i parameter coefficient values? The answer is most definitely "no," as unlike the b_i coefficients these r_i coefficients are not all mutually independent. For example, consider an invariance operation between, say, binary inputs x_1 and x_3, which in the first $n + 1$ coefficient values (the Chow parameters) would result in the interchange of the b_1 and b_3 coefficient values with no alteration to any of the other primary values. In the full 2^n set of coefficient values (the Rademacher–Walsh set) interchange of x_1 and x_3 inputs would again result in the interchange of the primary r_1 and r_3 coefficients, but in addition would interchange all the secondary coefficient values which contain a 1 and a 3 in their designation. So we cannot independently rearrange our primary and our secondary coefficient values freely and independently of each other to finish up with a classificiation entry necessarily in an all-positive, descending-magnitude order.

In Chapter 5 we shall be considering in detail the use of spectral data for logic synthesis. This will include a formal consideration of the invariance operations possible on the r_i spectral coefficients and their meaning in the corresponding Boolean domain. In the following, there-

The Classification of Logic Functions

fore, we shall merely state the invariance operations which are possible, the application of which will enable us to rearrange the spectral coefficient values for all possible functions $f(x)$ into the most compact numerical classification listing.

There are five useful invariance operations which result in reordering or sign changes of the full set of Rademacher–Walsh spectral coefficients. They are as follows.

(1) Interchange of any binary input variable x_i with $x_j, i \neq j \neq 0$; in the spectral domain this corresponds to the *interchange of all pairs of primary and secondary coefficients which contain i and j in their defining subscripts*, that is

$$r_i \leftrightarrow r_j$$

$$r_{ik} \leftrightarrow r_{jk} \quad \text{etc.}$$

Note, r_0, r_k, r_{ij}, etc. remain the same.

(2) Negation of any input variable x_i, $i \neq 0$; this results in the *change of sign of all primary and secondary coefficients which contain i in their subscript*, that is r_i, r_{ij}, r_{ijk}, etc.

(3) Negation of the complete function $f(x)$; this results in the *change of sign of all primary and secondary coefficients*, including r_0.

So far these three operations are the normal NPN invariance operations, as we have already considered in previous contexts. Their application to the reduced set of $n + 1$ Chow-parameter coefficient values was identical. However, we continue further.

(4) Replacement of any input variable x_i by the exclusive-OR signal $[x_i \oplus x_j], i \neq j \neq 0$; this operation results in the interchange of all pairs of primary and secondary coefficient values which contain i in their subscript as follows:

$$r_i \leftrightarrow r_{ij}$$

$$r_{ik} \leftrightarrow r_{ijk}, \quad \text{etc.}$$

with r_0 and all other coefficients which do not contain i in their subscript remaining unchanged. This may otherwise be stated as: *in all coefficients which contain the subscript i delete j if it also appears and append it if it does not.*

(5) Modification of the output of the network to $f^*(x)$, where $f^*(x) = [f(x) \oplus x_i]$; this operation results in the interchange of pairs of primary and secondary coefficient values as follows:

$$r_i \leftrightarrow r_0$$

$$r_{ij} \leftrightarrow r_j$$

$$r_{ijk} \leftrightarrow r_{jk}, \quad \text{etc.}$$

This may otherwise be stated as: *in all 2^n coefficients append i if it does not appear and delete it if it is already present.*

We shall consider the significance and use of these five invariance operations further in Chapter 5. Notice that exclusive-OR operations are very prominent in these operations, as might be expected from the logical significance of the full Rademacher–Walsh matrix. However, with their statement here we may now define the method of generating the canonic function classifications, applicable to any binary function $f(x)$.

Taking any given binary function $f(x)$ and its full 2^n spectral coefficient values, we may now re-order the coefficient values by the following operations.

(a) If the maximum-value coefficient lies in the secondary set, translate it into the primary set r_1 to r_n by an appropriate operation (4), repeated if necessary.

(b) When the maximum-value coefficient lies in the primary set r_1 to r_n, that is any $|r_i|$, $i = 1 \ldots n$, $> |r_0|$, generate the maximum value for r_0 by operation (5).

(c) Generate the maximum value for all primary coefficients r_1 to r_n (leaving r_0 maximally valued) by further operations (4) as necessary.

(d) Permutate the maximum-valued primary coefficients r_1 to r_n into descending order of magnitude by applications of operation (1) as necessary.

(e) Render all primary coefficients r_0 to r_n positive by the application of operation (3) followed by (2) as necessary.

The result of these five operations is to re-order the spectral coefficient values into a positive canonic order which represents the classification entry of the given binary function $f(x)$. Notice, however, two points: firstly, in the final re-ordering, whilst all the primary coefficients r_0 to r_n will be positive, not all the secondary coefficients will necessarily be positive values; secondly, the invariance operations can

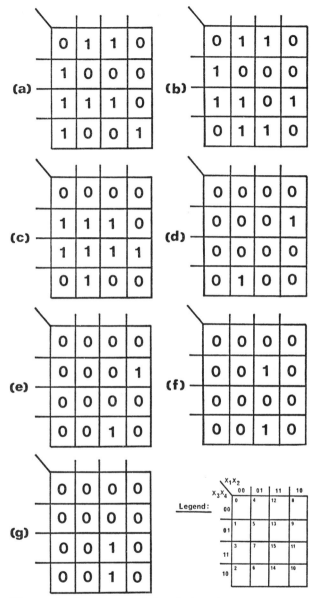

Fig. 2.10. The result of each step of the six invariance operations detailed in Table 2.11: (a) given function $f(x)$; (b) function $f^1(x)$; (c) function $f^2(x)$; (d) function $f^3(x)$; (e) function $f^4(x)$; (f) function $f^5(x)$; (g) function $f^6(x)$.

be undertaken in any order, although the order detailed above is possibly the most sensible.

Let us illustrate this re-ordering procedure with a four-variable non-linearly separable function, say

$$f(x) = [\bar{x}_1\bar{x}_2(x_3 + x_4) + x_2(\bar{x}_3\bar{x}_4 + x_3x_4) + \bar{x}_2x_3\bar{x}_4].$$

This function, shown plotted in Fig. 2.10a, possesses the following Rademacher–Walsh spectrum.

r_0	r_1	r_2	r_3	r_4	r_{12}	r_{13}	r_{14}	r_{23}	r_{24}	r_{34}	r_{123}	r_{124}	r_{134}	r_{234}	r_{1234}
0	−4	0	4	0	−4	0	4	4	0	−4	0	4	0	12	0

Notice that the Chow parameters for this function would be 0, −4, 0, 4, 0 which in positive canonic order is 4, 4, 0, 0, 0. This is not a listed Chow classification (see Appendix A), thus confirming that the example function is not linearly separable.

Performing the necessary invariance operations on this spectrum we have the following developments.

(a) Translate r_{234} coefficient value down to r_{24} by applying invariance operation (4) between inputs x_2 and $[x_2 \oplus x_3]$. This will interchange the coefficient values

$r_2 \leftrightarrow r_{23}$
$r_{12} \leftrightarrow r_{123}$
$r_{24} \leftrightarrow r_{234}$
$r_{124} \leftrightarrow r_{1234}$.

The result is shown in the second line of the complete table of coefficient value changes detailed in Table 2.11.

(b) Repeat operation (4) to move the coefficient value r_{24} down into the primary set by the invariance operation between x_4 and $[x_4 \oplus x_2]$. This will interchange the coefficient values

$r_4 \leftrightarrow r_{24}$
$r_{14} \leftrightarrow r_{124}$
$r_{34} \leftrightarrow r_{234}$
$r_{134} \leftrightarrow r_{1234}$

as detailed on the third line of Table 2.11.

(c) Apply operation (5) to maximize r_0 by the operation of exclusive-OR'ing the function with x_4. This will interchange all the coefficient values

$r_0 \leftrightarrow r_4$
$r_1 \leftrightarrow r_{14}$
$r_2 \leftrightarrow r_{24}$
$r_3 \leftrightarrow r_{34}$
$r_{12} \leftrightarrow r_{124}$

and so on, as detailed on the fourth line of Table 2.11.

The Classification of Logic Functions

(d) Begin to maximize the primary coefficient values r_1 to r_4 by operation (4). An operation between x_2 and $[x_2 \oplus x_4]$ will bring the r_{24} value down to position r_2, as shown on the fifth line of Table 2.11, whilst an operation between x_3 and $[x_3 \oplus x_4]$ will bring the r_{34} value down to position r_3. This is the move shown on the sixth line of Table 2.11. However, this is as far as we can go to maximize the primary set as no more invariance operations are possible to bring a further 4 into the primary set.

(e) Operations to positivize the primary coefficient values, operations (3) and (2), happen to be unnecessary with this particular example. Hence the final line of Table 2.11 represents the Rademacher–Walsh (or the spectral) canonic classification entry for the original function $f(x)$ and all functions we have generated in reaching this final classification entry, together of course with all other possible invariants on this set of numbers.

Table 2.11 The invariance operations starting with a given function

$$f(x) = [\bar{x}_1\bar{x}_2(x_3 + x_4) + x_2(\bar{x}_3\bar{x}_4 + x_3x_4) + \bar{x}_2x_3\bar{x}_4],$$

ending with the Rademacher–Walsh canonic classification entry

Spectral coefficients	r_0	r_1	r_2	r_3	r_4	r_{12}	r_{13}	r_{14}	r_{23}	r_{24}	r_{34}	r_{123}	r_{124}	r_{134}	r_{234}	r_{1234}
Given function $f(x)$	0	−4	0	4	0	−4	0	4	4	0	−4	0	4	0	12	0
Function $f^1(x)$, replacing x_2 by $[x_2 \oplus x_3]$	0	−4	4	4	0	0	0	4	0	12	−4	−4	0	0	0	4
Function $f^2(x)$, replacing x_4 by $[x_4 \oplus x_2]$	0	−4	4	4	12	0	0	0	0	0	0	−4	4	4	−4	0
Function $f^3(x) = [f^2(x) \oplus x_4]$	12	0	0	0	0	4	4	−4	−4	4	4	0	0	0	0	−4
Function $f^4(x)$, replacing x_1 by $[x_1 \oplus x_2]$	12	4	0	0	0	0	0	0	−4	4	4	4	−4	−4	0	0
Function $f^5(x)$, replacing x_2 by $[x_2 \oplus x_4]$	12	4	4	0	0	−4	0	0	0	0	4	0	0	−4	−4	4
Function $f^6(x)$, replacing x_3 by $[x_3 \oplus x_4]$	12	4	4	4	0	−4	−4	0	−4	0	0	4	0	0	0	0

What these particular invariance operations have done at each stage is illustrated in Figs. 2.10*b–g* respectively. It is fascinating that the trivial function of Fig. 2.10*g* is the canonic classification function for all the previous apparently dissimilar functions and the many other functions of this classification entry. Notice also two other points, namely (*a*) if we had performed the invariance operations in some other order from the above, we would finally have arrived at the same final canonic spectra, *but the intermediate functions would not have been those illustrated in Figs. 2.10b–g,* and (*b*) it is possible to combine more than one of these invariance operations at a time, if desired[35]. However, such combined operations are a mere linking of the separate steps we have illustrated here, and do not involve any new basic invariance operation or principle.

2.4.5. The Rademacher–Walsh Classifications

The classification for all binary functions which we have developed above from the Rademacher–Walsh coefficients may also be generated in dissimilar ways, but which finally yield an equivalent set of classification numbers. Earlier work of Slepian, Golomb, Ninomiya, Dertouzos, Berlekamp, and Lechner may be cited.[2,11,12,6,36–38]

However, in considering alternative developments, one must be extremely careful to appreciate fully the definitions of the resulting coefficients, particularly as it affects their signs. For example Dertouzos employs a classification set where the r_0 coefficient (b_0 in his terminology) is signwise identical to the Chow b_0 parameter, compared with the opposite sign which directly results from the Rademacher–Walsh orthogonal transform. Further, his invariance operations involve the exclusive-NOR operator rather than the exclusive-OR, and as a result his even-number coefficients $r_{12}, r_{13}, \ldots, r_{1234}$, etc. (his designation $b_{12}, b_{13}, \ldots, b_{1234}$, etc.) are also opposite in sign from our examples, but the odd-number coefficients $r_{123}, r_{124}, \ldots$, etc. (his designation $b_{123}, b_{124}, \ldots$, etc.) are the same sign.

A further problem with sign convention is of course that a same-number classification in two dissimilar conventions relates (in general) to two dissimilar functions. Hence it is very time consuming to draw a parallel with the functions which result in applying the relevant invariance operations in one convention with those that result in an alternative convention. With completely dissimilar sign conventions the corresponding invariance operations involving the movement of coeffi-

The Classification of Logic Functions

cient values from the primary set to the secondary set clearly involves different operations in the Boolean domain.

However, how compact is the spectral coefficient classification method? The answer is: extremely compact. If all the positive canonic NPN functions are reclassified under the additional spectral invariance operations (4) and (5) of §2.4.4, then for $n \leq 4$ we have a set of only eight entries for all binary 65,536 different functions, as detailed in Table 2.12.

Table 2.12. The canonic spectral classification of all binary functions of $n \leq 4$ under the full Rademacher–Walsh invariance operations

Canonic function	Spectral coefficient														
	r_0	r_1	r_2	r_3	r_4	r_{12}	r_{13}	r_{14}	r_{23}	r_{24}	r_{34}	r_{123}	r_{124}	r_{134}	r_{234}	r_{1234}
1	16	0	0	0	0	0	0	0	0	0	0	0	0	0	0	0
2	14	2	2	2	2	−2	−2	−2	−2	−2	−2	2	2	2	2	−2
3	12	4	4	4	0	−4	−4	0	−4	0	0	4	0	0	0	0
4	10	6	6	2	2	−6	−2	−2	−2	−2	2	2	2	−2	−2	2
5	8	8	8	0	0	−8	0	0	0	0	0	0	0	0	0	0
6	8	8	4	4	4	−4	−4	−4	0	0	0	0	0	0	−4	4
7	6	6	6	6	6	−2	−2	−2	−2	−2	−2	−2	−2	−2	−2	6
8	4	4	4	4	4	4	4	−4	−4	4	−4	−4	−4	4	4	−4

Note: Entries 1, 3, and 5 are for functions of $n < 4$ variables, whilst entries 2, 4, 6, 7, and 8, are for functions of exactly $n = 4$ variables.

The corresponding table for $n \leq 5$ will be found in Appendix B. This consists of 47 entries to cover the (approximately) 4.3×10^9 possible different Boolean functions of five or fewer variables.

The entries shown for $n \leq 4$ in Table 2.12 are fully detailed, with all 2^4 coefficient values tabulated. Notice, however, that the same set of primary coefficient values r_0 to r_4 never appears twice, and hence as far as an identification is concerned it would be sufficient to list the primary set only. Indeed, if merely r_0, r_1, and r_2 are listed, this subgroup forms a unique set for each $n \leq 4$ classification. However, the value of the higher-order coefficients must be available as soon as invariance operations are undertaken to generate other functions of the same classification entry, as all such invariance operations other than the simple NPN operations involve some interchange of primary and secondary coefficient values, as we have already illustrated. Similar remarks apply to the $n \leq 5$ classification tables given in Appendix B.

Another remarkable feature may be seen to emerge from this classification process. This is that although the majority of the binary functions of $n \leq 4$ are not linearly separable functions, seven out of the eight $n \leq 4$ spectral classification entries are threshold-logic functions. This may easily be confirmed by comparing the first $n + 1$ of the full 2^n spectral coefficient values with the Chow-parameter classification (see Appendix A). The first seven of the eight functions listed in Table 2.12 are threshold-logic functions, leaving only the eighth "all-fours" spectral classification as a non-linearly separable function.

What this means in practice is that all $n \leq 4$ functions, except those that fall into the all-fours classification, may be realized by one threshold-logic gate, appropriately prefaced and/or followed by some configuration of inverter and exclusive-OR gates. The exclusive-OR gates are required when the given function is not directly a threshold-logic function. Fig. 2.11 illustrates this principle.

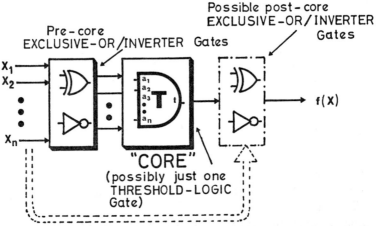

Fig. 2.11. A logic-design concept arising from spectral translation techniques. (Note: the "core" of the resulting network is essentially a simple function.)

What this finally leads on to, therefore, are various possibilities of synthesis using threshold-logic gates and exclusive-OR gates as well as conventional vertex gates, together with the possibilities of standardizing on some minimum set of gates from which all binary functions of a given $\leq n$ can be realized. This will be part of the substance of our subsequent Chapter 5.

There is still a great deal of interesting work to be pursued in this classification area. For example, what is the full significance of the all-fours function in the $n \leq 4$ classification, and what is the correlation

The Classification of Logic Functions

between the spectral classification procedure as examined here using the Rademacher–Walsh transform and classification procedures using other transforms or group theory? These and other interesting points we must, however, leave, and proceed with further known areas of information.

2.5. Enumeration and classification of ternary functions

The enumeration and classification of ternary (and higher-ordered) functions has not yet reached the stage of development of the binary classification area; indeed there are no pressing practical reasons to pursue higher-ordered classifications as there are for binary functions, and hence the subject is currently one of more fundamental and academic interest rather than practical. However, often the study of a more general case gives insight into the full significance of a particular or more simple situation.

When considering systems of higher order than the binary $R = 2$ case, the problems of function classifications are compounded owing to the increasing number of possible functions as R increases. For $R = 3$ we have 3^{3^n} possible functions of n variables, that is 27 single-variable functions, 19 683 two-variable functions, and so on. However, a start has been made on the classification for the ternary case along lines which show a close parallel with binary function classification. This extension of binary classification theories to the ternary case is based upon fundamentals, and therefore can be further extended to cover any multi-valued case ($R > 3$) if the need arises.

In the ternary case, however, the two main areas in which information may be found published are, not surprisingly (i) the canonic classification of linearly separable (threshold-logic) ternary functions, and (ii) work leading to the canonic classification of all ternary functions, whether linearly separable or not, by the application of orthogonal transforms. Notice the parallel with our previous binary developments in §§2.3 and 2.4 of this chapter. Let us, therefore, take a brief look at this further work of classifying three-valued functions, as so far developed.

2.5.1 Ternary Threshold-Logic Function Classification

In Chapter 1 we did not spend time considering higher-ordered threshold-logic functions. However, from the details covered on the more common binary-valued threshold-logic functions the concept of a three-valued threshold-logic function should readily be appreciated.

A ternary threshold-logic function $f(X)$ with input and output signals taking the signal values $-1, 0, +1$ may be defined as

$$f(X) = +1 \text{ if } \sum_{i=1}^{n} w_i X_i \geq t_H$$
$$ = 0 \text{ if } t_L < \sum_{i=1}^{n} w_i X_i < t_H$$
$$ = -1 \text{ if } \sum_{i=1}^{n} w_i X_i \leq t_L$$

where $w_i, i = 1 \ldots n$, are the weighting factors associated with each ternary input $X_i, i = 1 \ldots n$, respectively (cf. the binary threshold-logic equation given in §1.4, where w_i in this case corresponds to a_i in the binary case). This is shown in Fig. 2.12a with the concept of ternary linear separability illustrated in Fig. 2.12b.

Now closely corresponding to the b_i Chow parameters which were defined in §2.3.2 for linearly separable binary functions we may define a similar series of "Chow-type" parameters for the ternary case as follows[39,40]:

$$m_t \triangleq -\left\{ \sum_{f(X) \in \{-1\}} f(X) \right\}$$

that is {the number of minterms where output $f(X) = -1$}

$$p_t \triangleq +\left\{ \sum_{f(X) \in \{+1\}} f(X) \right\}$$

that is {the number of minterms where output $f(X) = +1$}

$$c_i, i = 1 \ldots n, \triangleq +\left\{ \sum_{f(X) \in \{-1,0,+1\}} \left(f(X) \cdot X_i \right) \right\}$$
$$ \equiv +\left\{ \sum_{f(X) \in \{\pm 1\}} \left(f(X) \cdot X_i \right) \right\}$$

that is {(the number of agreements between $f(X)$ and X_i taken over all minterms except where X_i or $f(X)$ or both are 0) − (the number of disagreements)}

It may be shown that such parameters (or their equivalent by some redefinition) uniquely charactize all linearly separable ternary

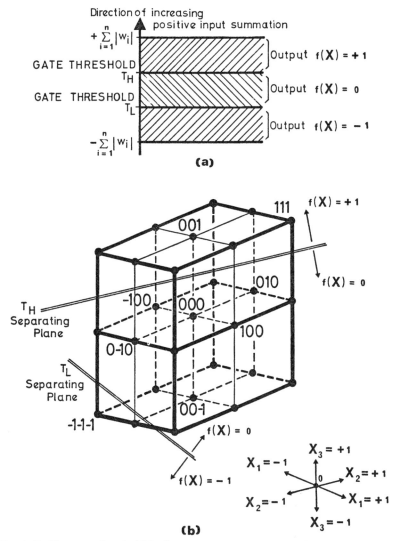

Fig. 2.12. Ternary threshold-logic concepts: (*a*) the gate threshold detection levels; (*b*) illustration of linear separability for a three-variable ternary function.

functions, as do the Chow b_i parameters for the binary case. Further there is a corresponding magnitude-order correspondence between the c_i's and the realizing weights w_i, as there is between the Chow b_i's and the binary realizing weights a_i[39-41].

If now the c_i parameter values for all linearly separable ternary functions are re-ordered in (conventionally) decreasing magnitude order, the positive canonic classification for the ternary functions results. Tables for $n = 3$ which also list the minimum-integer realizing weights w_i and thresholds t_H and t_L have been published[40]; the list for $n = 3$ contains 528 entries to cover the 85629 three-variable linearly separable ternary functions. Clearly, therefore, the increase from a binary to a ternary situation has vastly increased the length of such tabulation and classification procedures, although the basic concepts and principles of the classification and listing remain straightforward.

2.5.2. Ternary Non-Linearly Separable Function Classification

No classification of non-linearly separable ternary functions has been published to date. If and when published, it will almost certainly be based upon orthogonal transform methods, generating and then ordering the spectral coefficient values into some canonic ordering. Work on orthogonal transforms for higher than the binary case has, however, been considered by several authorities.

One possible approach to an orthogonal transform for $R > 2$ is to extend the orthogonal function system of Coleman[42]. As shown and proved by Kitahashi and Tanaka[43], the following orthogonality result holds for any R-valued system of n input variables X_1 to X_n:

$$\sum_{p=0}^{p=R^n-1} \left\{ \left(\exp(j\theta_R \vec{a}_k \vec{X}_p)\right) \cdot \left(\exp(-j\theta_R \vec{a}_l \vec{X}_p)\right) \right\}$$
$$= 0 \text{ when } k \neq l$$
$$= R^n \text{ when } k = l$$

where $j = \sqrt{-1}$, $\theta = 2\pi/R$, $\vec{a}_k$, $\vec{a}_l$ are the kth and l^{th} members of the set of R^n R-valued n-dimensional vectors $\vec{a} = (a_1, a_2, \ldots, a_n)$, $\vec{X}_p$ is the pth member of the set of R^n R-valued n-dimensional vectors $\vec{X} = (X_1, X_2, \ldots, X_n)$ and $\vec{a}_k \vec{X}_p$, $\vec{a}_l \vec{X}_p$ are the inner vector products of the two vector components.

This approach, however, leads on to trigonometrical functions in the subsequent development. Whilst mathematically correct complete orthogonal transforms can be developed for $R = 2$, $R = 3$, etc., it may be more direct to build up our orthogonal $R = 3$ transform from appropriate three-valued functions, in a manner similar to the way the two-valued Walsh-function transform was developed from the orthogonal but incomplete two-valued Rademacher functions.

The Classification of Logic Functions

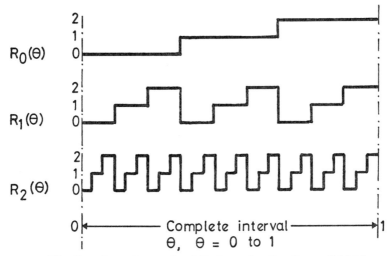

Fig. 2.13. The first three three-valued Rademacher functions of Liebler and Roesser.

This is the approach of Liebler and Roesser[44] following earlier work of Chrestenson[45]. Chrestenson defined functions which took the R values equal to the Rth roots of unity; the later work of Leibler and Roesser, however, is more familiar in a logic context, as the values 0, 1, and 2 are used in their development. The generation of their orthogonal set for $R = 3$ is based upon an incomplete set of functions which for similarity purposes are termed the "Rademacher functions of three values." The first three such functions are illustrated in Fig. 2.13. The primary "Rademacher" functions are then combined under modulo-3 addition to form a larger set, the "pre-Walsh function" set. Finally, in order to achieve orthogonality of the complete set, appropriate scaling factors on the magnitudes of these functions are imposed to give the final generalized Walsh set.

The result of this development is that the conventional two-valued Walsh functions have been generalized to a set of modified Walsh functions which may be applied to any $R \geq 2$ situation. As they form a complete orthogonal set, they may be used as a transform to convert uniquely, say, ternary data $f(X)$ into a spectral domain, the spectral coefficients in this domain uniquely defining the original function $f(X)$.

A considerable amount of continuing basic research work is anticipated in this area, including the possibility of alternative orthogonal transforms. Hence the short comments here should be regarded as a brief introduction to continuing and future work in this area.

2.6. Chapter Summary

In this chapter we have looked at many ways of classifying together dissimilar logic functions which possess certain common characteristics. A practical objective of such work is the possibility of being able to specify a small set of "standard" or "prototype" functions, from which all other functions can be realized by appropriate input interconnections and with possibly the use of addition inverter and exclusive-OR gates. In general algebraic classification methods have not proved to be the most useful.

The classification and listing of threshold-logic functions has had the greatest volume of research effort expended upon it. At first this may seem strange, seeing that threshold-logic gates are not in common usage to date alongside vertex gates. However, the classification of linearly separable functions and the listing of the resultant minimum-integer threshold-logic solutions was necessary in order to solve the problem of producing a threshold-logic realization for any given linearly separable function. It has been noted that it is not a direct procedure to generate the minimum-integer threshold-logic a_i values from the Chow-parameter b_i values, and so whilst the latter can be evaluated rapidly from the given binary data the former cannot. (We shall return to this awkward point in Chapter 4.)

However, this classification work on linearly separable functions and threshold-logic gates has by no means been wasted. The subsequent evolvement of the Rademacher–Walsh spectral coefficients, applicable to all functions, has direct ties with this work, leading to the very interesting current possibilities as suggested in Fig. 2.11.

The Rademacher–Walsh techniques introduced in this chapter represent possibly the most significant advance in logic theory and application for many years. It could well significantly alter many of our present techniques of logic design, fault diagnosis, and associated aspects. Chapter 5 will continue with the start we have now made with these non-Boolean methods of handling our digital data.

References

1. Staff of the Harvard Computational Laboratory *Synthesis of Electronic Computing Circuits.* Harvard University Press, Cambridge, Mass. 1951.

2. Slepian, D. On the number of symmetry types of Boolean functions on n variables. *Can. J. Math.* 5(2), 185–93, 1954.

3. Elspas, B. Self-complementary symmetry types of Boolean functions. *Trans. IRE* **EC9** 264–70, 1960.

4. Goto, E., and Takahasi, H. Some theorems useful in threshold-logic for enumerating Boolean functions. *Proc. IFIP Congress, Aug. 1962*, 747–51.

5. Harrison, M.A. *Combinatorial Problems in Boolean Algebras and Applications to the Theory of Switching,* Ph.D. Thesis, Electrical Engineering Dept., University of Michigan, 1963.

6. Dertouzos, M.L. *Threshold Logic: A Synthesis Approach.* MIT Press, Cambridge, Mass., 1965.

7. Muroga, S. *Threshold Logic and its Applications* John Wiley Interscience, New York, 1971.

8. Karnaugh, M. Pulse switching circuits using magnetic cores. *Proc. IRE* **43,** 570–84, May 1955.

9. Hurst, S.L. *Threshold-Logic*–An Engineering Survey. Mills and Boon, London, 1971.

10. Chow, C.K. On the characterisation of threshold functions. *Switching Theory and Logical Design,* 34–8. IEEE Special Publ. No. S.134, Sept. 1961.

11. Golomb, S.W. On the classification of Boolean functions. *Trans. IRE* **CT6** (Suppl) 176–86, May 1959.

12. Ninomiya, I. A study of the structures of Boolean functions and its application to the synthesis of switching circuits. Mem. Fac. Engng Nagoya Univ. **13**(2), 149–363, Nov. 1961.

13. Winder, R.O. Enumeration of seven-argument threshold functions. *Trans. IEEE* **EC14,** 315–25, June 1965.

14. Winder, R.O. *Threshold Functions through $n = 7$.* Sci. Rep. No. 7, AFCRL Contract AF 19(604)—8423, Oct. 1964.

15. Muroga, S., Tsuboi, T., and Baugh, C.R. *Enumeration of Threshold Functions of Eight Variables.* Rep. No. 245, Univ. of Illinois, August 1967. Abridged in *Trans. IEEE* **C19,** 815–25, Sept. 1970.

16. Winder, R.O. Chow parameters in threshold logic. *J. Ass. Computing Mach.* **18** (2), 265–89, Apr. 1971.

17. Bellman, R. *Introduction to Matrix Analysis.* McGraw-Hill, New York, 1960.

18. Harmuth, H.F. *Transmission of Information by Orthogonal Functions.* Springer, New York, 1969.

19. Andrews, H.C., and Caspari, K.L. A generalised technique for spectral analysis. *Trans IEEE* **C19**, 16–25, Jan. 1970.

20. Pratt, W.K., Kane, J., and Andrews, H.C. Hadamard transform image coding. *Proc. IEEE* **57**, 58–68, Jan. 1969.

21. Harr, A. Zur Theorie der orthogonalen Funktioneasysteme. *Math. Annen* **69**, 331–71, 1910.

22. Paley, R.E.A.C., On othogonal matrices. *J. Math. Phys.* **12** 311–20, 1933.

23. Cooley, J.W., and Tukey, J.W. An algorithm for the machine computation of complex Fourier series. *Math. Computation,* **19**, 297–301, Apr. 1956.

24. *Proc. Symp. on Theory and Application of Walsh Functions, June 1971, Hatfield, U.K.*

25. Walsh, J.L. A closed set of orthogonal functions. *Am. J. Math.* **45**, 5–24, 1923.

26. Fine, N.J. On the Walsh functions. *Trans. Am. Math. Soc.* **65**, 372–414, 1949.

27. Harmuth, H.F. A generalised concept of frequency and some applications. *Trans IEEE* **IT14**, 375–82, May 1969.

28. Paley, R.E.A.C. A remarkable series of orthogonal functions. *Proc. Lond. Math. Soc.* **34**, 241–79, 1932.

29. Levey, P. Sur une generalisation des functions orthogonales de Rademacher. *Comment Math. Helv.,* **16**, 146–52, 1944.

30. *Proc. Symp. on Theory and Application of Walsh and other Non-Sinusoidal Functions, June 1973, Hatfield, U.K.*

31. Rademacher, H. Einige Sätze über reihen von allgemeinen Orthogonalfunktionen. *Math. Annen* **87**, 112–38, 1922.

32. Shanks, J.L. Computation of the fast Fourier transform. *Trans. IEEE* **EC18**, 457–9, May 1969.

33. Ulman, L.J. Computation of the Hadamard transform and the R transform in ordered form. *Trans IEEE* **EC19**, 359–60, Apr. 1970.

34. Hurst, S.L. Application of Chow parameters and Rademacher–Walsh matrices in the synthesis of binary functions, *Computer J.* **16**, 165–73, May 1973.

35. Edwards, C.R. The application of the Rademacher–Walsh transform to Boolean function classification and threshold-logic synthesis. *Trans IEEE* **C24**, 48–62, Jan. 1975.

36. Berlekamp, E.R. Some mathematical properties of a scheme for reducing the bandwidth of matrix pictures by Hadamard smearing. *B.S.T.J.* **49**, 969–86, July/Aug. 1970.
37. Lechner, R.J. A transform approach to logic design. *Trans IEEE* **C19**, 627–40, July 1970.
38. Lechner, R.J. Harmonic analysis of switching functions. In *Recent Developments in Switching Theory* (Ed. A. Mukhopadhyay). Academic Press, New York, 1971.
39. Aibara, T., and Akagi, M. Enumeration of ternary threshold functions of three variables. *Trans. IEEE* **C21**, 402–7, Apr. 1972.
40. Moraga, C., and Nazarala, J. Minimum realization of ternary threshold functions. *Proc. 1974 Symp on Multivalued Logic, West Virginia, May 1974*, 347–58.
41. Kitahashi, T., Tezuka, Y., Kasahari, Y., and Nomura, H. Characterizing parameters of ternary logic functions. *J. IECE Japan* **52C**, 641–8, Oct. 1969 (in Japanese).
42. Coleman, R.P. Orthogonal functions for the logical design of switching circuits. *Trans. IEEE* **EC10**, 379–83, Sept. 1961.
43. Kitahasi, T., and Tanaka, A. Orthogonal expansion of many-valued logical functions and its application to their realization with single-threshold element. *Trans. IEEE* **C21**, 211–8, Feb. 1972.
44. Chrestenson, H.E. A class of generalised Walsh functions. *Pacif. J. Math.* **5**, 17–31, 1955.
45. Liebler, M.E., and Roesser, R.P. Multiple real-valued Walsh functions. *Conf. Rec. Symp. on the Theory and Application of Multiple-Valued Logic Design, Buffalo, N.Y., May 1971*, 84–102.

Chapter 3

Geometric Constructions for Binary and Non-Binary Functions

Introduction

In the preceding two chapters we have already used certain geometric constructions to illustrate aspects of digital functions. In particular the Karnaugh map construction has been used in several instances to show features of concern.

An important point has, however, been made in the Introduction at the beginning of this book which it is worth re-emphasizing here, this being that geometric constructions are limited in their practical application. Nevertheless they form an essential part in logic learning and development. The problem of course is that the layout of the various constructions becomes too complex to draw and interpret when the number of input variables becomes large. Above, say, $n = 4$ for binary systems or $n = 3$ for ternary systems, such constructions begin to fail to be useful and meaningful.

However, where they can be used and interpreted, they often offer a far superior means of analysis or synthesis compared with purely algebraic or mathematical means. The classic situation of course is in the minimization of combinatorial functions of, say, four input variables, which may readily be mapped on an $n = 4$ Karnaugh map. Minimization of the function from the completed map is immediate, as the adjacencies of the minterms and hence their assembly into a minimum irredundant sum-of-products result is immediately apparent[1,2]. This ready minimization may be contrasted with the classic Quine–McClusky minimization algorithm, which involves expansion of the given function into all possible pairings of minterms before final contraction down to the final minimized sum-of-products solution[1-4]. The large number of intermediate terms which are generated and have to be manipulated in the Quine–McClusky technique is a great drawback, but nevertheless is almost unavoidable where a geometric representation of the problem is not practical. A further awkward point is how to incorporate "don't-care" input conditions into an algebraic

Geometric Constructions for Binary and Non-Binary Functions 143

technique; with mapping or similar geometric representations it is usually immediately obvious how to use the "don't-care" minterms to best advantage.

The other supreme advantage which geometric constructions can offer is to illustrate the properties of and interrelationships within functions. Once such relationships are recognized and understood then mathematical relationships may be developed, which of course may be extended to situations where the number of variables is too great to allow geometric constructions to be practical. Hence the constructions which we shall look at in the subsequent pages of this chapter will be as much if not more for educational purposes as for practical usage. To see graphically the significance of a problem or feature is to be well on the way to understanding its full significance and potential application.

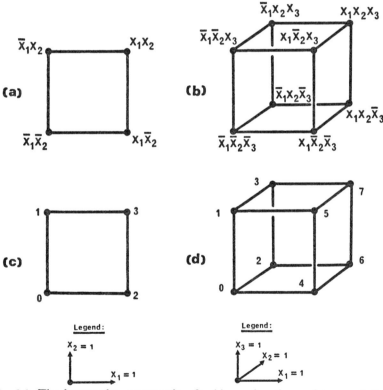

Fig. 3.1. The hypercube construction for binary functions of two and three variables: (a) $n = 2$, 4 nodes, one face; (b) $n = 3$, 8 nodes, six faces; (c) $n = 2$, with denary identification of the nodes; (d) $n = 3$, with denary identification of the nodes.

144   *The Logical Processing of Digital Signals*

3.1. The Hypercube Constructions

We have already encountered simple examples of the hypercube construction for binary functions in Chapter 1 and for a ternary function in Chapter 2, both in connection with linearly separable functions. However, let us here build up the picture in a slightly more comprehensive manner, and extend it beyond that which has previously been considered.

To recapitulate on previous information, Figs. 3.1a, b show the hypercube construction for two-variable and three-variable binary functions, respectively. As explained in Chapter 1, each node on such constructions uniquely represents one minterm, and hence each may conveniently be labeled by the denary number which corresponds to the binary minterm designation, i.e. 0 for minterm ...0, 1 for minterm ...01, 2 for minterm ...10, and so on. This is illustrated in the re-

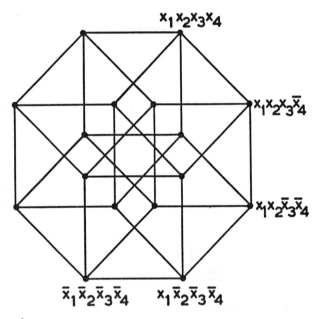

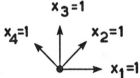

Fig. 3.2. The binary hypercube construction for $n = 4$.

Geometric Constructions for Binary and Non-Binary Functions 145

labeled Figs. 3c, d. Notice also that the tie-lines from any one node to all adjacent nodes are at right-angles in n-space.

For $n = 4$ the binary hypercube construction becomes as shown in Fig. 3.2. The 16 nodes of this figure form an interlocking assembly of three-dimensional cubes, but to illustrate the four right-angles in n-space which now exist between the tie-lines from each node is becoming increasingly difficult to represent realistically. For $n = 5$ the situation becomes impossible! Although more difficult to draw, label, and interpret easily, the $n = 4$ hypercube picture begins to illustrate more clearly than the smaller-dimensional versions the concept of all minterms of any binary function of n variables forming equispaced points on an n-dimensional sphere. With the binary values of 0 and 1, the center of gravity of the sphere has no obvious significance. However, let us convert our binary numbers from the 0, 1 domain into the $-1, +1$ domain, and see what results.

Starting with the simple $n = 2$ situation, the hypercube picture of Fig. 3.1a becomes relabeled as shown in Fig. 3.3a. Notice now that the origin is at the center of the four nodes, the nodes themselves lying on a circle (two-dimensional sphere) of radius $\sqrt{2}$.

For $n = 3$ the picture becomes one of a three-dimensional sphere, again with its center at the origin, but now with a radius of $\sqrt{3}$ from the origin to the surface of the sphere upon which all the 2^3 minterms lie. This is illustrated in Fig. 3.3b. Hence we may generalize this concept by stating that for any n-variable binary system with binary values -1, $+1$, all 2^n minterms will lie on the surface of an n-dimensional sphere of radius $\sqrt{n}$. The center of the sphere may also be referred to as the "center of gravity" of all the 2^n minterms.

Most of the features of binary functions may be illustrated on the hypercube construction, although it is possibly true to say that with the exception of linear separability most features can equally if not better be presented in alternative ways. We shall look at the exception of linear-separability as a separate immediately following section of this chapter.

Minimization of any given binary function may be illustrated and performed on the hypercube by identifying each node with the function output value 0 or 1 (false or true) and grouping true (or false) terms together in some optimum choice. Notice that minimization involves pairs of adjacent like nodes (two minterms) at the lowest level of minimization, faces of the hypercube (four minterms) at the next level of minimization, cubes of like nodes (eight minterms) at the third level of minimization, and so on. However, as it is difficult to disentangle

146   The Logical Processing of Digital Signals

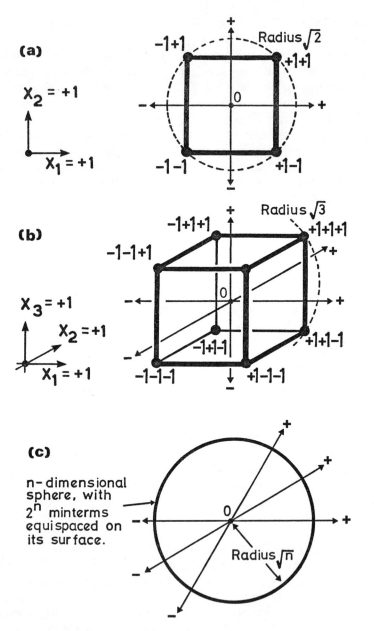

Fig. 3.3. The binary hypercube construction with truth values of $-1, +1$: (*a*) $n = 2$ case; (*b*) $n = 3$ case; (*c*) generalized case for any n.

Geometric Constructions for Binary and Non-Binary Functions   147

and see the various faces and cubes for above $n = 3$, constructions such as the Karnaugh map prove far more satisfactory for function minimization purposes.

Hamming distances are illustrated on the hypercube construction by considering how many tie-lines have to be transversed to go from one to another node in question. This concept was introduced in §1.5.3 of Chapter 1. Hamming distances of 1 correspond to adjacent nodes separated by merely one tie-line, Hamming distances of 2 by two tie-lines, and so on. Again, the main use of the hypercube format is to illustrate the principle and not to provide a ready means for the synthesis of codes etc. which require the property of exact Hamming distances. Other approaches are available for such synthesis purposes[5-7], this being, however, an area which we shall not be considering in this text.

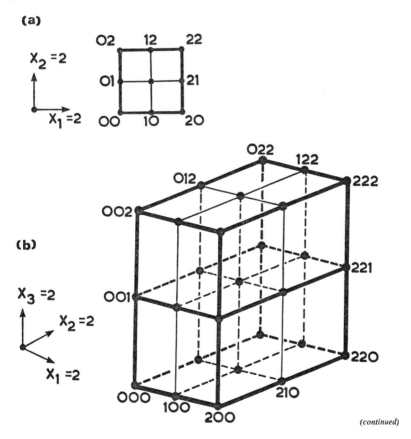

(continued)

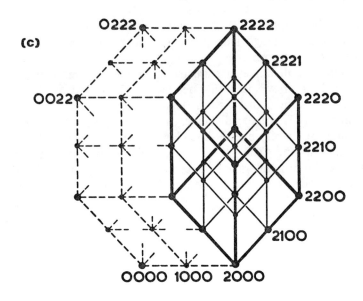

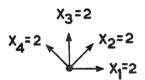

Fig. 3.4. The ternary hypercube construction for ternary functions with truth values of 0, 1, 2: (a) $n = 2$, $3^2 = 9$ nodes, one face: (b) $n = 3$, $3^3 = 27$ nodes, six faces; (c) $n = 4$, not fully detailed.

Hypercube representations for ternary (and higher-valued) functions become increasingly difficult to draw and interpret, as Fig. 2.12 of Chapter 2 may have already shown. Using the ternary values of 0, 1, 2 instead of −1, 0, +1, as was appropriate for the special application of Chapter 2, then the ternary hypercube formats for two and three variables are as illustrated in Figs. 3.4a, b respectively. The full hypercube for four variables becomes too complex to detail fully, and hence Fig. 3.4c illustrates this case in outline only. It is, however, not feasible to renumber the ternary truth values so that all 3^n ternary minterms lie on the surface of a sphere, with center at the origin, as we saw was possible in the binary case. Examination of the construction of Fig. 3.4b, for example, will show why this is not possible, bearing in mind that all the tie-lines present in Fig. 3.4 must remain unchanged in any

Geometric Constructions for Binary and Non-Binary Functions

renumbering (geometric shift) of the format. So the interesting n-dimensional sphere of the binary case, with equispaced nodes on its surface representing the minterms, is not possible in other than the binary case.

The use of the ternary hypersphere construction is mainly confined to simple ternary minimization exercises. Unlike the binary case, where the binary hypersphere is not the best possible construction for minimization purposes, the ternary hypersphere represents about the best three-valued construction that is possible. This is because the alternative Karnaugh map-type formats for three-valued functions prove most inconvenient, as we shall shortly see when we consider map-type constructions. Therefore we shall come back again to the ternary hypercube pictures in Chapter 6 when we shall be looking at the design of simple ternary combinatorial networks.

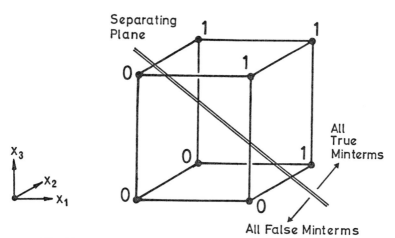

Fig. 3.5. The binary $n = 3$ hypercube construction and linear separability (cf. Fig. 1.2 of Chapter 1). Function $f(x) = [x_1 x_2 + x_1 x_3 + x_2 x_3]$ illustrated.

3.2. Linear Separability and the Hypercubes

The concept of linear separability of binary functions $f(x)$ using the hypercube representation was first illustrated in Fig. 1.2 of Chapter 1. By definition the set of all true and the set of all false minterms of any linearly separable function $f(x)$ may be separated by a simple separating plane within the n-dimensional hypercube. This feature is illustrated here again in Fig. 3.5.

Now the equation for any hyperplane in n-dimensional Euclidean space with axes $x_1, x_2, \ldots, x_n$ is given by the linear equation

$$w_1 x_1 + w_2 x_2 + \ldots + w_n x_n = k$$

where $w_1, w_2, \ldots, w_n$ and k are constants. Hence for any point

$$\vec{x} = x_1, x_2, \ldots, x_n \text{ in } n\text{-space, if}$$

$$w_1 x_1 + w_2 x_2 + \ldots + w_n x_n \geq k$$

then the point $\vec{x}$ lies on or to the non-origin side of the hyperplane, whilst if

$$w_1 x_1 + w_2 x_2 + \ldots + w_n x_n \leq k$$

then the point $\vec{x}$ lies on or to the origin side of the hyperplane.

Clearly, therefore, if we consider our discrete points in n-space which represent the minterms of a binary function $f(x)$ and the separating plane which divides them into true and false minterms in the case of a linearly separable (threshold) function, we have that the equation for such a separating plane gives the threshold realization for the function $f(x)$. In particular if the separating plane just touches the lowest-summation true minterms of $f(x)$, we have for all true minterms that

$$w_1 x_1 + w_2 x_2 + \ldots + w_n x_n \geq k$$

and for all false minterms that

$$w_1 x_1 + w_2 x_2 + \ldots + w_n x_n < k,$$

that is the separating-plane constants $w_1, w_2, \ldots, w_n$ and k are numerically equal to the threshold-logic realizing weights a_i and threshold t in the general equation

$$f(x) = 1 \text{ if } \langle a_1 x_1 + a_2 x_2 + \ldots + a_n x_n \rangle \geq t$$
$$= 0 \text{ if } \langle a_1 x_1 + a_2 x_2 + \ldots + a_n x_n \rangle < t.$$

Whilst this n-dimensional approach provides an interesting geometric interpretation of binary functions, in the end it does not offer any worthwhile avenues for everyday use. Documentation of this area will be found published in several sources[8-11]. However, let us proceed a little further in order to illustrate one other facet within this area which will fit in neatly with certain ideas covered in Chapter 2.

Converting our binary data from the 0, 1 domain into $-1, +1$ converts, as we have previously seen, the hypercube construction into a hypercube (or hypersphere) model of radius $\sqrt{n}$, the center of this hypersphere being at the origin. The separating plane which divides all

Geometric Constructions for Binary and Non-Binary Functions   151

true minterms from all false minterms for any given linearly separable function will not in general pass through the origin of this hypersphere. However, the linearly separable functions which do possess an equal number of true and false minterms *will* be divided by a separating plane passing through the origin of this hypersphere. Such functions are the *self-dual* linearly separable functions, whose property is that the value of their b_0 term in the Chow (or Rademacher–Walsh) classification is zero, and hence their Chow a_0 weighting is also zero. For example, in the $n = 3$ Chow classification 4 4 4 0 (see Appendix A), if we permutate so that

$$b_0 = 0 \qquad a_0 = 0$$
$$b_1 = 4 \qquad a_1 = 1$$
$$b_2 = 4 \qquad a_2 = 1$$
$$b_3 = 4 \qquad a_3 = 1$$

this defines the self-dual linearly separable function

$$f(x) = \langle x_1 + x_2 + x_3 \rangle_t$$

where $t = \tfrac{1}{2}\{(1+1+1) - 0 + 1\} = 2$, giving the Boolean function

$$f(x) = [x_1 x_2 + x_1 x_3 + x_2 x_3].$$

Proof that this particular function is self-dual has already been shown in §2.3 of Chapter 2; any other linearly separable function with $b_0 = 0$ likewise will be found to be self-dual as it possesses an equal number of true and false minterms.

However, let us now consider the $(n + 1)$-variable hyperfunction $f^h(x)$ of any binary function $f(x)$, where as defined in §2.3 of Chapter 2 $f^h(x)$ is

$$f^h(x) \triangleq [x_{n+1} f(x) + \bar{x}_{n+1} f^d(x)].$$

Now if we consider the case of a separating plane passing through the origin of the $(n + 1)$-dimensional sphere, then swinging this separating plane through all possible solid angles will produce *all possible linearly separable self-dual functions of the $n + 1$ variables*. However, all possible linearly separable self-dual functions of $n + 1$ variables correspond precisely to all the possible linearly separable functions of n variables, whether self-dual or not[1]. We therefore now have a possible

---

[1.] That this is so may be shown from Appendix A. Looking at, say, the $n \leq 4$ entries it will be seen that there are only three entries which contain $b = 0$ values, i.e. are self-dual. These entries correspond exactly to the full $n \leq 3$ classification when scaling by 2 is made to take into account the reduction in dimensions between $n = 4$ and $n = 3$. Similarly in $n \leq 5$ the seven entries with $b = 0$ values correspond to the full $n \leq 4$ classification, and so on for higher n.

geometric construction, the analysis of which may enable us to generate all possible linearly separable functions of n variables.

This interesting technique was pursued by Kaszerman[12]. His development also provided for dividing the radius of the $(n + 1)$ dimensional hypersphere by $(n + 1)^{1/2}$, thus ending up with a unit-radius $(n + 1)$-dimensional hypersphere which was precisely bisected by all possible separating planes to divide the representation minterms on the hypersphere surface.

Other related multi-dimensional concepts have also been pursued, including convex and non-convex space considerations[8,11]. Whilst providing useful insight into the basic problems of linear separability, in general no lasting usage for these techniques and concepts has emerged.

For ternary and higher-valued logic, the concept of linear separability and separating planes becomes correspondingly more difficult to consider geometrically. The underlying principles remain unchanged, but the conceptual and computational procedures are clearly more complex. To date, therefore, there appears to have been little work done or pending in this area.

3.3. Map Constructions for Binary Functions

The principal map format for binary work of course is the familiar Karnaugh map layout[13]. Veitch and other possible diagrams[14,15] provide no practical advantages, and indeed are usually disadvantageous for engineering purposes.

Familiarity with the Karnaugh map construction and its interpretation has already been assumed in the previous chapters of this book, as it nowadays forms a basic tool in logic design. What we shall do here, however, is to examine the map usage in illustrating or undertaking slightly more uncommon features and problems in binary logic.

3.3.1. Exclusive-OR Relationship

The normal grouping of like-valued minterms on a Karnaugh map layout, including the incorporation of any don't care minterms, produces the conventional prime-implicant product (AND) terms. The joint coverage from several prime implicants is the subsequent OR operation. The resultant two-level sum-of-products minimum solution may then be manipulated if necessary, say into an all-NAND or all-NOR form as required. Non-mapping algebraic algorithms are available as an alternative design procedure, not necessarily constrained to two-level

Geometric Constructions for Binary and Non-Binary Functions

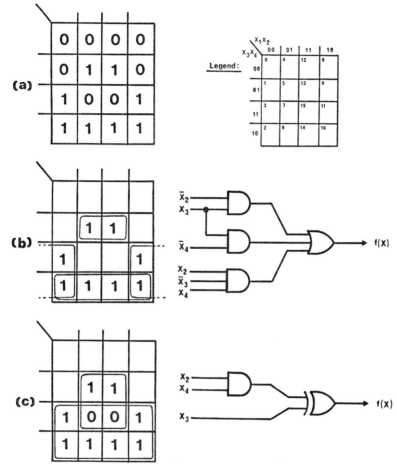

Fig. 3.6. A given function $f(x) = [\bar{x}_2 x_3 + x_3 \bar{x}_4 + x_2 \bar{x}_3 x_4]$: (a) the given function map (b) conventional AND/OR realization; (c) exclusive-OR realization.

networks[16], but frequently such algorithms do not generate a minimum solution to the given design problem.

The exclusive-OR relationships, however, may be extracted from a Karnaugh map by looking for overlapping prime-implicant terms, but where the overlap of any two terms takes the opposite truth value from that of the individual terms themselves. Initially this is not as easy to recognize on the maps as the more simple grouping of terms with or without overlap, but with a little practice considerable skill may be acquired in such procedures.

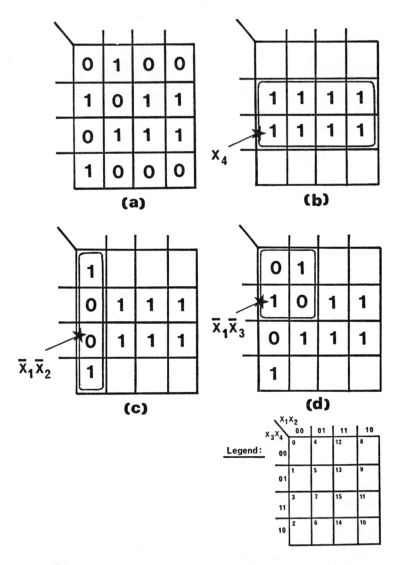

Fig. 3.7. A given function

$$f(x) = [x_1x_4 + x_2x_3x_4 + \bar{x}_1x_2\bar{x}_3\bar{x}_4 + \bar{x}_1\bar{x}_2(\bar{x}_3x_4 + x_3\bar{x}_4)]:$$

(a) the given function map; (b) the initial choice of prime implicant x_4; (c) exclusive-OR addition of prime implicant $\bar{x}_1\bar{x}_2$ to (b); (d) exclusive-OR addition of prime implicant $\bar{x}_1\bar{x}_3$ to (c) to complete the synthesis

$$f(x) = [x_4 \oplus \bar{x}_1\bar{x}_2 \oplus \bar{x}_1\bar{x}_3].$$

Geometric Constructions for Binary and Non-Binary Functions

As an example, consider the function illustrated in Fig. 3.6a. The conventional minimum sum-of-products synthesis of this function is the three-term solution in Fig. 3.6b. However, the prominent "hole" in the pattern of 1-valued minterms indicates the possibility of some exclusive-OR relationship, and indeed the function is readily realized by the exclusive-OR realization shown in Fig. 3.6c, namely

$$f(x) = [x_3 \oplus x_2 x_4].$$

Notice that no inversion of any input variable is required in this solution, compared with three inversions necessary in the sum-of-products solution. Fig. 3.7 illustrates a further function which may profitably use exclusive-OR relationships in its synthesis. Notice here the exclusive-OR overlapping of three prime-implicant terms in the chosen solution.

The search for a possible exclusive-OR solution to a given combinatorial problem will be seen to involve trying to take the largest-area prime-implicant coverage available or almost available on the map layout and then superimposing additional terms so as to add minterms to or amputate minterms from the initially chosen cover. For example, in Fig. 3.7 prime implicant x_4 is almost present and is therefore chosen as a starting cover. The subsequent prime-implicant terms serve to modify this x_4 coverage and to add the additional minterms required outside the x_4 area of the map.

When the map format cannot be used, owing to the number of input variables being greater than, say, 5 ($n > 5$), then network design using exclusive-OR or exclusive-NOR relationships becomes very difficult by any algebraic means. Indeed we then have to resort to spectral design methods, which is a subject which we shall be considering in Chapter 5 of this book.

3.3.2. Symmetries in Binary Functions

Symmetry considerations of individual logic gates were briefly considered in §1.5.2 of Chapter 1. Symmetries, however, are a prominent feature of many functions, but often the problem is to recognize which particular symmetry or symmetries are present, particularly when the function is complex. The symmetry of, say, an arithmetic function is usually clear from the definition of the function; for example, the sum or the carry output of a full adder circuit is completely symmetric in its three binary inputs as each input has exactly the same significance in determining the output state.

If the function can be plotted on a Karnaugh map layout, however, symmetries in the true and false minterm patterns may be found, thus identifying whatever symmetries may be present in the given function. However, in addition these map constructions may be used to classify different types of symmetry, which when analysed can form the basis for synthesis techniques. Such techniques will be introduced in Chapter 5.

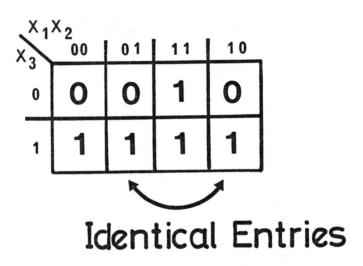

Fig. 3.8. The simple function $f(x) = [x_1 x_2 + x_3]$ which possesses symmetry between the two input variables x_1 and x_2.

Let us therefore look at symmetries which may be identified on a Karnaugh map and their underlying meaning.

Consider first the trivial function $f(x) = [x_1 x_2 + x_3]$. From this Boolean expression there is clearly a symmetry in the input variables x_1 and x_2, as interchanging these input signals leaves the function invariant. The Karnaugh map of this function is shown in Fig. 3.8, from which it will be seen that the two columns $\bar{x}_1 x_2$ and $x_1 \bar{x}_2$ have identical entries. Identical rows or columns or groupings on the Karnaugh map therefore are the key to the recognition of symmetries in any given function.

For functions of four variables the range of input symmetries similar to that of Fig. 3.8 which can be present is illustrated in Fig. 3.9, and will be seen to correspond to every possible different pairing of the input variables x_1, x_2, x_3, and x_4. More than one of these symmetries

Geometric Constructions for Binary and Non-Binary Functions 157

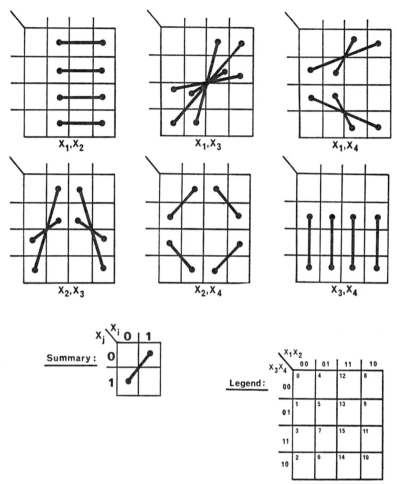

Fig. 3.9. The six non-equivalence symmetries, any of which may exist in a four-variable function $f(x_1, x_2, x_3, x_4)$. The links indicate like-valued minterms; outside the paired minterms any patterns of 0 and 1 may exist.

may be present in any given function. For example, the function plotted in Fig. 3.10 is symmetric in both x_1, x_3 and x_3, x_4.

This particular type of symmetry between two input variables x_i and x_j means that the function output is the same whether $x_i = 0$ and $x_j = 1$ or $x_i = 1$ and $x_j = 0$ is present, that is when the two input signals in question are dissimilar, or "non equivalent." For this reason this type of symmetry has been termed "non-equivalent" or "non-equivalence" symmetry. Where two or more such symmetries are present, as in Fig. 3.10, then multiple non-equivalence symmetry exists.

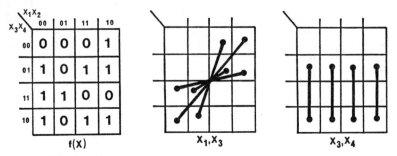

Fig. 3.10. The function

$$f(x) = [\bar{x}_1(\bar{x}_2 x_3 + \bar{x}_2 x_4 + x_3 x_4) + x_1(\bar{x}_2 \bar{x}_3 + \bar{x}_3 x_4 + x_3 \bar{x}_4)]$$

which is non-equivalence symmetric in both x_1, x_3 and in x_3, x_4.

Now suppose, say, the first and third columns in a four-variable Karnaugh map plot are the same, that is the $\bar{x}_1 \bar{x}_2$ and $x_1 x_2$ column entries. This agreement must also be some form of symmetry in the input variables x_1 and x_2, but not the non-equivalence symmetry condition previously illustrated. Instead this is another family of possible symmetries, as illustrated in Fig. 3.11. Notice that Fig. 3.11 again involves all possible pairs of input variables of the four inputs x_1 to x_4. However, what we are now illustrating is a form of symmetry between inputs x_i and x_j which makes the function output invariant whether x_i and x_j are both 0 or whether x_i and x_j are both 1. This type of symmetry has therefore been termed "equivalent" or "equivalence" symmetry. Like non-equivalence symmetry, two or more equivalence-symmetry conditions may be present in a given function.

Finally, both equivalence and non-equivalence symmetry may be simultaneously present between inputs x_i and x_j. This is effectively the simultaneous presence of these relationships separately illustrated in Figs. 3.9 and 3.11, this combined type of symmetry being termed "multiform" symmetry. Figure 3.12 illustrates the possible multi-form symmetries in a four-variable function. Recognition of such symmetries in a given function may aid the logic design of the function. Fundamentally a symmetry involves some repeated feature in the overall build-up of the function, and therefore it may be possible to synthesize the repeated feature only once in the complete function synthesis. This in its turn may produce an overall economy in the complete design.

Geometric Constructions for Binary and Non-Binary Functions 159

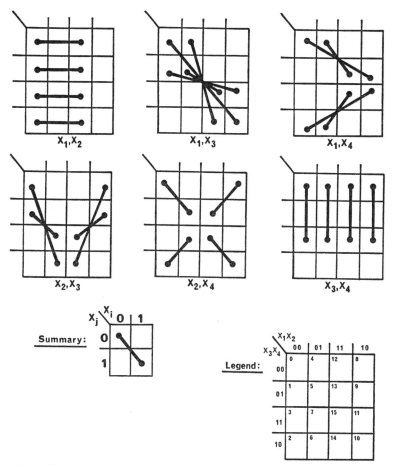

Fig. 3.11. The six equivalence symmetries, any of which may exist in a four-variable function $f(x_1, x_2, x_3, x_4)$.

This particular feature will be considered in Chapter 5 when dealing with logic design techniques. At this stage, however, the Karnaugh map has been used to introduce and illustrate the ideas of symmetries, and reference back to the concepts illustrated here will be made in the later chapter. It may be noticed that features such as symmetries are not always obvious in the conventional Boolean algebra and truth-table data; the Karnaugh map, on the other hand, is a very clear data format, but unfortunately becomes difficult to use for, say, $n \geq 5$. Rademacher–Walsh spectral techniques, however, may provide the means for

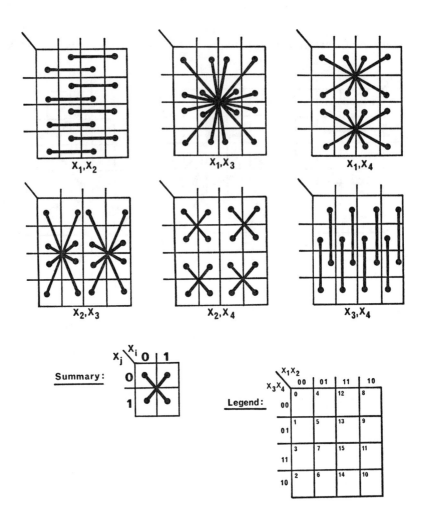

Fig. 3.12. The six multi-form symmetries, any of which may exist in a four-variable function $f(x_1,x_2,x_3,x_4)$.

handling large functions, but all the principles used for large n are extensions of the basic principles and concepts which may be visually developed on the Karnaugh map layouts.

3.3.3. Linearly-Separable Map Patterns

The coverage provided on a Karnaugh map by any vertex gate is a simple rectangular group of like-valued minterms. Threshold-logic gates, however, do not always provide such a trivial covering pattern, as Fig. 1.17 of Chapter 1 has already illustrated. Majority gates, being a special case of threshold-logic gates, likewise do not provide such simple covering patterns.

Geometric Constructions for Binary and Non-Binary Functions 161

For synthesis purposes recognition of rectangular coverage patterns on Karnaugh map layouts forms an essential part of the normal vertex gate design and minimization procedure. To extend the same technique to synthesis using threshold-logic (or majority) gates requires some *a priori* knowledge of what covering patterns are available on the Karnaugh map layout. However, as it is clearly impossible to list or draw all the possible covering patterns corresponding to every possible linearly separable function[2], some form of classification is essential in order to make the problem manageable.

Fortunately we have readily available the Chow-parameter numerical classification of linearly separable functions, which compresses into three entries all $n \leq 3$ functions, into seven entries all $n \leq 4$ functions, and so on as fully detailed in Appendix A. Each entry in these classification tables corresponds to the type-SD positive canonic threshold-logic function which is the representative of a large number of individual like functions.

Let us examine the $n \leq 3$ Chow classification to see if we can derive a corresponding map-pattern classification listing. The three $n \leq 3$ Chow classification entries may be expanded into the six $n \leq 3$ NPN positive canonic functions as shown in Table 3.1. These six functions of Table 3.1 are the positive canonic functions following:

(i) $f(x) = \langle x_1 + 0 + 0 \rangle_1$ (threshold)
      $= [x_1]$ (Boolean)

(ii) $f(x) = \langle 0 + 0 + 0 \rangle_0$ (threshold)
      $= [1]$ (Boolean)
      (a trivial, entirely degenerate case)

(iii) $f(x) = \langle 2x_1 + x_2 + x_3 \rangle_2$ (threshold)
      $= [x_1 + x_2 x_3]$ (Boolean)

(iv) $f(x) = \langle x_1 + x_2 + x_3 \rangle_1$ (threshold)
      $= [x_1 + x_2 + x_3]$ (Boolean)

(v) $f(x) = \langle x_1 + x_2 + x_3 \rangle_2$ (threshold)
      $= [x_1 x_2 + x_1 x_3 + x_2 x_3]$ (Boolean)

(vi) $f(x) = \langle x_1 + x_2 + 0 \rangle_1$ (threshold)
      $= [x_1 + x_2]$ (Boolean).

---

[2] There are 104 linearly separable functions for $n \leq 3$, 1882 for $n \leq 4$, and so on as was enumerated in Table 2.7 of Chapter 2.

Table 3.1. Chow-parameter classification for $n \leq 3$

Chow classification		Possible input weights a_1, a_2, a_3	Corresponding gate threshold t
$\lvert b_i \rvert$	$\lvert a_i \rvert$		
8, 0, 0, 0	1, 0, 0, 0	or 1, 0, 0 0, 0, 0	1 0
6, 2, 2, 2	2, 1, 1, 1	or 2, 1, 1 1, 1, 1	2 1
4, 4, 4, 0	1, 1, 1, 0	or 1, 1, 1 1, 1, 0	2 1

The Karnaugh map patterns of each of these functions is now illustrated in Fig. 3.13. These patterns constitute the NPN positive canonic covering patterns for all possible $n \leq 3$ linearly separable functions.

The 104 different individual linearly separable functions of $n \leq 3$ correspond to all different possible positions and orientations of these six covering patterns or their complements within the $n = 3$ Karnaugh map layout. The pattern shapes remain invariant, but their positions within the map change, corresponding precisely to all the possible NPN permutations of each canonic classification[8,17]. To confirm that this is so Table 3.2 gives the full statistics of the situation. It is an interesting and informative exercise for the reader to confirm these figures.

Table 3.2. The map-pattern classification details for $n \leq 3$

Canonic function and map pattern	Equivalent hypercube division	Number of different map positions & functions
Function (i), pattern (a)	Any one side of the hypercube	6 possible positions = 6 possible Boolean functions (pattern & complement of pattern give identical functions)
Function (ii), pattern (b)	Whole hypercube	1 position = 2 possible Boolean functions, either $f(x) = 0$ or $f(x) + 1$ (trivial case)

(continued)

Function (iii), pattern (c)	Any two meeting edges of the hypercube	24 possible positions = 48 possible Boolean functions (pattern & complement of pattern are dissimilar functions)
Function (iv), pattern (d)	Any one node of the hypercube	8 possible positions = 16 possible Boolean functions (pattern & complement of pattern are dissimilar functions)
Function (v), pattern (e)	Any three meeting edges of the hypercube	8 possible positions = 8 possible Boolean functions (pattern & complement of pattern give identical functions)
Function (vi) pattern (f)	Any two adjacent nodes of the hypercube	12 possible positions = 24 possible Boolean functions (pattern & complement of pattern are dissimilar functions)
		Total 104 Boolean functions

If we extend this concept to list the $n \leq 4$ linearly separable map patterns, then the seven $n \leq 4$ Chow-parameter classificaiton entries give 15 NPN map patterns. These will be found fully detailed in Appendix D. Together they encompass the full 1882 different linearly separable functions of $n \leq 4$ variables[18].

If we briefly examine the $n \leq 4$ map patterns detailed in Appendix D, we shall find that, if we divide any map into two halves, bisecting it along any axis which represents the change of any x_i from logic 0 to logic 1 the map patterns which result in each half of the full $n \leq 4$ map will be one of the $n \leq 3$ map patterns. This must be the case, since if any linearly separable function $f(x)$, $f(x) = f(x_1, \ldots, x_i, \ldots, x_n)$, is divided into

$$f(x) = [\bar{x}_i f'(x_1, \ldots, -, \ldots, x_n) + x_i f''(x_1, \ldots, -, \ldots, x_n)]$$

then each function $f'(\ldots)$ and $f''(\ldots)$ *is itself always a linearly separable function*. This is what the bisection of the map patterns confirms. However, what does *not* follow is that we can take the map patterns of, say, $n-1$ variables and combine them in any random manner so as to generate the n-variable map patterns; this is not so since the combination of map patterns into larger groupings has to obey certain constraints such as unateness, monotonicity, and asummability, these

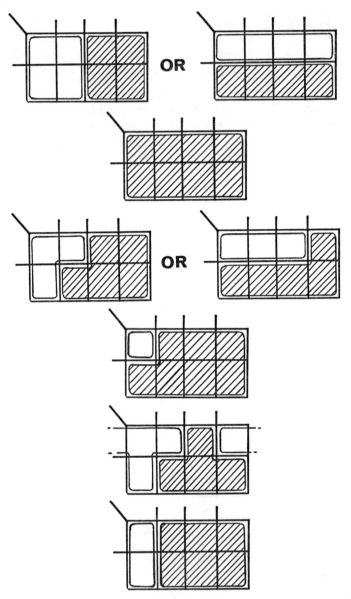

Fig. 3.13. The positive canonic linearly separable map patterns of $n \leq 3$. (Note that the shaded area or the non-shaded area may be the area of the true minterms of a given function. The position may be anywhere and with any orientation within the map layout to represent the particular functions.)

Geometric Constructions for Binary and Non-Binary Functions 165

being increasingly restrictive constraints which all linearly separable functions must obey[8-11]. We shall be briefly looking at these constraints in the chapter following.

The use of these classified map patterns for threshold-logic synthesis will also be a topic covered in the next chapter. From this subsequent work it will be appreciated that these map patterns form a very important place in the field of threshold-logic synthesis.

3.3.4. Rademacher–Walsh Functions and Spectral Translations

The orthogonal Rademacher–Walsh functions developed in Chapter 2 were shown to be equivalent to the binary input variables x_i, $i = 1 \ldots n$, of a system plus all the different exclusive-OR's of the x_i's. In addition a constant-value term was necessary in order to complete the set.

In order to determine the Rademacher–Walsh spectral coefficients for any given Boolean function $f(x)$, an {agreement $-$ disagreement} count was developed. In Fig. 2.8 of Chapter 2 it was suggested that this count could be made by effectively overlapping the map of the given function $f(x)$ with each of the 2^n maps of the Rademacher–Walsh functions. This forms a very easy method of hand calculation of the spectral coefficient values for $f(x)$, provided the Rademacher–Walsh function maps are available. Fig. 2.8 gave these maps for the $n = 3$ case. For $n = 4$ the maps are as illustrated in Fig. 3.14. For $n = 5$ a total of 32 maps are necessary, of which the first four and the last two only are illustrated in Fig. 3.15. The remaining ones may be very quickly prepared if necessary. However, it is in general true to say that hand computation of Rademacher–Walsh coefficient values becomes tedious for $n \geq 5$, say, and computer computation using fast transform methods becomes much more sensible. Nevertheless for up to say $n = 4$ the map-pattern correlation technique is adequate and very convenient.

Also in Chapter 2, the various invariance operations in the spectral domain which may be used to derive the spectral classification tables were detailed. These invariance operations will also be used in Chapter 5. However, for up to $n = 4$ these translations may be shown and explained very neatly on Karnaugh map layouts. They are as follows.

(i) Firstly, interchange of any binary input x_i with input x_j, $i \neq j \neq 0$, as depicted in Fig. 3.16a. Such interchange corresponds to

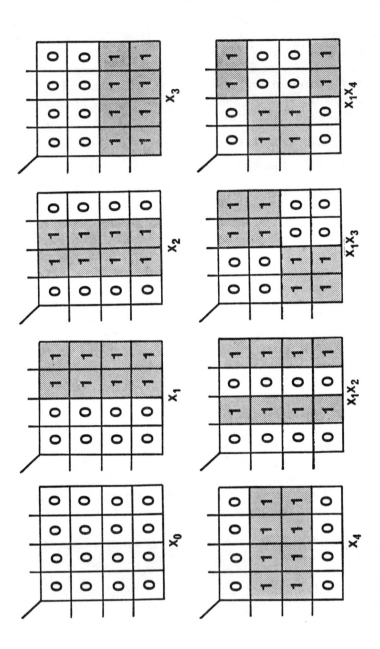

Geometric Constructions for Binary and Non-Binary Functions 167

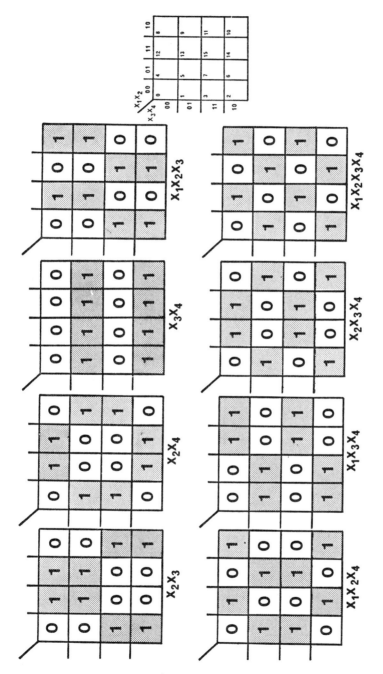

Fig. 3.14. The 16 Rademacher–Walsh functions of $n = 4$.

(continued)

Geometric Constructions for Binary and Non-Binary Functions 169

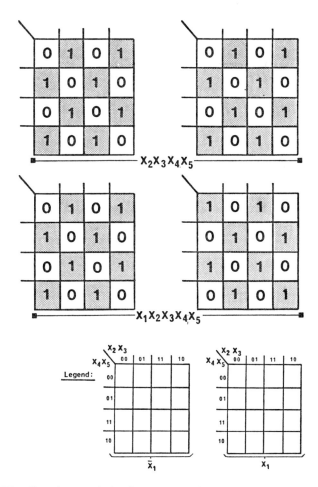

Fig. 3.15. The first four and the last two Rademacher–Walsh functions of $n = 5$.

the two groups of minterm values $\bar{x}_i x_j$ and $x_i \bar{x}_j$ being interchanged on the map layout as shown in Fig. 3.16b–g respectively. Notice that should the two groups of minterms in question be identical, then the function is non-equivalence symmetric in inputs x_i and x_j, as was illustrated in Fig. 3.9.

(ii) Negation of any input variable x_i consists of interchanging the entries in one half of the map layout with those in the other half. Fig. 3.17 illustrates the four such invariance operations for a $n = 4$ function.

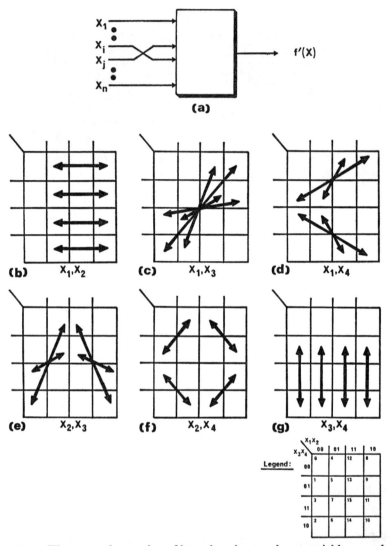

Fig. 3.16. The spectral operation of interchanging two input variables x_i and x_j: (a) the input interchange; (b)–(g) the six possible interchanges with four input variables.

(iii) Negation of the complete function $f(x)$, the third of the basic NPN invariance operations, merely consists of reversing the value of all the individual 0- and 1-valued minterms in the particular map of a function $f(x)$. It does not involve any regular interchange of groups of minterms on the map layout.

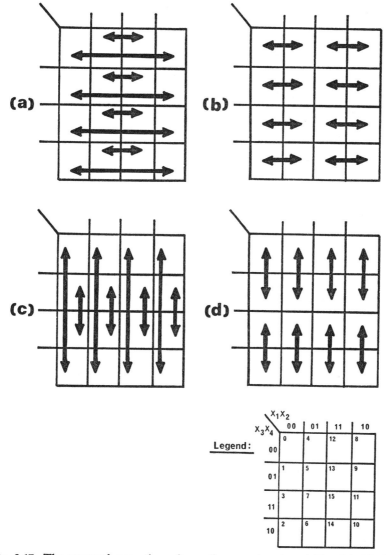

Fig. 3.17. The spectral operation of negating one input variable: (a)–(d) the negation of x_1, x_2, x_3, and x_4, respectively.

(iv) Replacement of any input variable x_i by the exclusive-OR signal $[x_i \oplus x_j]$, $x_i \neq x_j$, is the fourth spectral invariance operation. The effect of this operation is the interchange of minterms as shown in Fig. 3.18.

(v) The final spectral invariance operation is that of modifying the function output from $f(x)$ to $[f(x) \oplus x_i]$. This operation, like that of negating the complete function, is unique to a given function $f(x)$, and therefore no regular interchange of groups of minterms on the map layout is involved. Figure 3.19, however, illustrates this particular operation with an arbitrary function $f(x)$, from which the usefulness of the Karnaugh map in illustrating the operation is clear.

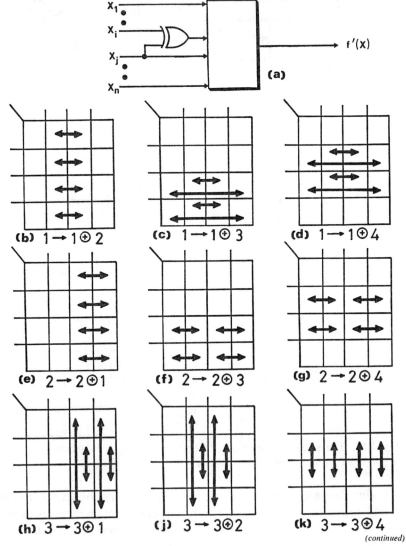

(continued)

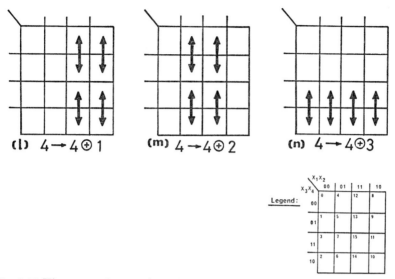

(l) $4 \rightarrow 4 \oplus 1$ (m) $4 \rightarrow 4 \oplus 2$ (n) $4 \rightarrow 4 \oplus 3$

Fig. 3.18 The spectral operation of replacing any input variable x_i with the exclusive-OR signal $[x_i \oplus x_j]$: (a) the input interchange; (b)–(n) the 12 possible operations with four input variables.

Fig. 3.19. The spectral operation of replacing the function output $f(x)$ with $[f(x) \oplus x_i]$: (a) the output interchange; (b) an example function $f(x)$; (c) choose $x_i = x_4$; (d) the resultant function $[f(x) \oplus x_4]$.

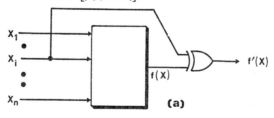

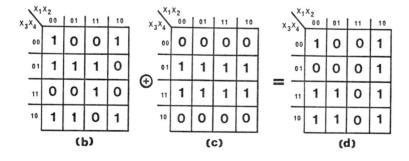

To summarize this particular section, the Karnaugh map gives an invaluable insight into the various operations in the spectral domain. In particular the reader may now be interested to go back to Fig. 2.10 of Chapter 2, and apply the mapping interpretations which we have now covered to the evolutions shown in Figs. 2.10a–g.

3.4. Map Constructions for Ternary and Higher-Valued Functions

The basic principle of the two-dimensional Karnaugh map layout for binary functions is that only one variable changes value when crossing any horizontal or vertical dividing line on the map. Without this fundamental principle the whole usefulness and application of the map collapses.

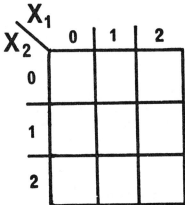

Fig. 3.20. A Karnaugh-type map format for ternary functions of two variables.

Let us attempt to extend the principle to ternary functions. Consider first a two-variable map $f(X_1, X_2)$, which therefore involves nine combinations of truth values. The simple Karnaugh-type map layout shown in Fig. 3.20 is clearly a possible layout, smooth continuity of the truth values being maintained across each horizontal and vertical dividing line, including edge-to-edge and top-to-bottom adjacencies.

To extend this concept to three variables X_1, X_2, X_3 immediately presents problems. It now proves impossible to preserve the continuity of all three variables across each horizontal and vertical dividing line of the map. Figure 3.21 shows three possible attempts to achieve a satisfactory two-dimensional format, but in all cases certain discontinuities always arise. In Fig. 3.21a adjacency is clearly lost between the

Geometric Constructions for Binary and Non-Binary Functions 175

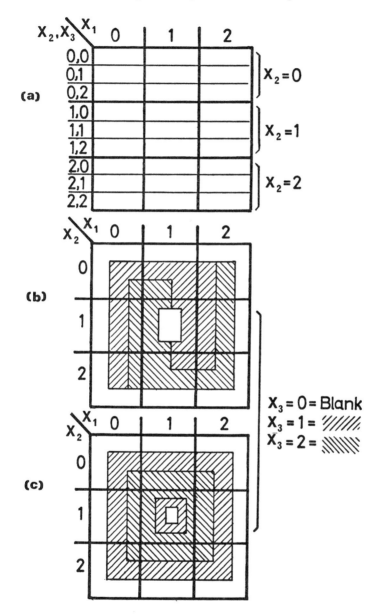

Fig. 3.21. Possible Karnaugh-type map format for ternary functions of three variables: (*a*) a simple non-cyclic format; (*b*) a partitioned format; (*c*) an alternative partitioned format.

$X_2 = 0, 1$, and 2 sections of the map, for example going from 000 to 010, from 001 to 011, and so on. Figure 3.21b is an improvement on this situation, but discontinuity is still introduced by the unavoidable central $X_3 = 0$ area which produces discontinuities between minterms 010, 110, and 210 and between minterms 100, 110, and 120. This particular format is something similar to certain formats suggested for binary functions of more than four variables[19], though in these binary diagrams the cross-hatched areas extend to the edges of the map which aids their interpretation to some degree. In the third suggested map format a more recognizable adjacency pattern for the majority of the terms is achieved, but at the loss of adjacency for both the central $X_3 = 0$ and the $X_3 = 1$ regions.

For four ternary variables X_1, X_2, X_3, X_4 it proves completely unsatisfactory to attempt to extend the cross-hatched-areas idea of

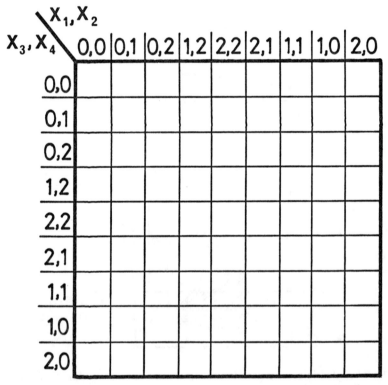

Fig. 3.22. A cyclic Karnaugh-map format for ternary functions of four variables.

Figs. 3.21b, c. Instead one is forced back into a plain map layout similar to Fig. 3.21a, with no possibility of maintaining the basic principles of adjacency between all truth values of the variables. About the best format which can be proposed is illustrated in Fig. 3.22.

To use any of the above diagrams for minimization purposes, triplets of adjacent minterms with one variable taking all three truth values have to be sought. (This compares with the search for pairs of 0- and 1-valued minterms in a binary minimization procedure.) For example the ternary expression

$$[X_1^1 X_2^0 X_3^0 + X_1^0 X_2^1 X_3^1 + X_1^1 X_2^1 X_3^0 + X_1^0 X_2^2 X_3^1 + X_1^1 X_2^2 X_3^0 + X_1^0 X_2^0 X_3^1]$$

will be found to minimize to

$$[X_1^1 X_3^0 + X_1^0 X_3^1]$$

where X_1^0 = input X_1 with value 0, X_1^1 = input X_1 with value 1, and so on. However, the lack of complete adjacency in any of the suggested map formats for $n \leq 3$ will be found to be a very serious inconvenience and the visual recognition of ternary prime implicants greatly impaired by this imperfection in the formats.

For $n \leq 3$ the hypercube picture of Fig. 3.4 proves to be a far superior construction for illustrative and minimization purposes[20]. We shall be using this construction in preference to any of these two-dimensional map formats in Chapter 6 of this book. For still higher-valued digital systems than ternary ($R > 3$) the pictorial situation becomes even more impossible to resolve for any practical purpose. No really satisfactory two-dimensional map formats can be proposed, and it is therefore possibly fortunate that no foreseeable need for considering digital systems of $R > 3$ is likely to be encountered!

3.5. Chapter Summary

Of all the geometric constructions reviewed in this chapter, it is undoubtedly the two-dimensional Karnaugh map for representing binary functions which possess the greatest significance. The various patterns of true and false minterms which may arise on the map layouts have a particular advantage in that the human eye and brain can identify patterns in a way which it is very difficult to quantify and mechanize—pattern recognition is a particularly strong human ability upon which our existence, relationships, and culture largely depend.

To extend map formats to more than, say five binary variables, or to adapt it to ternary and higher-valued digital systems, proves increasingly difficult and finally provides little if any practical significance. Subsequent chapters of this book, therefore, will often refer to the Karnaugh map to illustrate binary features of direct concern, but not very great use will be made in future pages of the other geometric constructions which for completeness have been reviewed in this particular chapter.

References

1. Lewin, D. *Logical Design of Switching Circuits*. Elsevier/North Holland, New York; Nelson, London, 1974.

2. Barna, A., and Porat, D.I. *Integrated Circuits in Digital Electronics*. John Wiley Interscience, New York, 1973.

3. Quine, W.V. The problem of simplifying truth functions. *Am. Math. Mon.* **59**, 521–31, 1952.

4. McClusky, E.J. Minimisation of Boolean functions. *B.S.T J.* **35**, 1417–44, 1956.

5. Blake, I.F. *Algebraic Coding Theory* Dowden, Hutchinson, and Ross, Stroudsburg, Pa. 1973.

6. Peterson, W.W., and Weldon, E.J. *Error Correcting Codes* MIT Press, Cambridge, Mass., 1972.

7. Mukhopadhyay, A. Lupanov decoding networks. *Recent Developments in Switching Theory*. Academic Press, New York, 1971.

8. Dertouzos, M.L. *Threshold Logic: A Synthesis Approach*. MIT Press, Cambridge, Mass., 1965.

9. Winder, R.O. *Fundamentals of Threshold-Logic*. Air Force Cambridge Sci. Rep. No. 1, AFCLR-68-0086, 1968.

10. Hu, S.T. *Threshold Logic*. University of California Press, Berkeley, Calif., 1965.

11. Lewis, P.M., and Coates, C.L. *Threshold-logic* John Wiley, New York, 1967.

12. Kaszerman, P. A geometric test-synthesis procedure for a threshold device. *Inform. Control* **6**, 381–98, Dec. 1963.

13. Karnaugh, M. A map method for synthesis of combinational logic circuits. *Trans. AIEE* **72**, 593–99, 1953.

14. Veitch, E.W. A chart method of simplifying truth functions. *Proc. Association for Computing Machinery Conf., May 1952,* 127–33.

15. Flegg, H.G. *Boolean Algebra and its Application.* Blackie, London, 1964.

16. Zissos, D. *Logic Design Algorithms.* Oxford University Press, Oxford, U.K., 1972.

17. Hurst, S.L. The correlation between Karnaugh map patterns and Chow parameters for 3-variable linearly-separable functions. *Electron. Letter.* **6**, 182–3, March 1970.

18. Hurst, S.L. Synthesis of threshold-logic networks using Karnaugh-mapping techniques. *Proc. IEE* **119**, 1119–28, Aug. 1972.

19. Anderson, D.E., and Cleaver, F.L. Venn-type diagrams for arguments of N terms. *J. Symbol. Logic* **30**, 113–18, June 1965.

20. Hurst, S.L. An extension of binary minimisation techniques to ternary equations. *Computer J.* **11**, 277–86, Nov. 1968.

Chapter 4

The Design of Binary Logic Networks Using Principally Threshold-Logic Gates

Introduction

In this chapter we shall be leading on to the use of threshold-logic gates to realize random-logic combinatorial networks, followed by a consideration of the possibilities of their use in simple sequential applications.

Currently this work may be viewed as a mainly theoretical exercise in the handling of digital signals, owing to the present lack of threshold-logic gates on the commercial market. This situation, however, may change if and when integrated circuit manufacturers begin to look for further products for the small-scale integrated circuit (s.s.i.) market[1,2,3,], or more effective monolithic circuit layouts for medium-scale (m.s.i.) or large-scale (l.s.i.) applications[4].

To date the two main difficulties of threshold logic, namely (i) network synthesis methods to exploit efficiently the increased logical power of threshold relationships and (ii) the circuit design of the threshold-logic gates themselves, have together precluded the committed industrial pursuit of the use of such gates alongside of or as an alternative to conventional vertex gates. Further, the vast investment and successful expansion in integrated circuit technology has so far meant that chip size has not been a serious limitation—chip areas have increased and active device areas have decreased such that in general the l.s.i. designer has succeeded in laying out his requirements on a chip without finding the need for new design methods or logically more efficient gate configurations. This situation may, however, change, and

Design of Threshold-Logic Networks 181

hence the concepts which we shall cover in this chapter may be an aspect of future network design. In a later chapter we shall be looking at the parallel problem of the circuit design of the more complex logic gates themselves.

4.1. Direct Threshold Synthesis of Given Binary Functions

It would be ideal if some universally applicable technique could be formulated whereby any given combinational function $f(x)$ could be readily translated, preferably by hand, into a threshold-logic realization. This objective, however, is fundamentally unattainable owing to the non-linear relationships that exist between the Boolean domain and the threshold domain. For very simple functions a quick trial-and-error translation is usually possible, for example

$$f(x) = [x_1 + x_2 x_3] \quad \text{(Boolean)}$$
$$= \langle 2x_1 + x_2 + x_3 \rangle_2 \quad \text{(threshold)}.$$

Beyond such trivial examples more scientific techniques are desirable, such as are involved in the developments following.

The three basic questions which arise when a threshold-logic realization of any given binary function $f(x)$ is being considered are as follows.

(*a*) Is the given function a linearly separable function; that is, can it be realized by the use of one single threshold-logic gate?

(*b*) If the function *is* linearly separable, then what are the optimum input weighting values and output threshold value required on the gate?

(*c*) If the function *is not* linearly separable, then how best can it be decomposed so that threshold-logic gates may be used to best advantage in its realization?

Let us therefore consider each of these areas to see in more detail the problems involved.

4.1.1. Linear Separability; Unateness, Monotonicity, Asummability

All linearly separable binary functions have to satisfy three increasingly restrictive criteria, namely unateness, monotonicity, and asummability. Functions which fail to satisfy all three are not linearly separable, and therefore do not have a single threshold-logic gate realization.

Unateness is the weakest of these three criteria. A unate function is one in which *no input variable x_i, $i = 1 \ldots n$, appears in both uncomplemented and complemented form in the minimized Boolean expression for the given function.*[1] For example the function

$$f(x) = [x_1\bar{x}_2 + \bar{x}_2 x_3]$$

is a unate function, but

$$f(x) = [x_1\bar{x}_2 + x_2 x_3]$$

is a non-unate function. The properties of unate functions have been extensively investigated and have been shown to be a necessary but not sufficient test for linear separability[5-8]. Thus all threshold-logic functions are unate, but not all unate functions are threshold-logic functions. Necessity is readily shown by the following development, which is simpler than the more formal algebraic proofs given by several authorities.

Suppose in a given function $f(x)$, given in a minimized sum-of-products form, that the prime-implicant product term $x_1 x_2 x_3$ is present. This requires that, if a threshold-logic realization is being sought, the arithmetic sum of the weights $a_1 + a_2 + a_3$ is greater than or equal to the gate threshold value t in order to realize the term $x_1 x_2 x_3$. If a_1 (say) was zero or negative, then $a_2 + a_3$ would equal or exceed the value t, but this would require the product term $x_2 x_3$ rather than $x_1 x_2 x_3$ to appear in the minimized expression for $f(x)$. Hence a_1 must be non-zero and positive, and similarly for a_2 and a_3. Now by a similar reasoning, should a product term such as $\bar{x}_1 x_4 x_5$ also appear in the minimized expression for $f(x)$, then the weight a_1 associated with x_1 should now be non-zero and negative as, numerically, $\bar{x}_1 x_4 x_5 = \{1 - x_1\} x_4 x_5$. Clearly the weight a_1 associated with x_1 cannot be both non-zero positive and non-zero negative, and hence terms such as $x_1 x_2 x_3$ and $\bar{x}_1 x_4 x_5$ cannot simultaneously arise in a linearly separable function.

Unateness therefore is the most simple but weakest test which may be applied to a given Boolean function to check for linear separability.

If the given function is unate, the next stronger test is *monotonicity*. Monotonicity first requires the definition of function comparability which is defined as follows. Two functions $f_1(x)$ and $f_2(x)$ are said to be *comparable* if wherever the value of $f_1(x) = 1$ then the value of $f_2(x)$

---

[1] In all these classification criteria it is a prerequisite that the Boolean function shall be in a minimized form; the minimized sum-of-products form is normally employed.

is also 1, or *vice versa*. Specifically, if $f_1(x) = 1$ whenever $f_2(x) = 1$, then

$$f_1(x) \supseteq f_2(x)$$

or if $f_2(x) = 1$ whenever $f_1(x) = 1$ then

$$f_2(x) \supseteq f_1(x).$$

A simple case is AND and OR comparability; if $f_1(x) = [x_1 + x_2 + x_3]$ and $f_2(x) = [x_1 x_2 x_3]$, then clearly $f_1(x) = 1$ whenever $f_2(x) = 1$, that is

$$f_1(x) \supseteq f_2(x).$$

Reverting to monotonicity, let any function

$$f(x) = f(x_1, \ldots, x_i, \ldots, x_n)$$

be decomposed about x_i into two disjoint functions

$$f(x_1, \ldots, 0, \ldots, x_n) \quad \text{and} \quad f(x_1, \ldots, 1, \ldots, x_n).$$

Then the given function $f(x)$ is said to be 1-monotonic if the two $n - 1$ variable disjoint functions per decomposition *are comparable for all x_i decompositions*, $i = 1 \ldots n$.

Similarly, if $f(x)$ is still further decomposed about a second variable $x_j, j \neq i$, to give a total of four disjoint $n - 2$ variable functions per decomposition, then $f(x)$ is said to be 2-monotonic if all possible pairs of functions decomposed about each x_i, $x_j = 00, 01, 10$, and 11 are comparable (see Fig. 4.1, page 184).

Continuing further, a function $f(x)$ is said to be k-monotonic if comparability holds for all possible pairs in each set of $(n - k)$-variable reduced functions, and finally $f(x)$ is said to be *completely monotonic* if comparability holds for $k = n - 1$, that is no matter how far the original function $f(x)$ is decomposed into all possible disjoint functions, at all stages of decomposition the corresponding reduced functions are comparable.

Monotonicity is a very powerful check of linear separability, and has been extensively investigated[6-11]. All linearly separable functions are completely monotonic, 1-monotonic, 2-monotonic, etc., providing correspondingly more selective checks for linear separability. One-monotonicity corresponds precisely to unateness, and is therefore the weakest check for linear separability. It was originally hoped that complete monotonicity would be a necessary and sufficient test of linear separability for any given function[8], but sufficiency was dis-

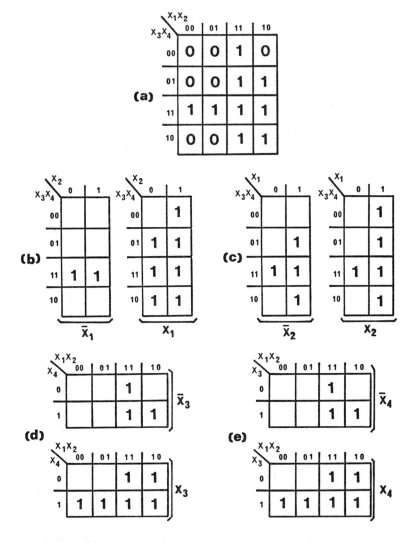

Fig. 4.1. Decomposition and comparability of the function

$$f(x) = [x_1(x_2 + x_3 + x_4) + x_3 x_4],$$

proving 1-monotonicity: (a) the given function; (b) decomposition about x_1, giving the reduced functions $f(x_2,x_3,x_4)$; (c) decomposition about x_2, giving the reduced functions $f(x_1,x_3,x_4)$; (d) decomposition about x_3, giving the reduced functions $f(x_1,x_2,x_4)$; (e) decomposition about x_4, giving the reduced functions $f(x_1,x_2,x_3)$.

proved by a counter-example function of $n = 9$ disclosed by Gabelman[12]. Muroga et al., however, have exhaustively proved that complete monotonicity is a sufficient test for linear separability for all functions of $n \leq 8$[13].

As an illustration of monotonicity, consider a known linearly separable function

$$f(x) = [x_1(x_2 + x_3 + x_4) + x_3x_4].$$

Its Karnaugh map plot is shown in Fig. 4.1.a. The first decomposition of $f(x)$ about each of the four input variables in turn produces the four pairs of maps shown in Figs. 4.1.b–e. The given function will now be seen to be 1-monotonic, as all pairs of functions of three variables are comparable. Decomposing the function still further will demonstrate the 2-monotonicity and the 3-monotonicity checks. Figure 4.2.a shows the further decomposition into sets of four two-variable functions, with Fig. 4.2b showing the further final decomposition into sets of eight single-variable functions. All decompositions will be seen to show comparability and hence the given function is proved to be completely monotonic.

To contrast these illustrations of complete monotonicity, the function $f(x) = [x_1x_2 + x_3x_4]$ is a simple unate function but not a linearly separable one. Figure 4.3 shows its progressive decomposition into disjoint functions of fewer variables, from which it will be seen that 1-monotonicity (unateness) is present but 2-monotonicity is absent as comparability fails in four of the six reduced functions of two variables. The function therefore is not completely monotonic, and therefore cannot be linearly separable.

The failure of complete monotonicity to be a sufficient test for linear separability for functions of greater than $n = 9$ has meant that a third yet stronger test for linear separability has been proposed. This is *asummability*. Asummability, however, first requires that we define the opposite property, that of summability. Any given function $f(x)$ of n variables is said to be "k-summable," k being any integer number $2 \leq k \leq \frac{1}{2}2^n$, if a set of k true minterms and a set of k false minterms can be found such that the vector addition of the k true minterms is equal to the vector addition of the k false minterms; that is

$$\sum_1^k \vec{x}_{\text{true}} = \sum_1^k \vec{x}_{\text{false}}.$$

If any such set of minterms can be found, then the function $f(x)$ is said to be k-summable. For example, suppose a function $f(x)$ contains min-

186 The Logical Processing of Digital Signals

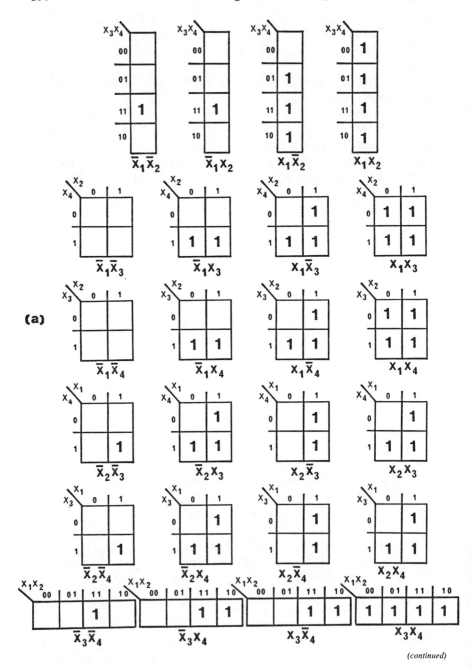

(a)

(continued)

Design of Threshold-Logic Networks   187

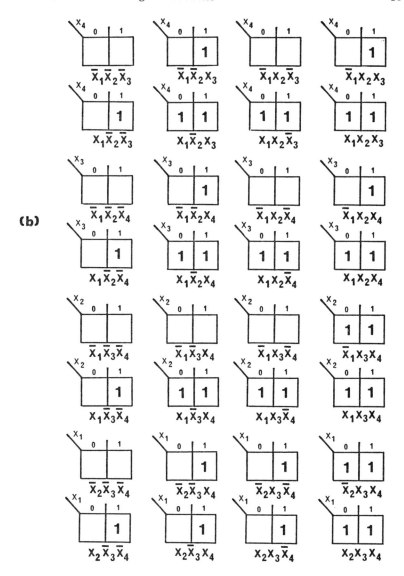

Fig. 4.2. Further decomposition and comparability of the function of Fig. 4.1, proving 2-monotonicity and 3-monotonicity. (*a*) decompositions about x_1x_2, x_1x_3, x_1x_4, x_2x_3, x_2x_4 and x_3x_4, for proof of 2-monotonicity. (*b*) decompositions about $x_1x_2x_3$, $x_1x_2x_4$, $x_1x_3x_4$ and $x_2x_3x_4$, for proof of 3-monotonicity, which in this case is complete monotonicity.

terms 001 and 111 amongst its true minterms, and 011 and 101 amongst its false minterms, then the vector sum of these two pairs gives

$$\sum \vec{x}_{true} = \{(0,0,1) + (1,1,1)\} = 1,1,2$$

and

$$\sum \vec{x}_{false} = \{(0,1,1) + (1,0,1)\} = 1,1,2,$$

$$= \sum \vec{x}_{true}.$$

(a)

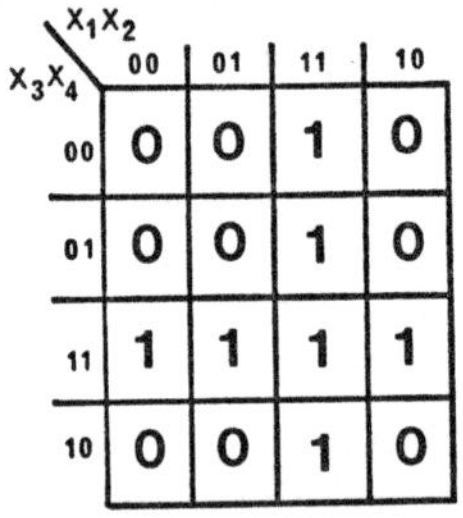

x_3x_4 \ x_1x_2	00	01	11	10
00	0	0	1	0
01	0	0	1	0
11	1	1	1	1
10	0	0	1	0

(b)

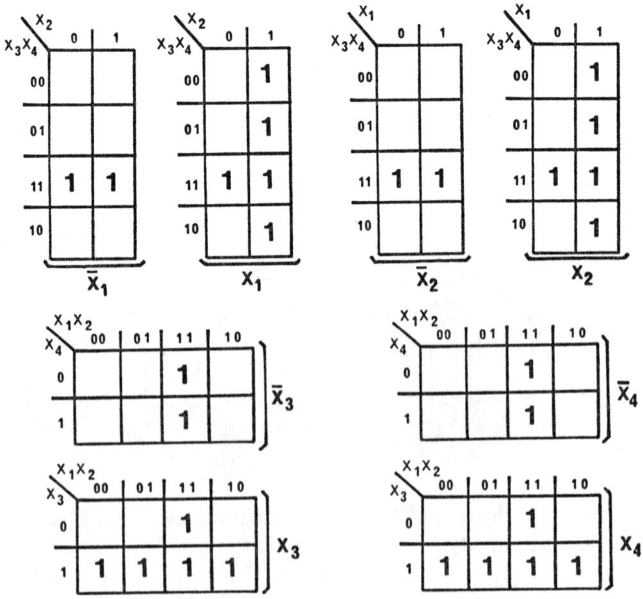

(continued)

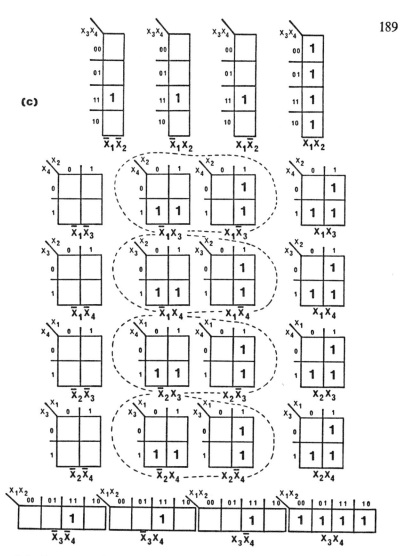

Fig. 4.3. Decomposition and comparability checks on the function $f(x) = [x_1x_2 + x_3x_4]$: (a) the given function; (b) first decompositions, comparability present in all cases; (c) second decompositions, comparability fails in the ringed x_1x_3, x_1x_4, x_2x_3, and x_2x_4 decompositions.

Thus the given function is shown to be 2-summable. If no such set of k true and k false minterms can be found, then the function is said to be *k-asummable*.

Now, if a threshold-logic realization is being considered, the input weighting summation of each true minterm will be equal to or greater than the gate threshold t. Thus if the input weighting summation of k separate true minterms are added together, this total will be $\geq k \cdot t$. In a similar manner, the addition of the input summation weighting of k separate false minterms must be $< k \cdot t$. Thus in a threshold-logic realization, which implies linear separability, the total input summation of any k true and any k false minterms can never be equal, but if a function is shown to be k-summable then, no matter what input weighting values are given to the individual x_i inputs, there will be an equal input summation for k true and k false minterms. Thus a k-summable function cannot be a linearly separable function, and hence the corollary follows that all linearly separable functions must be k-asummable for all possible values of k. Formal proof that complete asummability is a necessary and a sufficient condition for linear separability may be found in Elgot[14] and elsewhere[15,16].

It can be shown that 2-asummability is exactly equivalent to complete monotonicity. However, as a test for linear separability of any given function $f(x)$, as a preliminary to see whether a single threshold-logic realization is possible, complete asummability can be a very lengthy test. By hand it is very tedious for n greater than, say, 4, although it is a relatively simple digital computer exercise provided sufficient computing time and storage capacity is available.

To summarize, therefore, if a function is given in a minimum sum-of-products Boolean form, then unateness is a simple first check to apply, as it will show immediately if the function is not linearly separable. Monotonicity checks are convenient if the function is small enough to be readily plotted on a Karnaugh map layout, that is for functions of, say, $n \leq 5$, and if complete monotonicity is found for such small functions then linear separability is proved. Indeed the 2-monotonic check is an adequate check for a very high proportion of given functions. Asummability checks, whilst guaranteeing to prove or disprove linear separability, are tedious and not at all convenient for hand calculation.

4.1.2. Input Weights and Gate Threshold Values

Given that a function is linearly separable and therefore has a single threshold logic-gate realization, the next problem is to determine the set of input weights a_i, $i = 1 \ldots n$, and the gate-threshold value necessary in order to realize the given function. For any linearly separable

Design of Threshold-Logic Networks

function there are theoretically an infinite set of a_i and t values which will realize the function, but minimum-integer values are the values desired in all normal circumstances.

Now the threshold realization of any function $f(x)$ of n input variables must obey the series of 2^n independent inequalities

$$a_1 x_1 + a_2 x_2 + \ldots + a_n x_n \geq t \text{ or } < t$$

where $\geq t$ is obeyed wherever $f(x) = 1$ and $< t$ is obeyed wherever $f(x) = 0$. If an arithmetic solution for a_1 to a_n and t can be found then this gives a threshold realization for the given function. However, blindly to try numbers in the full list of 2^n inequalities is obviously an inefficient procedure towards a final solution. Some guide towards a solution is clearly desirable.

One approach is to try to reduce the number of inequalities which have to be considered. Now it may be recalled from developments in Chapter 2 and the notes given in Appendix A, that when the magnitude of known minimum-integer realizing weights $|a_1|$ to $|a_n|$ of any linearly separable function are compared with the $|b_1|$ to $|b_n|$ values for the function, where

$$b_i, i = 1 \ldots n, \triangleq \sum_{p=0}^{2^n-1} \{f(y) \cdot y_i\}$$

then the maximum, co-equal, and minimum values of the $|a_i|$'s directly reflect the maximum, co-equal, and minimum values of the respective $|b_i|$'s. For example, should, say, $|b_2|$ be larger than any other $|b_i|$, then the input weight $|a_2|$ will be larger than any other $|a_i|$ input weight. There is, however, no direct arithmetic relationship between the $|b_i|$ and the $|a_i|$ values. Therefore if the given function is transformed into an equivalent positive canonic standard form by permutating and negating (NPN) operations, such that finally we obtain the order

$$x_1 \gtrsim x_2 \gtrsim \ldots \gtrsim x_n,$$

where $x_i \gtrsim x_{i+1}$ indicates that

$$\sum_{p=0}^{2^n-1} \{f(y) \cdot y_i\} \geq \sum_{p=0}^{2^n-1} \{f(y) \cdot y_{i+1}\}$$

then with this procedure we know that $a_1 \leq a_2 \leq \ldots a_n \leq 0$ in the final threshold realization, provided one exists. This will now enable many of the 2^n inequalities to be discarded in the subsequent search for the realizing a_i values.

Techniques from here on differ somewhat according to different authorities. All, however, produce an "initial tableau" or "initial matrix" of inequalities derived from the essential prime implicants of the standardized positive canonic function. For example, the standardized positive canonic function $f(x) = [x_1 (x_2 + x_3 + x_4 x_5) + x_2 x_3 x_4]$ gives the following inequalities:

$$a_1 + a_2 \geq t \qquad (4.1)$$
$$a_1 + a_3 \geq t \qquad (4.2)$$
$$a_1 + a_4 + a_5 \geq t \qquad (4.3)$$
$$a_2 + a_3 + a_4 \geq t \qquad (4.4)$$
$$a_1 + a_4 < t \qquad (4.5)$$
$$a_1 + a_5 < t \qquad (4.6)$$
$$a_2 + a_3 + a_5 < t \qquad (4.7)$$
$$a_2 + a_4 + a_5 < t \qquad (4.8)$$
$$a_3 + a_4 + a_5 < t \qquad (4.9)$$

However, because the a_i's have been arranged in a strict numerical value order, equation (4.1) may be disregarded in view of equation (4.2), equation (4.6) may be disregarded in view of equation (4.5), and so on. Finally a minimum-length matrix of inequalities is left for which a minimum-integer solution may be sought by linear programming or similar means. For small functions this can be done by hand, but for more complex functions computer aid is clearly necessary. Details of the considerable existing work in this area may be found published[7,9,12,17,18].

A slightly different approach to the determination of the realizing weights and threshold value is to choose initial values for the a_i's and t and then iteratively perturb the values chosen so as to obtain a final acceptable solution. This approach requires two desirable features, firstly that the "first guess" of the values shall be a reasonable guess and not completely random, and secondly that there shall be some converging mechanism to provide convergence towards a final satisfactory solution.

Considerable research in this area has been undertaken. Normally a "first-guess" solution for the a_i's is to take the directly corresponding b_i values, which of course may be calculated from the given function, this approach corresponding to an earlier but generally unacknowledged observation of Hawkins[19] who observed that an approximate solution for each input weight is frequently given "by the number of agreements of the variable value and the function value, minus the number of disagreements." This of course is numerically identical to the b_i values. The "first-guess" values are then tested to see if they

Design of Threshold-Logic Networks

provide a correct realization for the given function. This may be done by individually testing the realization on all 2^n minterms, or on a reduced set of inequalities such as previously mentioned, or by using the realizing equation (2.8) given in Chapter 2. If these "first-guess" values for the a_i's do not correctly realize the given function, then certain perturbation algorithms have been proposed which modify the originally chosen values such that eventually a satisfactory solution for the a_i's is reached, provided of course that the given function is linearly separable. There may, however, be several iterative perturbations of the values before a successful solution is found.

Yet other numerical solutions for the a_i values based upon the b_i values have been proposed by Dertouzos, Winder, and others[20-22]. These frequently involve equations based upon n-dimensional geometric considerations, and as a result do not directly give integer values for the a_i's. Such non-integer results then require trial-and-error leveling up or down to achieve a final integer result for the weights and threshold.

The theory why the b_i values sometimes fail to give a correct threshold realization will be found in Kaplan and Winder[23]. However, even when the b_i values do give a functionally correct result, this result is frequently not the minimum integer realization. As an example of this very interesting feature, let us consider a simple linearly separable function and work through the b_i and a_i values involved. Let us take the linearly separable function

$$f(x) = [x_1(x_2 + x_3) + x_2 x_3 x_4]$$

which almost intuitively can be seen to have a threshold realization of

$$f(x) = \langle 3x_1 + 2x_2 + 2x_3 + x_4 \rangle_5.$$

However, assuming for the moment we do not know this result, let us attempt a calculation for the a_i values based upon the b_i's. Evaluation by the usual agreement-disagreement count (see Appendix A) will yield the b_i values following:

b_0	b_1	b_2	b_3	b_4
-2	10	6	6	2

Now from equation (2.8) (Chapter 2) we have that, when the chosen a_i values correctly realize $f(x)$,

$$\sum_{i=0}^{n} a_i b_i = \sum_{p=0}^{2^n-1} \left\{ + \left| \sum_{i=0}^{n} a_i y_i \right| \right\}.$$

Taking the left-hand side of this equation we have

$$\{(a_0)(-2) + (a_1)(+10) + (a_2)(+6) + (a_3)(+6) + (a_4)(+2)\}$$

which taking $a_i = b_i$ gives the summation

$$\{(-2)(-2) + (+10)(+10) + (+6)(+6) + (+6)(+6) + (+2)(+2)\}$$
$$= \{4 + 100 + 36 + 36 + 4\} = 180.$$

Taking the right-hand side of the above equation and again taking $a_i = b_i$, we have the summation tabulated below.

Minterm p	$a_0 = -2$ y_0	$a_1 = 10$ y_1	$a_2 = 6$ y_2	$a_3 = 6$ y_3	$a_4 = 2$ y_4	$\sum_{i=0}^{4} a_i y_i$	$\left\|\sum_{i=0}^{4} a_i y_i\right\|$
0	+1	−1	−1	−1	−1	−26	26
1	+1	−1	−1	−1	+1	−22	22
2	+1	−1	−1	+1	−1	−14	14
3	+1	−1	−1	+1	+1	−10	10
4	+1	−1	+1	−1	−1	−14	14
5	+1	−1	+1	−1	+1	−10	10
6	+1	−1	+1	+1	−1	− 2	2
7	+1	−1	+1	+1	+1	+ 2	2
8	+1	+1	−1	−1	−1	− 6	6
9	+1	+1	−1	−1	+1	− 2	2
10	+1	+1	−1	+1	−1	+ 6	6
11	+1	+1	−1	+1	+1	+10	10
12	+1	+1	+1	−1	−1	+ 6	6
13	+1	+1	+1	−1	+1	+10	10
14	+1	+1	+1	+1	−1	+18	18
15	+1	+1	+1	+1	+1	+22	22

$$\therefore \sum\left\{\left|\sum_{i=0}^{4} a_i y_i\right|\right\} = 180$$

Thus the right-hand summation of the equation is numerically equal to the left-hand summation, and hence the values chosen for the a_i do provide a correct realization for the function $f(x)$. Therefore we have proved that the solution

$$\begin{array}{ccccc} a_0 & a_1 & a_2 & a_3 & a_4 \\ -2 & 10 & 6 & 6 & 2 \end{array}$$

Design of Threshold-Logic Networks

is satisfactory. However, we can clearly divide all these values by 2 to give the smaller-value realization of

$$\begin{array}{ccccc} a_0 & a_1 & a_2 & a_3 & a_4 \\ -1 & 5 & 3 & 3 & 1 \end{array}$$

which gives the gate threshold value t of

$$t = \tfrac{1}{2}\{(5 + 3 + 3 + 1) - (-1) + 1\}$$
$$= \tfrac{1}{2}\{12 + 1 + 1\} = 7$$

that is, a threshold-logic realization of

$$f(x) = \langle 5x_1 + 3x_2 + 3x_3 + x_4 \rangle_7.$$

However, now we notice that this is not the minimum-integer solution of

$$f(x) = \langle 3x_1 + 2x_2 + 2x_2 + x_4 \rangle_5$$

and that there is no mathematical path available to reduce the result based on the b_i values to this minimum-integer result! So generation of minimum-integer values cannot be guaranteed starting with the b_i values, even if the b_i values do correctly realize the given function. It is left for the reader to check that if the minimum-integer solution for the a_i's are used in equation (2.8) (Chapter 2) instead of the b_i-based ones, then the two sides of the equation are still numerically equal—their total is now each 58 compared with the 180 total in the above b_i-based summations.

Thus the problem of directly determining the a_i values, and in particular the minimum-integer values, for any given function $f(x)$, first knowing that $f(x)$ is linearly separable and therefore has single-gate realizability, is seen to be a particularly difficult problem with no direct mathematical means of solution. Fortunately, for $n \leq 7$ there is no need for such determination to be undertaken as the results have been freely published[24,25]; for $n \leq 6$ the values are tabulated in Appendix A. The same information has also been evaluated for $n \leq 8$, but the length of the tabulation is now becoming too great for general publication. The statistics and particular comments on the $n \leq 8$ results, however, have had separate publication[26].

We shall return to consider these published Chow-parameter look-up tabulations in more detail in §4.2.

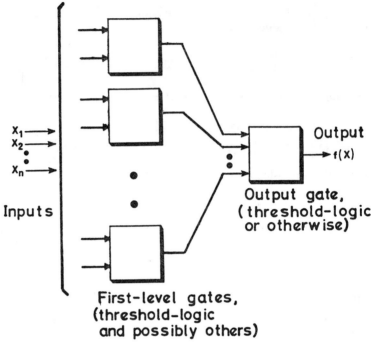

Fig. 4.4. The schematic two-level network realization of a non-linearly separable function $f(x)$, incorporating threshold-logic gates.

4.1.3. Non-Linearly Separable Functions

If a given function $f(x)$ is a non-linearly separable function, shown by being non-unate, or not completely monotonic, or by being k-summable, then fundamentally no single threshold-logic realization for $f(x)$ is possible. How then may threshold-logic gates be used to advantage in such situations?

Unfortunately there is no simple synthesis method available which will directly lead to a "best" solution, where "best" may be defined as using the fewest number of gates, or the fewest number of gate interconnections, or any other goodness criteria. The basic problem is to break the given problem down into two or more reduced functions, each of which shall be a simple threshold-logic or vertex function, or possibly an exclusive-OR function, these separate functions then being appropriately combined to produce the required function output $f(x)$.

Design of Threshold-Logic Networks

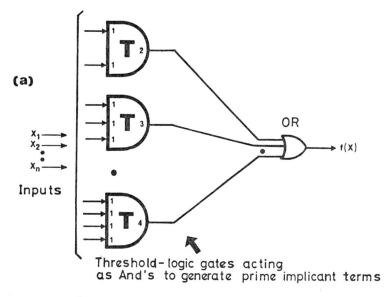

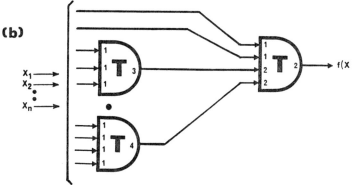

Fig. 4.5. Threshold-logic realization of a non-linearly separable function $f(x)$, based upon the minimum sum-of-products prime implicants: (a) straightforward use of threshold-logic gates for each sum-of-products term; (b) a slight improvement to incorporate the second-level OR relationship within a threshold-logic gate.

This build-up of $f(x)$ is illustrated in Fig. 4.4, and is of course identical to our normal form of network realization if vertex gates only are being used.

The Boolean data for $f(x)$ in truth-table or algebraic form does not provide a good basis for this synthesis. If $f(x)$ is given in a minimum sum-of-products form, then a direct but most inefficient form of threshold-logic realization is to realize each product (AND) term and the overall sum (OR) term by a separable threshold-logic gate, remembering that all vertex-gate relationships may be generated by single gates with simple weighting/gate threshold values. This is as shown in Fig. 4.5a, and is clearly a completely unjustified use of threshold-logic gates. A slight improvement can be suggested by combining the final OR connective with one of the proceeding AND's, as shown in Fig. 4.5b but even so we have done very little to exploit fully the power of the threshold-logic relationships.

A trial-and-error technique starting with the minimized sum-of-products expression for $f(x)$ which may yield a more efficient threshold-logic solution is as follows. The basic concept is to try to amalgamate as many prime-implicant terms together into one threshold relationship as possible, as distinct from the realizations shown in Fig. 4.5 which basically kept each prime-implicant term as a separate entity.

Let us take an example function to illustrate this possible technique. Take the function

$$f(x) = [x_1 x_2 + x_1 \bar{x}_3 + x_1 x_4 + x_2 x_3 + \bar{x}_1 x_3 \bar{x}_4]$$

which is in minimum form. This function is clearly not linearly separable, as both true and complemented variables appear in its equation. Let us, however, collect the terms together into unate groups which we can then separately test for linear separability. One grouping for this function is

$$f(x) = [\{x_1(x_2 + \bar{x}_3 + x_4)\} + \{x_3(x_2 + \bar{x}_1 \bar{x}_4)\}].$$

Each of the terms within the { } braces is unate, and being in this case simple functions have an obvious threshold realization of

$$\langle 3x_1 + x_2 + \bar{x}_3 + x_4 \rangle_4$$

and

$$\langle 3x_3 + 2x_2 + \bar{x}_1 + \bar{x}_4 \rangle_5$$

respectively. Hence the realization shown in Fig. 4.6a is produced. The OR gate of this realization may be incorporated in one of the threshold-logic gates, for example as indicated in Fig. 4.6b.

This method of synthesis is largely based upon recognition of unateness, but as we have seen unateness is not an absolute guarantee of linear separability. Hence rearrangement of the prime implicants into a particular unate grouping will not always provide a solution, and trial-and-error regrouping may become necessary.

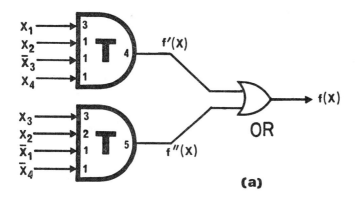

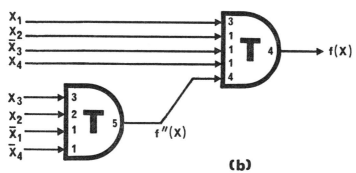

Fig. 4.6. Threshold-logic realization of the non-linearly separable function
$$f(x) = [x_1x_2 + x_1\bar{x}_3 + x_1x_4 + x_2x_3 + \bar{x}_1x_3\bar{x}_4]$$
based upon unate groupings of prime-implicant terms: (a) threshold-logic/OR realization; (b) elimination of the second-level OR gate.

A further shortcoming of this method is that the prime-implicant division may not be the optimum division of the given function into linearly separable terms. Indeed a better result for the function illustrated in Fig. 4.6 is the OR of the two threshold-logic functions

$$\langle 2x_1 + \bar{x}_3 + x_4 \rangle_3$$

and

$$\langle 2x_3 + \bar{x}_1 + x_2 + \bar{x}_4 \rangle_4$$

a result which can be found by mapping techniques which we shall consider later in this chapter. Nevertheless, in spite of shortcomings this prime-implicant grouping approach can conveniently produce reasonably efficient realizations for many non-linearly separable problems.

To summarize the sections of this chapter so far, it is not straightforward to generate directly by hand an efficient threshold-logic synthesis of a function $f(x)$ from given binary data, unless the given function is simple. The published Chow-parameter tabulations, however, provide a much more relevant basis upon which to proceed, with techniques such as we will consider in the following sections.

4.2. Chow-parameter tabulations and single-gate realizability

4.2.1. The Chow Parameters

The definition of the b_i Chow parameters has already been covered. The numerical order of the b_i magnitudes consitutes a type SD numerical classification for all functions, although as we have just seen the b_i values themselves do not always lead to the minimum-integer a_i realizing weights for each classification entry.

Generation of all the canonic classification entries for any given value n may be approached in two ways. The first way is to generate all n-valued linearly separable functions from the known complete set of $(n-1)$-variable linearly separable functions, by pairing and testing all possible resulting functions:

$$f(x) = [\bar{x}_n f_1(x_{n-1}) + x_n f_2(x_{n-1})]$$

where $f(x)$ is the sought n-variable function, $f_1(x_{n-1})$ and $f_2(x_{n-1})$ are all possible pairs of the known $(n-1)$-variable functions, and x_n is the additional n^{th} variable. Note that $f_1(x_{n-1})$ may be the same function as $f_2(x_{n-1})$ in this pairing. This is the same operation as was mentioned towards the end of §3.3.3 in connection with the build-up of n-variable map patterns from known $(n-1)$-variable patterns. The n-variable functions generated in this manner have to be tested for unateness, monotonicity, and asummability to ensure linear separability of $f(x)$, and discarded if they fail this test.

Alternatively, and found to be the preferable method where $n \geq 6$, the standard linearly separable functions of n variables may be generated by determining all the self-dual linearly separable functions of $n+1$ variables. As we have already seen in Chapter 2, the self-dual linearly separable functions of $n+1$ variables precisely encompass all the possible linearly separable functions of n variables.

Details of these generation techniques will be found well documented by Dertouzos[19,28], Winder[24,25], and Muroga et al.[26,27,29].

However, having determined all the positive canonic linearly separable functions of n variables under this b_i classification procedure, there remains the equally formidable problem of generating the optimal a_i values for each entry, where optimality is initially defined as minimizing the total summation A, where

$$A = \sum_{i=0}^{n} a_i$$

subject to the constraint at all 2^n minterms that

$$|a_0 y_0 + a_1 y_1 + \ldots + a_n y_n| \geq 1.0.$$

What this latter constraint effectively means is that fractional values between 0 and 1 will not be allowed, and that (in general) all a_i values will be integer values.

Details of the iterative linear programming techniques used to derive the optimal a_i values for each b_i classification entry will be found in Winder[25], Muroga et al.[17,29] and elsewhere. A number of interesting features of linearly-separable functions emerge from the final results, amongst them being the following.

(a) For the 2470 linearly separable canonic functions of $n \leq 7$, the optimal realizing weights and threshold are all without exception integer values. This is a very remarkable result. For the 175 428 canonic functions of $n \leq 8$, however, there are just two $n = 8$ functions which have fractional weights under the above constraint, these being the following two entries.

Chow $|b_i|$	Weights $|a_i|$
66, 54, 40, 30, 24, 16, 16, 6, 6	14.5, 12.5, 9.5, 7.5, 6, 4, 4, 1.5, 1.5,
65, 55, 39, 31, 25, 17, 15, 5, 5	16.5, 14.5, 10.5, 8.5, 7,5, 4, 1.5, 1.5

(b) There is no completely monotonic function of ≤ 8 variables which is not also a linearly separable function. Thus complete monotonicity is a sufficient test for linear separability for up to $n \leq 8$, but not for $n = 9$, as demonstrated by Gableman[12].

(c) No function of $n \leq 7$ has more than one possible solution for the optimal a_i values. For $n = 8$, however, 12 entries have two possible solutions for the a_i values.

These features and further statistical information will be found more fully documented in the literature, particularly by Muroga[26,29].

4.2.2. Use of the Chow Parameters for Single-Gate Realizability

If we are considering a given function $f(x)$ of n inputs, where n lies within the range of the available Chow-parameter tables, then the tabulated b_i and a_i values provide a ready look-up means for checking whether $f(x)$ is linearly separable, and if so what are the required gate input weight and threshold values. If n lies outside the published tables, then we have a less-direct procedure available, this being a problem which we shall mention in §4.2.3.

Now the Chow b_i parameters are listed in a positive canonic order, with $b_0 \leq b_1 \leq \ldots \leq b_n$. For any given function $f(x)$ we can transform $f(x)$ into its representative positive canonic Boolean function by performing the necessary invariance operations, from which the final representative b_i values can be determined. These values may then be looked up in the Chow tabulations to see if they appear, and if they do then the given function $f(x)$ is linearly separable.

However, as first pointed out by Dertouzos[19,28], algebraic transformation of the given function $f(x)$ into its representative Boolean form before determination of the canonic b_i values is unnecessary. Instead, and far more convenient, the b_i values for $f(x)$ itself may be evaluated *and then these values rearranged in descending-magnitude order* which will always correspond to the required positive canonic ordering. Let us take a simple function to illustrate this procedure.

Consider the arbitrary function

$$f(x) = [x_1\bar{x}_2 + \bar{x}_2\bar{x}_3 + \bar{x}_2 x_4 + x_1\bar{x}_3 x_4].$$

Calculating the b_i values for this function using, say, the counting procedure given in Chapter 2 or Appendix A, we have the values

b_0	b_1	b_2	b_3	b_4
0	+4	−12	−4	+4.

Rearranging in descending magnitude order:

b_2	b_1	b_3	b_4	b_0
−12	+4	−4	+4	0.

Taking magnitude values only we have

$\|b_2\|$	$\|b_1\|$	$\|b_3\|$	$\|b_4\|$	$\|b_0\|$
12	4	4	4	0.

Design of Threshold-Logic Networks

Now this sequence of numbers is listed in the Chow-parameter tabulations (see Appendix A, $n \leq 4$). Therefore the given function $f(x)$ is a linearly separable function.

However, the Chow-parameter tables also give us the a_i magnitude values for each $|b_i|$ entry, which we may proceed to employ as follows. From the tables we see that the minimum-integer $|a_i|$ values associated with the above $|b_i|$ entry are

$$2 \quad 1 \quad 1 \quad 1 \quad 0.$$

Thus the actual weights a_1 to a_4 required for the given function $f(x)$ are given by these values, allocated according to the rearrangement of the original b_i values for $f(x)$, and with the sign of the original b_i values. Thus we have for this particular function that

$$\begin{array}{ccccc} a_2 & a_1 & a_3 & a_4 & a_0 \\ -2 & +1 & -1 & +1 & 0 \end{array}$$

giving us the threshold-logic realization

$$f(x) = \langle x_1 - 2x_2 - x_3 + x_4 \rangle_t.$$

Two points require further attention, firstly to determine the required gate-threshold value t and secondly to eliminate any negative weights which are present in our initial realization. Calculating t from the above a_i values we have

$$t = \tfrac{1}{2}\left\{\left(\sum_{i=1}^{n} a_i\right) - a_0 + 1\right\}$$
$$= \tfrac{1}{2}\{(-2 + 1 - 1 + 1) - (0) + 1\}$$
$$= 0.$$

Therefore $f(x) = \langle x_1 - 2x_2 - x_3 + x_4 \rangle_0.$

The negative weighting factors can be eliminated by substituting $\{1 - \bar{x}_i\}$ for each x_i which has a negative weighting factor, therefore giving

$$f(x) = \langle x_1 - 2\{1 - \bar{x}_2\} - \{1 - \bar{x}_3\} + x_4 \rangle_0$$
$$= \langle x_1 + 2\bar{x}_2 + \bar{x}_3 + x_4 - 3 \rangle_0$$
$$= \langle x_1 + 2\bar{x}_2 + \bar{x}_3 + x_4 \rangle_3$$

where the -3 inside the threshold summation has been taken outside to form a revised threshold value in the final realization.

This shows that the original negative weights can therefore *be replaced immediately by positive weights if the input variable x_i associated with each negative weight is complemented.* If we had done this before calculating t, we would have had

$$f(x) = \langle x_1 + 2\bar{x}_2 + \bar{x}_3 + x_4 \rangle_t$$

with t now given by

$$\tfrac{1}{2}\{(2 + 1 + 1 + 1) - (0) + 1\}$$
$$= 3$$

which is precisely the same as the previous final result. Indeed it is usual practice to follow the latter steps rather than the ones which we initially detailed.

Further examples of this simple look-up technique for producing the optimum threshold-logic gate realization for any linearly separable function may be found in most texts on threshold-logic.

4.2.3. Large-n Networks.

Should a given function $f(x)$ lie outside the range of n listed in the available Chow-parameter tables, then the problem of checking $f(x)$ for linear separability and determining the appropriate a_i values remains. Assuming that the $n \leq 7$ tables are available, this becomes necessary for eight or more input variables.

The size of problem, however, is now too great for hand calculation. CAD techniques must be invoked unless the function has a particularly well-defined structure, such as a word-recognition problem or an arithmetic requirement. In the latter cases, however, the strong structure with relatively few minterms necessary to define fully the required function is frequently amenable to an heuristic solution without complex design algorithms.

Indeed we may observe that one seldom encounters a complex random-logic design requirement with a large number of input variables per network. There may be a large number of input variables involved in the complete system, but the basic specification of the system tends to group them into manageable proportions in the majority of cases. However, it may well be that a more efficient system realization could be produced if the designer could handle more input variables at a time, but no general design concepts have yet evolved for this, for either vertex or threshold-logic realizations.

Design of Threshold-Logic Networks

The maximally efficient design of large-n networks therefore remains an area of research, but is bound to be computer based in order to handle the volume of data involved.

4.2.4. Gate-Sensitivity Considerations

So far in our considerations of threshold-logic gates, we have not had occasion to consider any possible limitations of circuit performance which may impose a limit on the input weights or gate-threshold value. However, in the same way that there may be, say, a fan-in limitation with conventional vertex gates, so there may be some limitation present in the availability of threshold-logic gates.

In Chapter 7 we shall be particularly concerned with the circuit design of non-vertex gates, and we shall encounter gate-sensitivity constraints which can limit gate specifications. Thus we may have a situation similar to that which can exist in normal vertex-gate networks, namely that two (or more) gates may have to be used in practice to do the duty of a single larger gate. Let us here consider the problem of how a single threshold-logic realization may be split into two (or more) threshold-logic gates.

This problem is, fortunately, simple. No matter how a single linearly-separable function $f(x)$ of n variables is divided into two disjoint reduced functions about any input variable, both reduced functions of $n-1$ variables are themselves always linearly separable. This feature was noted and illustrated in §3.3.3. of Chapter 3.

As an example of disjoint synthesis, consider the linearly separable function

$$f(x) = [x_1(\bar{x}_2 + x_3 + \bar{x}_4) + \bar{x}_2(x_3 + \bar{x}_4) + x_3\bar{x}_4 x_5]$$
$$= \langle 3x_1 + 3\bar{x}_2 + 2x_3 + 2\bar{x}_4 + x_5 \rangle_5.$$

Suppose for certain reasons we cannot use the latter single-gate realization, but still require a threshold-logic solution using smaller threshold-logic gates. Let us choose to divide $f(x)$ about, say, x_1. This will give us the two disjoint functions

$$[(\bar{x}_2 + x_3 + \bar{x}_4) + \bar{x}_2(x_3 + \bar{x}_4) + x_3\bar{x}_4 x_5]$$
$$= [(\bar{x}_2 + x_3 + \bar{x}_4)] \quad \text{when } x_1 = 1$$

and

$$[\bar{x}_2(x_3 + \bar{x}_4) + x_3\bar{x}_4 x_5] \quad \text{when } x_1 = 0.$$

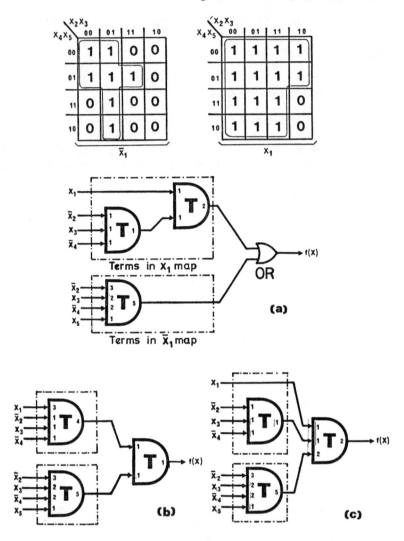

Fig. 4.7. The disjoint synthesis of the linearly separable function

$$f(x) = [x_1(\bar{x}_2 + x_3 + \bar{x}_4) + \bar{x}_2(x_3 + \bar{x}_4) + x_3\bar{x}_4x_5]:$$

(a) the basic decomposition about input x_1; (b) one possible realization; the output gate may of course be a simple OR vertex gate) (c) another possible realization, still using threshold-logic gates with not more than four inputs; the upper threshold-logic gate may of course be a simple OR vertex gate.

Now the threshold realization of these two disjoint functions may be seen to be

$$\langle \bar{x}_2 + x_3 + \bar{x}_4 \rangle_1$$

for the $x_1 = 1$ function, and

$$\langle 3\bar{x}_2 + 2x_3 + 2\bar{x}_4 + x_5 \rangle_5$$

for the $x_1 = 0$ function. Notice that we could have obtained these results directly from the original threshold realization for $f(x)$ without using the Boolean equations, by merely setting $x_1 = 1$ or 0 in the original threshold equation. This would have given the above $x_1 = 0$ realization immediately, but the $x_1 = 1$ realization would initially be $\langle 3\bar{x}_2 + 2x_3 + 2\bar{x}_4 + x_5 \rangle_2$, which clearly minimizes to $\langle \bar{x}_2 + x_3 + \bar{x}_4 \rangle_1$.

We may now use these reduced threshold-logic functions in several ways to generate the complete function $f(x)$. Figure 4.7 illustrates some of the possibilities for the final synthesis.

Thus the reduction of any large threshold-logic function into an assembly of smaller functions is a fairly trivial exercise. It is clearly advantageous to decompose the function about a variable with a high input weighting rather than one with a low input weighting, as maximum simplification of one of the two reduced functions is then achieved. The reduction process may of course be applied more than once if necessary.

4.3. Threshold Synthesis of Non-Linearly Separable Functions

The synthesis of a non-linearly separable function cannot use the Chow-parameter tables to give a direct solution. If the computed b_i values of the function are not a listed entry, then no clear indication is given by these values how best to proceed. Looking ahead to the next chapter, we shall there be concerned with the use of the comprehensive Rademacher–Walsh parameters and we shall then see a more powerful and general approach to this problem.

However, to continue here with Chow-parameter considerations, our problem with any non-linearly separable function $f(x)$ is how best to split $f(x)$ into parts which have a "best" threshold-logic realization. We have already seen that the prime-implicant approach is not necessarily efficient, and some other method for choosing more efficient grouping of the minterms of $f(x)$ is required.

The most direct method available for this purpose is the Karnaugh map pattern technique. This technique may be readily applied to all functions of $n \leq 5$, with some tedium to $n = 6$, but so tedious as to be generally impractical for $n \geq 7$. However, within these limits the technique offers an unsurpassed method of threshold-logic synthesis. The basic concept is to plot the given function $f(x)$ on a conventional Karnaugh map layout. All true (or all false) minterms are then covered by choosing appropriate covering patterns from the range of standard linearly separable map patterns. It will be recalled from Chapter 3 that these standard patterns are based upon the Chow-parameter entries, and therefore constitute a *geometric classification* for all possible linearly separable functions. Figure 3.13 detailed the available patterns for $n \leq 3$, whilst Appendix D details and classifies them for $n \leq 4$.

The great advantage of this design technique is that a visual build-up of the synthesis is available. Its feasibility relies upon the human ability to recognize patterns and shapes, regardless of orientation or extraneous information. A further advantage in terms of the synthesis is that it readily enables specific threshold-logic gates, that is specific linearly separable patterns, to be incorporated in the synthesis procedure. Finally "don't care" conditions in the given function may readily be used to best advantage.

4.3.1. Map Synthesis for Functions of n ≤ 4

The general synthesis procedure for $n \leq 4$ is as follows.

(i) If the map pattern for $f(x)$ is clearly in two (or more) blocks of true (or false) minterms, then check whether each block is a linearly separable pattern, using the D_k ratio (see Appendix D) as a guide if necessary.

(ii) If no obvious separate groupings are present, determine the D_k ratio of the complete map. Using this ratio as a guide see whether any standard linearly separable pattern of about the same D_k value can be found which almost fits the $f(x)$ plot[2].

(iii) Choosing a reasonably good "first fit," proceed to make up the exact cover requirements of $f(x)$ by using further covering patterns as necessary. Note that the additional covering pattern(s) may be OR-ed or AND-ed or exclusive OR-ed with the first cover, so as to add or amputate minterms from the first cover as necessary.

[2] Notice that if a standard map pattern can be found which exactly fits the plot of $f(x)$, then $f(x)$ must be a linearly separable function realizable by one threshold-logic gate.

Design of Threshold-Logic Networks

(iv) "Don't care" minterms in the given function $f(x)$ may be used to make up a standard map pattern as necessary.

(v) If specific threshold-logic gates have to be used in the synthesis, first determine and note all the possible patterns which these gates provide, and then use these patterns to best advantage in covering $f(x)$.

(vi) Finally, if two (or more) outputs $f_1(x)$ and $f_2(x)$ are required, then common covering patterns between the two functions should be considered.

Full details of these techniques have been published[30-32]. As an illustration of the results, Fig. 4.8 shows three typical syntheses; it is left as an exercise for the reader to see if alternative designs can be proposed taking different minterm groupings, so as to use gates with smaller fan-ins or input weightings, etc.

4.3.2. Map synthesis for Functions of $n > 4$

Above $n = 4$, pairs of 16-minterm Karnaugh maps have to be used. The problem which now arises is how the input variables shall best be allocated to the maps. For $n = 5$ we have five possibilities; for example $x_2, \ldots, x_5$ may be chosen as the map row and column designations, with $\bar{x}_1, x_1$ as the designation of the individual maps of the necessary pair. For $n = 6$ there are 15 possible allocations of the x_i's for the quadruple group of maps now involved.

Considering first a five-variable function $f(x)$, which is non-linearly separable, the ideal aim would be to split $f(x)$ into two disjoint halves about one of the x_i's, that is into one of the five pairs of four-variable maps covering $f(x)$, *each half of which is now linearly separable*. Unfortunately there is no known way of determining whether such a division is possible directly from $f(x)$ without testing the possibilities. Hence the suggested threshold-logic synthesis procedure for $f(x)$ is to complete all five possible pairs of maps for $f(x)$ and check with the standard four-variable map patterns whether pairs of maps are linearly separable patterns[3]. If one (or more) of such pairs of maps is found containing two standard map patterns, then a threshold/OR or a threshold/AND realization can be built up in the usual manner using these two reduced functions.

---

[3] This is quite a rapid process to do by hand provided blank maps with minterm identification in all squares of each map pair are available. Alternatively a trivial computer program will sort and print out the false and true minterm patterns.

210 The Logical Processing of Digital Signals

Details of this procedure may be found published[30,31]. The full procedure, including the possibility of not being able to find any pair of maps with standard map patterns, is briefly as follows.

(i) Plot the given function on all five pairs of Karnaugh maps.

(ii) If a pair of maps can be found in which *both* map patterns are standard linearly separable patterns, adopt these reduced functions for the realization. If more than one such pair of maps can be found, then

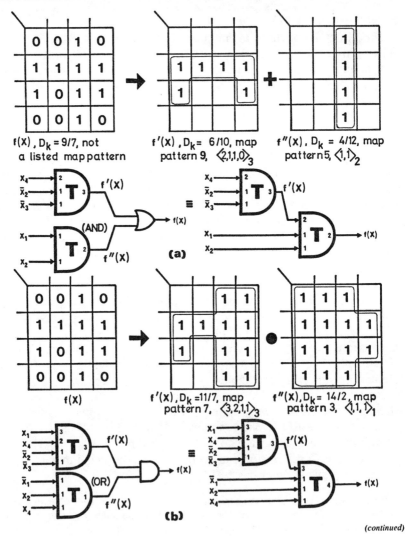

(continued)

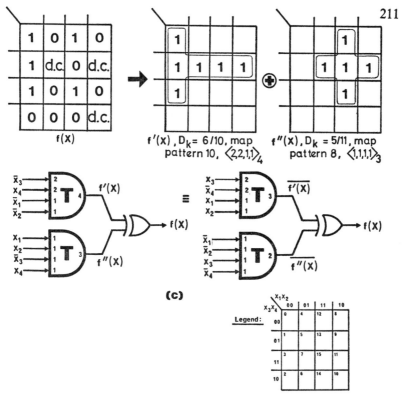

Fig. 4.8. Example $n = 4$ syntheses using Karnaugh map patterns (see Appendix D): (a) a two-level threshold/OR realization of

$$f(x) = [x_1(x_2 + x_4) + \bar{x}_2 x_4 + \bar{x}_3 x_4];$$

(b) a two-level threshold/AND realization for the same function; (c) a two-level threshold/exclusive-OR realization of a function with "don't care" conditions (cf. Figs. 3.6 and 3.7 of Chapter 3 for vertex/exclusive-OR constructions).

adopt the pair which give preferable gate specifications (e.g. lowest input summation, or fan-in, or any other desirable constraint) or comparability between the two reduced functions.

(iii) If condition (ii) cannot be found then choose any pair in which *one* of the two maps is a standard pattern. If more than one such pair can be found then in general adopt the one in which comparability exists between the two maps. The standard-pattern reduced function is of course realized with one threshold-logic gate, and an appropriate realization for the other non-linearly separable reduced function is built up by normal means.

212   The Logical Processing of Digital Signals

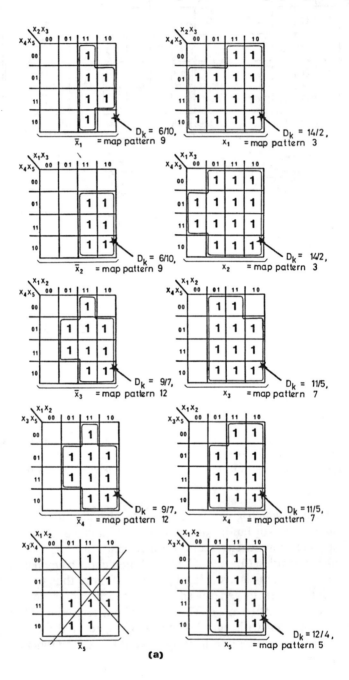

(a)

Design of Threshold-Logic Networks

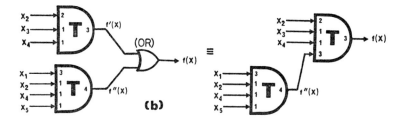

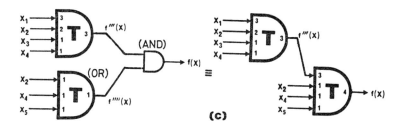

Fig. 4.9. Example $n = 5$ synthesis using Karnaugh map patterns of the unate but non-linearly separable function

$$f(x) = [x_1(x_2 + x_4 + x_5) + x_2(x_3 + x_5)]:$$

(a) the five possible pairs of Karnaugh maps, all of which except the $\bar{x}_5$ map are linearly separable patterns; notice comparability holds in all pairs of maps; (b) synthesis using the $\bar{x}_1, x_1$ patterns, involving the OR of the two linearly separable functions

$$f'(x) = [x_2(x_3 + x_4)]$$

and

$$f''(x) = [x_1(x_2 + x_4 + x_5)]$$

which have specifications of $\langle 2,1,1 \rangle_3$ and $\langle 3,1,1,1 \rangle_4$ respectively; (c) an alternative synthesis from the same pair of map patterns, involving the AND of the two larger linearly separable functions

$$f'''(x) = [x_1 + x_2(x_3 + x_4)]$$

and

$$f''''(x) = [x_2 + x_4 + x_5];$$

notice the use of comparability in producing this choice of AND coverage.

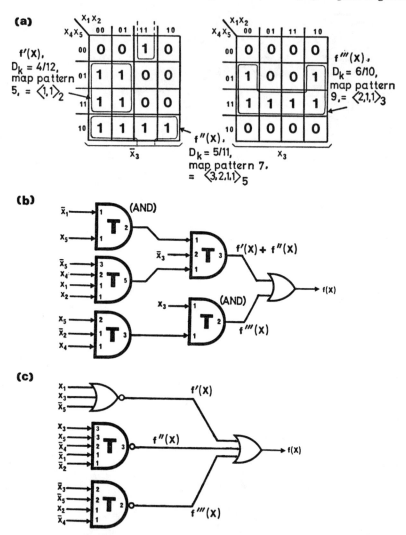

Fig. 4.10. Further example synthesis of an $n = 5$ non-linearly separable function $f(x) = [x_1x_2\bar{x}_3\bar{x}_5 + \bar{x}_1\bar{x}_3x_5 + \bar{x}_2x_3x_5 + x_3x_4x_5 + \bar{x}_3x_4\bar{x}_5]$ which fails to produce any pair of linearly separable maps: (a) the $\bar{x}_3, x_3$ pair of maps, in which x_3 but not $\bar{x}_3$ is a linearly separable pattern ($x_5, \bar{x}_5$ maps also have one but not two linearly-separable patterns); notice comparability is also absent in this function; (b) a non-minimal threshold-OR solution based upon the minterm groupings shown in (a); (c) an alternative solution for the reader to verify, using negated outputs on the threshold-logic gates.

Design of Threshold-Logic Networks

(iv) If no standard map pattern can be found in any map of any pair, choose a pair in which comparability holds between the two maps and proceed to synthesize each reduced function by normal means.

The object of comparability in the above choices is to eliminate the fifth variable from the smaller of the two four-variable functions. The fifth variable (the x_i, $\bar{x}_i$ map designation variable) is of course a redundant variable under such circumstances. The development of typical syntheses for five-variable non-linearly separable functions is shown in Figs. 4.9 and Fig. 4.10.

It may be thought that in doing this form of synthesis, the possibility of finding a five-input threshold-logic gate which provides a useful cover for some pattern of minterms in one map and some other pattern of minterms in the other map is lost. However, when the cover produced by individual five-variable threshold-logic gates is analysed, the majority of covering patterns consist of a standard four-variable map pattern in one map with *a full or almost-full*, or *an empty or almost-empty*, map pattern in the second map. Thus it is a viable procedure individually to synthesize each four-variable map pattern, with comparability being a useful constraint between the two map patterns. A more detailed exposition on this aspect will be found published[30,31].

For six-variable non-linearly separable functions, a similar threshold synthesis procedure may be followed. The given function should first be plotted on the appropriate sets of four-variable Karnaugh maps, looking for any set in which all four map patterns are standard linearly separable patterns. The best choice of decomposition of the given function follows the guide lines given above for the $n = 5$ function synthesis.

Further techniques of network synthesis using threshold-logic gates will be found published, for example by Sheng[33] under the heading of "compound synthesis," but without the map pattern techniques to aid the choice of optimal logic coverage.

4.3.3. Other Considerations

In all combinatorial networks the possibility of "logic hazards" may be encountered. Basically a hazard is the occurrence, or possibility of occurrence, of a momentary wrong output logic signal, or rapid sequence of wrong signals, caused by a change in the inputs applied to the network. Such hazards are caused by the different signal paths in the network from inputs to output, which can have slightly dissimilar response times to changing input signals.

In combinatorial networks two basic classifications of logic hazards may be defined. They are as follows.

(i) A *static 1 logic hazard* may occur between a pair of adjacent input minterms (minterms a Hamming distance of 1 apart), both of which produce a steady logic 1 output signal, but which during transition a momentary logic 0 output signal may be observed.

(ii) A *static 0 logic hazard* may occur between a pair of adjacent minterms, both of which produce a steady logic 0 output, but during transition a momentary logic 1 output signal may be observed.

It is also possible for multiple transitory outputs to occur between adjacent minterms, similar to contact bounce on mechanical contacts, this feature being termed a *dynamic logic hazard*.

The general theory of gate delays and logic hazards may be found in several readily available sources[34-36]. The general state of the art may be briefly summarized for both combinatorial and sequential networks as follows.

(*a*) If changes of only one input variable at a time to a combinatorial network are assumed to take place, then the elimination of all static logic hazards will also eliminate all dynamic logic hazards in the network.

(*b*) If simultaneous multiple-input changes to a combinatorial network take place, then it may be impossible to eliminate all logic hazards. Such hazards are now inherent in the function and cannot be removed if the inputs are allowed to change in any arbitrary order. This type of hazard may therefore be referred to as a *function hazard*.

(*c*) The elimination of logic hazards for single-input changes in a two-level network realization may be achieved by ensuring that the coverage produced by the first-level terms used in the realization overlap each other.

(*d*) There is no systematic method for the elimination of hazards in more complex situations; a general approach is to include additional delays in certain paths such that the network timing as a whole becomes determinable.

(*e*) Hazards in sequential networks depend upon whether the network is synchronously or asynchronously operating. In synchronously operating circuits the timing of the master clock should ensure correct operation, but in asynchronously operating circuits considerable care must be exercised in formulating the state assignment tables to attempt to ensure that no hazards in the state-assignment transitions occur.

The technique of (*c*) above, however, is one of the most common requirements for hazard-free networks. It is the one which we shall

Design of Threshold-Logic Networks

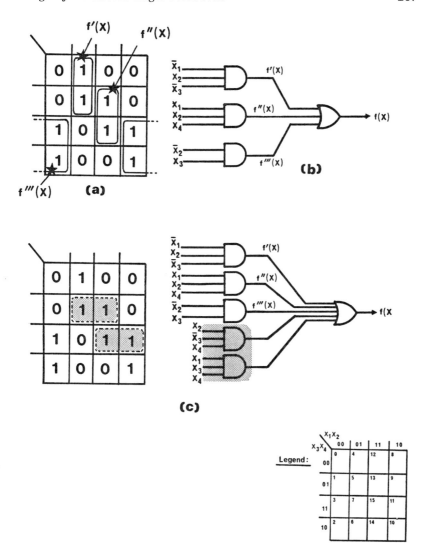

Fig. 4.11. Example hazard-free synthesis of

$f(x) = \sum 2,3,4,5,10,11,13,15$:

(a) plot of the function; (b) minimum Boolean realization; (c) hazard-free Boolean realization, requiring the addition of the first-level covering terms $x_2\bar{x}_3x_4$ and $x_1x_3x_4$.

briefly look at here, with threshold-logic realizations in mind.

Consider, for example, the four-variable function

$$f(x) = \sum 2, 3, 4, 5, 10, 11, 13, 15$$

which is shown plotted in Fig. 4.11a. The minimum two-level sum-of-products realization is shown in Fig. 4.11b. This realization, however, has two logic hazards present, namely when inputs change between input $x_1 x_2 \bar{x}_3 x_4$ and the adjacent input $\bar{x}_1 x_2 \bar{x}_3 x_4$, and similarly between input $x_1 x_2 x_3 x_4$ and $x_1 \bar{x}_2 x_3 x_4$. In both of these transitions the network output is controlled by dissimilar AND gates which may have different propagation delays, and hence give rise to momentary discontinuities in the network output. The elimination of these potential logic hazards is achieved by the incorporation of two additional covering terms, as illustrated in Fig. 4.11c.

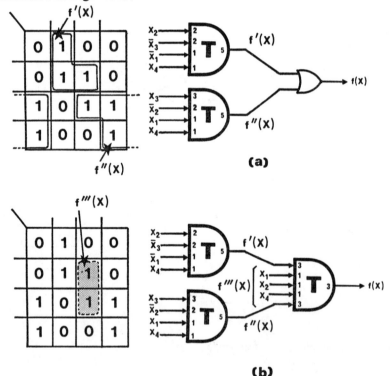

Fig. 4.12. The hazard-free threshold-logic synthesis of the same function of Fig. 4.11: (a) threshold realization, not hazard-free; (b) hazard-free realization with $f'''(x) = \langle x_1 + x_2 + x_4 \rangle_3$ added to achieve this requirement.

Design of Threshold-Logic Networks

However, if we consider a threshold-logic realization, then advantage may be taken of the non-rectangular cover which threshold-logic gates can provide. Figure 4.12 illustrates how the same principles as used in Fig. 4.11c may be applied to build up a hazard-free threshold-logic realization.

It may of course be possible to provide the "overlap" between the first-level gates without employing any additional gates for the hazard elimination. Figure 4.13 illustrates a simple problem in which appropriate choice of the first-level gate coverage immediately provides this requirement. Notice, however, that in all this development we have implicitly assumed that the output from a single threshold-logic gate is itself hazard free and that no transitory discontinuity in gate output occurs when the gate inputs are changed. As will be appreciated later when we consider possible circuits for such gates, this will in general be the case—it is a similar situation to considering the output from an OR gate, which should not change momentarily when the gate inputs are changed provided at least one of the gate inputs remains at logic 1.

Turning from logic hazards to another consideration, so far in this chapter we have not considered whether particular threshold-logic gates may be preferable or of more universal use than others. We have, however, indicated how the design techniques based upon standard linearly separable map patterns may be constrained to use preferred choices, but we have not looked at gate specifications which may be of most general use.

The choice of the most useful ("universal") threshold-logic gate is complicated by the factors mentioned in §2.3.1 of Chapter 2. Briefly, it will be recalled that if two (or more) gate inputs are strapped together, this effectively provides a single input x_i with an increased input weighting value a_i, whilst if unused gate inputs are connected to a permanent logic 1 this effectively reduces the gate-threshold value to the variable inputs from t to $\{t - \sum a_j\}$, where $\sum a_j$ is the sum of the input weights which are connected to the permanent logic 1 signal. A basic method of producing a single threshold-logic gate which could be used for a large number of specific requirements would of course be to have a gate with a very large number of unity-weighted inputs and a high gate-threshold value.

Supposing it was desired to provide a general-purpose gate with a maximum input summation $\sum_{i=1}^{n} a_i$ of, say, S. Then the useful range of gate-threshold value t for such a gate will lie in the range $1 \leq t \leq S$, the lower limit giving the OR of the gate inputs and the upper limit the AND of the gate inputs. Therefore if a gate with a threshold $t = S$ and

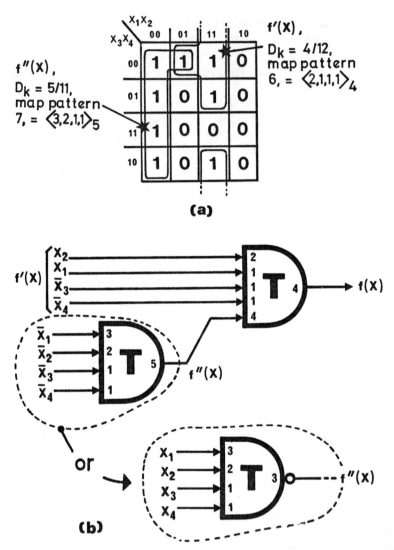

Fig. 4.13. Hazard-free synthesis of the function $f(x) = \sum 0,1,2,3,4,12,13,14$ which does not involve any additional first-level cover to achieve this requirement: (*a*) plot of the function; (*b*) threshold realization.

with $\{S + (S - 1)\}$ unity-weighted inputs was available, this would enable all possible gate specifications within the requirements to be provided. This concept is illustrated in Fig. 4.14.

Design of Threshold-Logic Networks

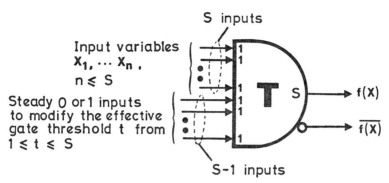

Fig. 4.14. The concept of a multi-input unity-weighted threshold-logic gate for universal usage.

However, intuitively this is not the most efficient way of providing an optimum general-purpose threshold-logic gate. How, therefore, can a more compact universal gate specification be derived? The answer is again to drawn upon the information contained in the canonic Chow-parameter classification tables, which list all the positive canonic threshold functions for any n. If we can specify a single threshold-logic gate which will realize all the entries of any chosen n, then this single gate will, by appropriate connections to its inputs, be able to realize all the possible linearly separable functions of $\leq n$.

Let us look at the simple case of the Chow parameters for $n \leq 3$. They are as tabulated in Table 4.1. If we expand each of the three Chow entries into the different possible combinations of the gate input weightings a_i, $i = 1$ to 3, and compute the corresponding gate-threshold values t for each set of input weights, we obtain the right-hand-column data in this tabulation. Notice one particular point concerning the gate-threshold values; this is that because t is given by $\frac{1}{2}\{(\sum_{i=1}^{n} a_i) - a_0 + 1\}$ we may include both signs of a_0 in a computation for t and hence obtain two gate-threshold values per set of input weights, except of course where a_0 is zero. From Table 4.1 we may conclude that the highest total input summation we need provide on an $n \leq 3$ gate is four. This can be provided by four unity-weighted inputs a_1 to a_4, from which all the particular entries in the center column of Table 4.1 can be generated. The highest gate-threshold value listed in the right-hand column of Table 4.1 is 3. However, if we try to standardize upon a "universal" gate specification of $\langle 1,1,1,1 \rangle_3$, it will be seen that the specification of $\langle 2,1,1 \rangle_2$ cannot be produced as we are using all the x_i inputs to generate the 2,1,1 input weighting requirements and have no further means of reducing the effective gate threshold from 3 to the required value of 2. Thus to be able to generate the $\langle 2,1,1\rangle_3$ and the $\langle 2,1,1\rangle_2$ gates we required a universal gate specification of $\langle 1,1,1,1,1\rangle_3$ to be available.

Table 4.1. Chow parameters and corresponding canonic gate specifications for $n \leq 3$

Chow $\|a_i\|, i = 0 \ldots n$	$\|a_i\|, i = 1$ to 3	Corresponding gate-threshold value t
1, 0, 0, 0	1, 0, 0 0, 0, 0	1 −1 or 0
2, 1, 1, 1	2, 1, 1 1, 1, 1	2 or 3 1 or 3
1, 1, 1, 0	1, 1, 1 1, 1, 0	2 1 or 2

Note: the second entry of the right-hand column is the trivial all-0 or all-1 output function, and is therefore of no practical significance.

However, what happens if we provide a universal threshold-logic gate with *true and complemented* outputs? Again looking at the $\langle 2,1,1\rangle_2$ and $\langle 2,1,1\rangle_3$ cases by way of illustration, these two gates realize the canonic Boolean functions

$$f(x) = [x_1 + x_2 x_3]$$

and

$$f(x) = [x_1(x_2 + x_3)]$$

respectively. However, if we complement the first function we obtain the function $[\bar{x}_1(\bar{x}_2 + \bar{x}_3)]$, which in positive canonic form is precisely the same as the second function. This concept is illustrated in Fig. 4.15. Therefore if we allow, as indeed we are doing, full invariance operations on the x_i inputs to our universal gate, it is unnecessary to provide directly the higher of the two gate-threshold values per entry classification as the lower value with output inversion will generate this function for us. Hence to realize all the possible different linearly separable functions of $n \leq 3$, the universal threshold-logic gate of

Design of Threshold-Logic Networks

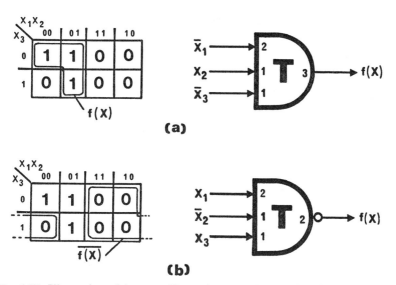

Fig. 4.15. Illustration of the use of inversion on a threshold-logic gate output in order to modify the gate-threshold value: (a) a given function

$$f(x) = [\bar{x}_1(x_2 + \bar{x}_3)] = \langle 2\bar{x}_1 + x_2 + \bar{x}_3 \rangle_3;$$

(b) the function

$$f(x) = \overline{[x_1 + \bar{x}_2 x_3]} = \overline{\langle 2x_1 + \bar{x}_2 + x_3 \rangle_2}$$
$$\equiv \langle 2\bar{x}_1 + x_2 + \bar{x}_3 \rangle_3.$$

specification $\langle 1,1,1,1 \rangle_2$ with true and complemented outputs will satisfy all our requirements.

Applying the same development to the $n \leq 4$ and $n \leq 5$ cases, the universal gate for both of these situations may be derived. It is left for the reader to confirm that they are as illustrated in Fig. 4.16. Notice that the $n \leq 4$ gate will of course generate all the functions which the $n \leq 3$ gate will cover, and similarly the $n \leq 5$ gate will cover all lower-value cases.

With the question of universal gates must also arise the question of the gate packages, with the attendant problems of pin limitations and pin utilization. If we consider what is currently the most popular form of packaging of logic gates for random-logic applications, then 14- or 16-pin dual-in-line packages are the universal standard, although there are trends to "stretch" this standard to 18 or 20 pins maintaining the

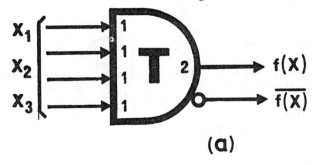

(a)

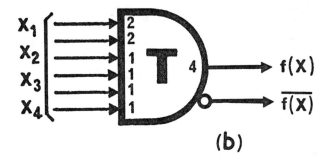

(b)

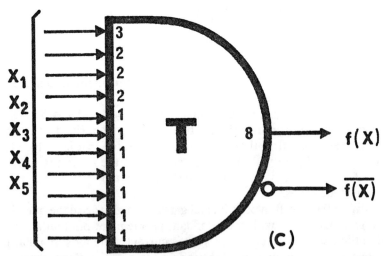

(c)

Fig. 4.16. Proposed universal threshold-logic gates based upon the Chow-parameter classifications for all linearly separable functions: (*a*) the universal $n \leq 3$ gate, $\sum a_i = 4$; (*b*) the universal $n \leq 4$ gate, $\sum a_i = 8$; (*c*) the universal $n \leq 5$ gate, $\sum a_i = 16$.

Design of Threshold-Logic Networks

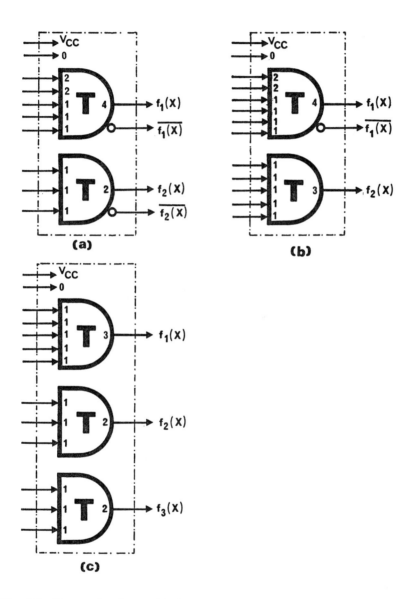

Fig. 4.17. Proposed universal 14-pin and 16-pin threshold-logic packages: (*a*) a 14-pin package of Amodei et al.; (*b*) a 16-pin two-gate package; (*c*) a 16-pin three-gate package.

same pin spacings. Let us consider how threshold-logic gates may fit into these pin-limited situations.

The first published disclosure in this area was the 14-pin package of Amodei et al.[37]. This was a two-gate package, as illustrated in Fig. 4.17a. This choice of gates was not based upon general-purpose usage, but for specific arithmetic operations. Nevertheless, such a package could be used quite effectively to realize random-logic requirements, as has been illustrated[31]. However, when we attempt to incorporate the more universal gates shown in Fig. 4.16 into 14- or 16- pin packages, multiples of these gates do not nicely fit into such pin constraints and certain compromises must be made.

Figure 4.17b and c illustrate two published possibilities. The former incorporates one $n \leq 4$ universal gate, plus one five-input majority gate which may be regarded as an alternative to the $n \leq 3$ universal gate of Fig. 4.16a. The other suggested 16-pin package will be seen to contain all majority gates, which individually are not as generally useful as the universal gates but which nevertheless collectively form a very powerful multi-gate package.

The logic coverage available from the gates shown in these packages together with other suggested possibilities is elaborated in Table 4.2. A discussion on these coverages may be found published[38], but in general it may be concluded that no one perfect solution for a general-purpose package can be found.

Table 4.2. Possible gate specifications for general-purpose use and their coverage

(a)

Gate description	Input weights	Gate threshold	True and complemented outputs provided	Total/gate input–output connections
(a) 3-input majority[37]	1,1,1	2	No	4
(b) 5-input majority	1,1,1,1,1	3	No	6
(c) $n \leq 3$ universal[38]	1,1,1,1	2	Yes	6
(d) Non-universal	2,1,1,1	3	Yes	6
(e) Non-universal[37,39]	2,2,1,1,1	4	Yes	7
(f) Non-universal	2,1,1,1,1	4	Yes	7
(g) 7-input majority[39]	1,1,1,1,1,1,1	4	No	8
(h) $n \leq 4$ universal[38]	2,2,1,1,1,1	4	Yes	8

(continued)

Design of Threshold-Logic Networks

(b)

Chow-parameter classifications		Coverage provided by gates (a) to (h)							
Weights a_i	Threshold t	(a)	(b)	(c)	(d)	(e)	(f)	(g)	(h)
$n \leq 3$:									
1,0,0	1	✓	✓	✓	✓	✓	✓	✓	✓
2,1,1	2 or 3	–	✓	✓	✓	✓	✓	✓	✓
1,1,1	1 or 3	–	✓	✓	✓	–	✓	✓	✓
1,1,1	2	✓	✓	✓	–	✓	✓	✓	✓
1,1,0	1 or 2	–	✓	✓	✓	✓	✓	✓	✓
$n \leq 4$:									
1,0,0,0	1	✓	✓	✓	✓	✓	✓	✓	✓
3,1,1,1	3 or 4	–	–	–	–	–	✓	✓	✓
1,1,1,1	1 or 4	–	–	–	–	–	✓	✓	✓
2,1,1,1	3	–	✓	–	✓	–	✓	✓	✓
2,1,1,0	2 or 3	–	✓	✓	✓	✓	✓	✓	✓
1,1,1,0	1 or 3	–	✓	✓	✓	–	✓	✓	✓
3,2,2,1	4 or 5	–	–	–	–	–	–	–	✓
3,2,1,1	3 or 5	–	–	–	–	–	–	–	✓
2,2,1,1	2 or 5	–	–	–	–	–	–	–	✓
1,1,1,0	2	✓	✓	✓	–	✓	✓	✓	✓
1,1,0,0,	1 or 2	✓	✓	✓	✓	✓	✓	✓	✓
2,2,1,1	3 or 4	–	–	–	–	✓	✓	✓	✓
2,1,1,1	2 or 4	–	–	–	–	✓	✓	✓	✓
1,1,1,1	2 or 3	–	✓	✓	–	–	✓	✓	✓

(a) Eight possible candidates and the total input–output connection count.
(b) The coverage of the standard canonic functions of $n \leq 3$ and 4 provided by the candidates (note, the trivial case of gate output always 0 or 1 has not been included).

There is of course no reason why suggested packages for general-purpose use should contain only threshold-logic gates; a mixture of threshold logic plus other types of gate may in practice prove very advantageous. A step in this direction has already been made with the commercial availability of a package containing majority plus exclusive-NOR gates[1]; threshold-logic plus exclusive-OR/NOR gates may provide even more interesting possibilities.

The provision of multi-outputs on a threshold-logic gate, that is a gate with one set of input weights a_i, $i = 1 \ldots n$, but with more than one output, each separate output being associated with a particular gate-threshold value t, increases the problem of choosing a "best" universal gate or package specification. In the next section of this chapter we shall be considering such gates, and it will then be seen that multi-output threshold-logic plus exclusive-OR gates are essential partners for certain duties. However, whilst we are at present considering standard packages, let us briefly consider this final aspect.

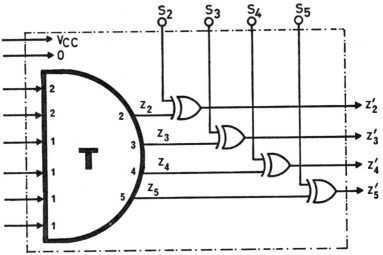

Fig. 4.18. A proposed 16-pin package containing multi-output threshold-logic plus exclusive-OR capability.

Restricting ourselves still to a 16-pin package, a multi-output threshold-logic package as illustrated in Fig. 4.18 has been suggested[2]. If steady logic 0 signals are applied to the exclusive-OR inputs S_2 to S_5, then the final package outputs z'_k, $k = 2\ldots 5$, are identical to the gate outputs z_k, $k = 2\ldots 5$. If any one or more input S is connected to a logical 1 signal, however, then the appropriate z'_k output becomes $z'_k = \bar{z}_k$. Finally if the z'_k outputs are fed back to the exclusive-OR inputs, then a range of output functions become available from the various z'_k outputs.

Clearly we now have a very versatile logic package with which to realize both linearly separable and non-linearly separable functions. The generally unsolved problem at present, though, is what design techniques can we use to exploit fully the considerable logic discrimination of such proposed assemblies. Also would it be more useful to have some exclusive-OR gates available on the package inputs *as well as or instead of* on the gate outputs as shown in Fig. 4.18? We shall look at these and other associated problems in subsequent sections of this book, but no firm guidelines are yet available in many of these areas.

Finally in this section on other aspects of the synthesis of threshold-logic networks, mention should be made of the *number of gate levels* in the network realizations. The basic techniques illustrated in the preceding sections have tended initially to produce two-level networks of the threshold/OR or threshold/AND form. Indeed conventional Boolean design techniques tend to be of this form also, unless or until further manipulation of the network design is undertaken. However, any two-level threshold-logic network with, say, N first-level threshold-logic gates can always be converted into an n-level cascade, as illustrated in Figs. 4.19a and b. Proof that this is so may be found in Dertouzos[28] and elsewhere, although it is a reasonably obvious result that any prime-implicant term P_i can be made to cascade through to the output $f(x)$ by appropriate selection of the input cascade weighting value on each level of gating. Indeed our previous Figs. 4.6b, 4.8a, 4.8b, and others were simple illustrations of this fact. Equally though, networks between the two extremes of Fig. 4.19a and b can be proposed, as indicated by Fig. 4.19c. What is not so readily possible with this latter topology is to design the "best" network, where "best" may be defined for example as using the fewest total number of gates in the multi-level realization. Algebraic factorization of the given function $f(x)$ into unate terms is one general approach, as may be found in Sheng[33], but no guarantee of the "best" solution can be given. If the threshold-logic gates are restricted to being simple majority gates, however, then more rigorous structures may be proposed; indeed we now find ourselves back into the area reviewed in Chapter 1, dealing with circuit networks of the form shown originally in Fig. 1.8.

4.4. Variable-threshold and multi-threshold gates

In the area of threshold logic which we shall now consider, we need to be careful to distinguish precisely which form of threshold-logic gate or network we are considering. There may be some confusion when reading published papers owing to lack of accepted formal definitions.

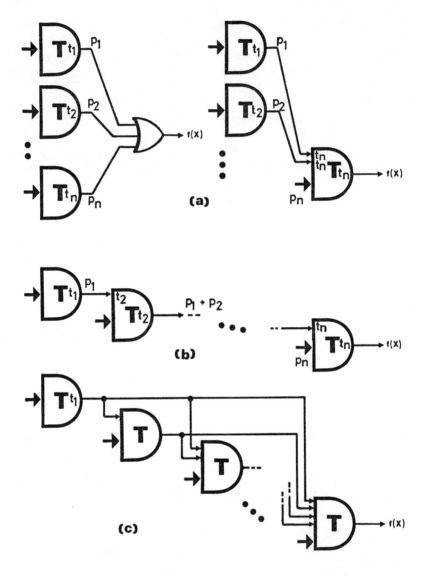

Fig. 4.19. The equivalence of multi-level network realizations for non-linearly separable functions: (*a*) two-level realizations; (*b*) the equivalent N-level realization; (*c*) the general multi-level network configuration, where some of the cascaded gate outputs may be zero weighted in the subsequent gate inputs; note that (*a*) and (*b*) are extreme cases of this general arrangement.

Design of Threshold-Logic Networks

In this section we shall be considering gates which have a fixed set of input weights a_i, $i = 1\ldots n$, but which have some form of gate threshold against which the input summation $\sum_{i=1}^{n} a_i x_i$ is equated *other than the single fixed value* which we have so far considered. The three types of gate with non-simplex thresholds which we shall briefly consider are as follows.

(i) *Single-output* gates with a single *variable-value* threshold t, output

$$f(x) = 1 \quad \text{if} \quad \sum_{i=1}^{n} a_i x_i \geq t$$

which we shall therefore term a "single-output variable-threshold" gate.

(ii) *Single-output* gates controlled by *several separate gate thresholds* $t_1, t_2, \ldots, t_k$, with output $f(x)$ defined for example by

$$f(x) = 1 \quad \text{if} \quad \sum_{i=1}^{n} a_i x_i \geq t_1, < t_2$$
$$\text{or} \quad \geq t_3, < t_4$$
$$\text{etc.}$$

$$= 0 \quad \text{if} \quad \sum_{i=1}^{n} a_i x_i < t_1$$
$$\text{or} \quad \geq t_2, < t_3$$
$$\text{or} \quad \geq t_4, < t_5$$
$$\text{etc.}$$

which type of gate we shall refer to as a "single-output multi-threshold" gate.

(iii) *Multiple-output* gates, with separate outputs $f_1(x), \ldots, f_k(x)$, each output corresponding to a *separate fixed gate-threshold value* $t_1, t_2, \ldots, t_k$, respectively, where

$$f_1(x) = 1 \quad \text{iff} \quad \sum_{i=1}^{n} a_i x_i \geq t_1$$

$$f_2(x) = 1 \quad \text{iff} \quad \sum_{i=1}^{n} a_i x_i \geq t_2$$

$$\vdots \qquad\qquad\qquad\qquad \vdots$$

$$f_k(x) = 1 \quad \text{iff} \quad \sum_{i=1}^{n} a_i x_i \geq t_k$$

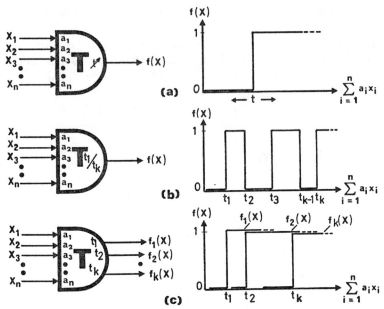

Fig. 4.20. Types of threshold-logic gate with a constant weight input but different forms of output threshold: (*a*) the single-output variable-threshold gate, where the gate-threshold value *t* may be varied; (*b*) the single-output multi-threshold gate, whose single output is controlled by several distinct threshold values; note that the gate output may be 0 or 1 at the maximum input summation and may be negated from that shown; (*c*) the multi-output threshold-logic gate, containing several distinct threshold values *t*.

which type of gate we shall refer to as a "multi-output threshold-logic" gate.

These three categories of gate are as illustrated in Fig. 4.20. Notice that they may also be found referred to as "constant-weight" gates in the literature because we are not modifying the input weighting factors in these considerations. We shall consider each of these categories in the sections following.

4.4.1. Single-Output Variable-Threshold Gates

When the threshold value of a threshold-logic gate is changed but the input weightings remain constant, a number of different functions of the input variables are realized.

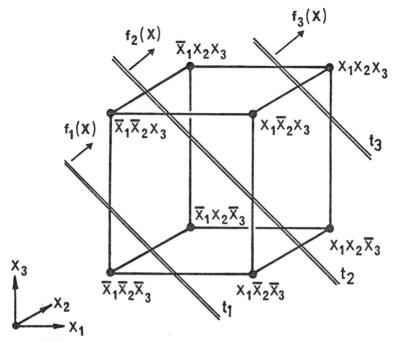

Fig. 4.21. The $n = 3$ hypercube construction divided by three parallel separating planes.

In the hypercube picture of linear separability, as discussed and illustrated in Chapter 3, the effect of a change in the gate threshold value t is to generate a new separating plane in the n-dimensional space, this new separating plane being parallel to the original separating plane. roof that all such planes produced by variation of t are parallel planes is a fairly trivial geometric exercise, but may be found detailed in Brown[40] and elsewhere. Fig. 4.21 illustrates an $n = 3$ hypercube construction, divided by three parallel separating planes.

The linearly separable functions which may be realized by a single threshold-logic gate with different threshold values are termed *isobaric* or *simultaneously realizable* functions[15,29]. It may be shown that the number of isobaric functions of n input variables cannot be greater than 2^n+1, including the trivial cases of $f(x)$ always 0 or always 1[7,29]. This can be informally visualized by considering the separating plane in the hypercube picture being moved slowly across the hypercube in a direction perpendicular to its plane, progressively moving through each of the 2^n nodes and thus generating a new function $f(x)$ as it passes each node in turn. Notice, however, that to achieve this maximum number of linearly separable functions requires that the various input weighting values a_i be such that they total to a different input summation value at each of the 2^n input minterms, for example as shown in Table 4.3 for the $n = 3$ case.

Table 4.3. Possible integer input weightings for the $n = 3$ case such that separation between any of the 2^n input minterms can be achieved by appropriate choice of the gate-threshold value t

Inputs and weights a_i			
$a_3 = 3$ x_3	$a_2 = 2$ x_2	$a_1 = 1$ x_1	$\sum_{i=1}^{n} a_i x_i$
0	0	0	0
0	0	1	1
0	1	0	2
0	1	1	3
1	0	0	4
1	0	1	5
1	1	0	6
1	1	1	7

Some further theoretical properties of single-output variable-threshold gates may also be cited. For example, for any given set of input weights a_i the perpendicular distance D between two adjacent separating planes in the hypercube construction is given by[41]

$$D_{t_{k-1}, t_k} = \left\{ \frac{t_{k-1} - t_k}{\sum_{i=1}^{n} a_i} \right\}$$

where t_{k-1} and t_k are the gate-threshold values of the two separating planes in question. This is not a particularly significant result, except for purely geometric considerations. A more significant property, however, is one which we have already encountered and used, but which has not been specifically stated or proved. It may be formally stated as follows. If a threshold-logic gate realizes any linearly separable function $f(x)$ at some threshold value t_1, then it also realizes the dual of $f(x)$, that is $f^d(x)$, at some other threshold value t_2. If $f(x)$ is a self-dual function, then of course t_1 will be equal to t_2. Proof of this will be found in Meisel[41,42], and is readily demonstrated by the following simple example:

given $f(x) = [x_1 + x_2x_3]$
$= \langle 2x_1 + x_2 + x_3 \rangle_2$ by inspection

then $f^d(x) \triangleq \overline{[\bar{x}_1 + \bar{x}_2\bar{x}_3]}$
$= [x_1(x_2 + x_3)]$
$= \langle 2x_1 + x_2 + x_3 \rangle_3$

which are the same input weightings, but modified gate-threshold value t. The function $f(x)$ and its dual $f^d(x)$ are therefore two of the possible isobaric functions which the particular gate input weights can realize.

Another property of single-output variable-threshold gates is as follows. If, for two distinct functions $f_1(x)$ and $f_2(x)$ of n variables realizable by threshold-logic gates with identical input weightings but dissimilar threshold values t_k and t_j, $t_k > t_j$, the $(n + 1)$-variable function $f(x)$ is proposed, where

$$f(x) = [\bar{x}_{n+1}f_1(x) + x_{n+1}f_2(x)]$$

then $f(x)$ is also a linearly separable function, realizable by the set of input weights and gate threshold

$$\langle a_1, a_2, \ldots, a_n, a_{n+1} \rangle_{t_k} \quad \text{where } a_{n+1} = t_k - t_j.$$

Proof of this may be found in Muroga and elsewhere[7,29,43]. For example, consider the two simple functions

$$f_1(x) = [\bar{x}_1(\bar{x}_2 + x_3)]$$

and

$$f_2(x) = [\bar{x}_1 + \bar{x}_2x_3]$$

as illustrated in Fig. 4.22a. The four-variable function which results from combining these two identical input-weighting functions is the function

$$f(x) = [\bar{x}_4\{\bar{x}_1(\bar{x}_2 + x_3)\} + x_4(\bar{x}_1 + \bar{x}_2x_3)]$$

which is shown in Fig. 4.22b. Notice that this function is linearly separable with weights and threshold given by the above expression. Also, as $t_k > t_j$, then $f_1(x) \subset f_2(x)$, and therefore the function $f(x)$ may be minimized to

$$f(x) = [f_1(x) + x_{n+1}f_2(x)].$$

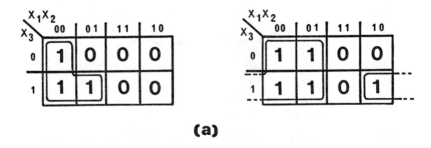

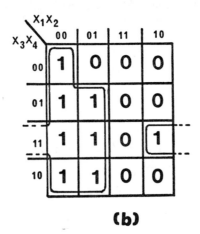

Fig. 4.22. The combining of two linearly separable functions of three variables with the same input weightings but different thresholds into one four-variable linearly separable function: (a) the two three-variable functions

$$f_1(x) = [\bar{x}_1(\bar{x}_2 + x_3)]$$
$$= \langle 2\bar{x}_1 + \bar{x}_2 + x_3 \rangle_3$$

and

$$f_2(x) = [\bar{x}_1 + \bar{x}_2 x_3]$$
$$= \langle 2\bar{x}_1 + \bar{x}_2 + x_3 \rangle_2;$$

(b) the four-variable function

$$f(x) = [\bar{x}_4\{\bar{x}_1(\bar{x}_2 + x_3)\} + x_4(\bar{x}_1 + \bar{x}_2 x_3)]$$
$$= [\bar{x}_1(\bar{x}_2 + x_3 + x_4) + \bar{x}_2 x_3 x_4]$$
$$= \langle 2\bar{x}_1 + \bar{x}_2 + x_3 + x_4 \rangle_3.$$

The converse of this expansion is equally true, namely given any linearly separable function $f(x)$ it may be decomposed into two linearly

Design of Threshold-Logic Networks

separable functions of one fewer variables, with the same input weighting values but with one different value threshold than that of the given function $f(x)$.

When areas of potential usefulness of single-output variable-threshold gates are considered, then two distinct areas may be found. The first of these is in that of adaptive (learning) networks for pattern recognition and artificial intelligence, wherein the value of the gate threshold may be adjusted to obtain correct system response. The second area is that of network design with standardization upon a particular specification of variable-threshold gate.

We shall not concern ourselves in this book with the many and varied aspects of pattern recognition, artificial intelligence, and learning machines. This is a vast subject area in its own right. Suffice here to say that it is usually too restrictive to consider threshold-logic gates with fixed input weightings and variable thresholds only for this duty; the greater flexibility of *variable input weights* as well as the threshold value is desirable in order to achieve useful theoretical results. This is undoubtedly a very important area of research and development, one which will continue to be increasingly important[40,44-48].

The second possible area of application of single-output variable-threshold gates is that of the network synthesis of non-linearly separable functions. This of course is the same area as we have already considered in previous sections of this chapter, but now with the constraint to use one specific specification of gate in the complete network realization. The situation is similar to the network arrangements previously illustrated in Fig. 4.19, but with all gates in the network having identical input weightings.

In practice this constraint proves too severe for economic network synthesis. The ability to be able to employ different input weighting on the various levels of gating is very necessary in order to achieve a good network realization. Hence we shall find that the variable-threshold facility by itself on a threshold-logic gate is not of great usefulness; of far greater potential are multi-threshold facilities which we shall consider in the following pages.

4.4.2. Single-Output Multi-Threshold Gates

The literature on single-output multi-threshold gates is at times confusing owing to dissimilar definitions of the multiple thresholds at which the gate output changes from 0 to 1 and *vice versa*. In the first classic

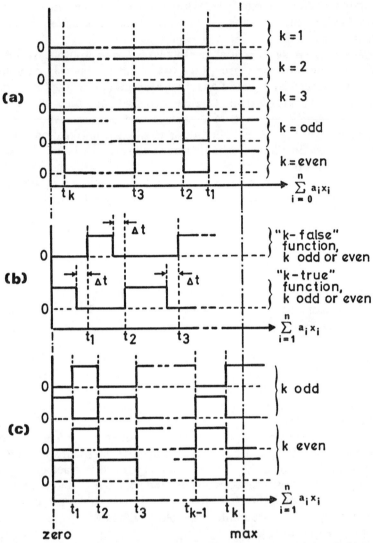

Fig. 4.23. The output characteristics of single-output multi-threshold gates with different possible definitions of the gate thresholds: (*a*) the definitions of Haring and others, with $t_k < t_{k-1}$, and where the values for t were taken as 0.5, 1.5, 2.5, etc. with integer input summation values; (*b*) the definitions of Muroga and others, where $t_k > t_{k-1}$, but with a margin Δt allowed for correct operation (note $\Delta t = 1$ in Muroga); (*c*) the definitions we shall here employ, which are similar to (*b*) but neglecting Δt.

Design of Threshold-Logic Networks

papers on this subject by Haring *et al.*[49-51] the definition of the threshold values $t_1, t_2, \ldots$ was $t_1 > t_2$ or in general $t_k < t_{k-1}$. This gives rise to output characteristics such as shown in Fig. 4.23a. In addition there is the question of defining precisely when the gate output switches; for example if the gate output is 0 at, say, an input summation $\sum_{i=1}^{n} a_i x_i$ of 3, and 1 at an input summation value of 4, then how should the transition point between output 0 and 1 be defined? Haring and others adopt fractional values for the effective gate thresholds, e.g. 1.5, 2.5, etc., which with integer gate input summations $\sum_{i=1}^{n} a_i x_i$ overcomes any ambiguity. Others, such as Muroga, adopt a "margin" Δt for correct operation, for example as illustrated in Fig. 4.23b, still with nominal integer input summations [29,52-54].

Here we shall adopt the following general conventions.

(a) We shall continue to assume the gate input summation $\sum_{i=1}^{n} a_i x_i$ is always an integer-value summation, so that fractional values are never encountered.

(b) t_1 is the lowest-value gate threshold and t_k the highest-value gate threshold, that is $t_k > t_{k-1}$.

(c) The gate output mechanism gives an unambiguous output 0 or 1 at every integer input summation.

(d) The gate output may be 0 or 1 at the minimum (= zero) input summation and 0 or 1 at the maximum (= $\sum_{i=1}^{n} a_i$) input summation, as illustrated in Fig. 4.23c.

(e) To distinguish between a gate threshold at which the output changes from 0 to 1 and a gate threshold at which the output changes from 1 to 0, the symbol t_j, $1 \leq j \leq k$, will be used for the former case and $\bar{t}_j$ for all the latter case. Hence the four situations shown in Fig. 4.23c are expressed in threshold terminology as follows for the order shown:

$$\langle a_1 x_1 + \ldots + a_n x_n \rangle_{t_1, \bar{t}_2, t_3, \ldots, t_k}$$

$$\langle a_1 x_1 + \ldots + a_n x_n \rangle_{\bar{t}_1, t_2, \bar{t}_3, \ldots, \bar{t}_k}$$

$$\langle a_1 x_1 + \ldots + a_n x_n \rangle_{t_1, \bar{t}_2, t_3, \ldots, \bar{t}_k}$$

and

$$\langle a_1 x_1 + \ldots + a_n x_n \rangle_{\bar{t}_1, t_2, \bar{t}_3, \ldots, t_k}.$$

Note that the threshold value symbols are alternatively barred and unbarred to represent the changes in the output from 1 to 0 and *vice versa*, respectively.

The general properties of functions realizable with single-output multi-threshold gates have been extensively investigated. Among these results the following points may be mentioned.

Firstly, any given function $f(x)$ of n input variables may be realized by a single single-output multi-threshold gate with at most 2^n-1 gate thresholds. This may readily be seen by considering an n-input gate with input weightings a_i, $i = 1 \ldots n$, which sum to a different value for each of the 2^n input minterms—this of course is the case if binary-ordered weighting values are used, as illustrated in Table 4.3. With such input weightings, then 2^n-1 thresholds can uniquely distinguish every input minterm and hence any output function $f(x)$ can be realized.

Secondly, if there exists a decomposition of a given function $f(x)$ of n variables into two functions

$$f(x) = [f_1(x) + f_2(x)]$$

such that

$$f'(x) = [\bar{x}_{n+1}f_1(x) + x_{n+1}\overline{f_2(x)}]$$

is a linearly separable function, then the given function $f(x)$ is realizable by a single-output two-threshold gate. This may be illustrated by the example following. Given

$$f(x) = \sum 0, 2, 5, 7, 11, 12, 13, 14, 15$$

which plots as shown in Fig. 4.24a. Dividing this function into

$$f_1(x) = \sum 5, 7, 11, 12, 13, 14, 15$$

and

$$f_2(x) = \sum 0, 2$$

we obtain the five-variable function $[\bar{x}_5 f_1(x) + x_5\overline{f_2(x)}]$ shown in Fig. 4.24b. From Appendix D, and noting that comparability holds between the two maps, it may be quickly confirmed that the function of Fig. 4.24b is linearly separable with the normal single-threshold realization of

$$f'(x) = \langle 2x_1 + 3x_2 + x_3 + 2x_4 + 3x_5 \rangle_5.$$

Design of Threshold-Logic Networks 241

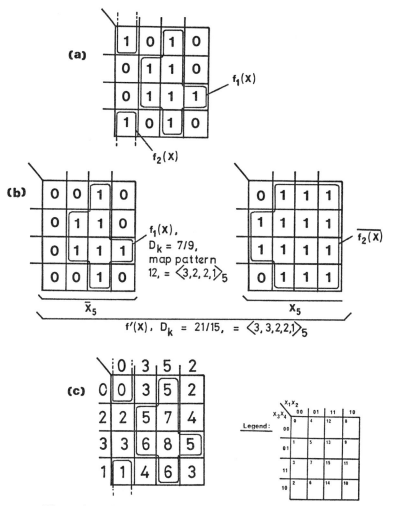

Fig. 4.24 Illustration of the theorem of two-threshold realizability: (*a*) plot of the given function $f(x)$, which we may divide into $f_1(x)$ and $f_2(x)$; (*b*) plots of $\bar{x}_5 f_1(x)$ and $x_5 \overline{f_2(x)}$, which jointly constitute the linearly separable function $f'(x)$; (*c*) the input summation value $\sum_{i=1}^{n} a_i x_i$ of each row and column and at every minterm with the weighting values necessary to realize $f'(x)$.

Thus $f_1(x)$ of our given function is clearly the threshold function

$$f_1(x) = \langle 2x_1 + 3x_2 + x_3 + 2x_4 \rangle_5.$$

If we put these values in the key Karnaugh map shown in Fig. 4.24c, we shall now see that $f_2(x)$ is realized with the same input weightings but with thresholds of 0 and 1 only. Thus the two-threshold realization of $f(x)$ is given by

$$f(x) = \langle 2x_1 + 3x_2 + x_3 + 2x_4 \rangle_{\bar{2},5}$$

where $\bar{2}, 5$ indicates that $f(x)$ becomes 0 when the input summation equals or exceeds 2, but becomes 1 again when the summation equals or exceeds 5. If we call these two thresholds t_1 and t_2, respectively, then it will be noted that $|t_1| = \{|t_2| - a_5\}$, where a_5 was the weighting associated with the additional input variable in making the linearly separable function $f'(x)$. The action of reducing the threshold value from 5, which realizes the function $f_1(x)$, can be clearly seen from Fig. 4.24c; in effect the function "expands" within the map, and at the threshold value of 2 has reached the exact perimeter of the function $f_2(x)$. Thus our original function

$$f(x) = [f_1(x) + f_2(x)]$$

is realizable with one set of input weights but two distinct output threshold values. Haring[49] has extended this concept to the case of a k-threshold function, where $k > 2$. No new fundamental principles are, however, involved.

Finally in this brief general review, we have a general algebraic relationship for multi-threshold functions. If $f(x)$ is any function realized by a single-output multi-threshold gate, say the k-odd realization

$$f(x) = \langle a_1 x_1 + \ldots + a_n x_n \rangle_{t_1, \bar{t_2}, \ldots, \bar{t_{k-1}}, t_k}$$

then the function is also given by the Boolean expression

$$f(x) = [f_1 \bar{f_2} + \bar{f_3} f_4 + \ldots + f_k]$$

where

$$f_1 = \langle a_1 x_1 + \ldots + a_n x_n \rangle_{t_1}$$
$$f_2 = \langle a_1 x_1 + \ldots + a_n x_n \rangle_{t_2}$$
$$\vdots \quad \vdots \quad \quad \vdots$$
$$f_k = \langle a_1 x_1 + \ldots + a_n x_n \rangle_{t_k}.$$

Design of Threshold-Logic Networks

It is left for the reader to derive the similar expression for the other three multi-threshold possibilities, as defined in Fig. 4.23c. The correctness of the Boolean expression may be readily demonstrated, for example by a Karnaugh mapping procedure. Further theoretical properties and concepts of multi-threshold functions may be found in the published work of Yen and elsewhere [54].

Network synthesis using single-output multi-threshold gates originally pursued the concept of considering the number of changes from 0 to 1 and *vice versa* in the truth table for a given function $f(x)$ and attempted to minimize this number of changes by invariance operations on the input variables. Longer "runs" of output 0 or 1 were sought, which therefore minimized the maximum number of threshold values necessary for a threshold realization. This approach will be found detailed by Haring and by Necula [49,55]. However, the algorithms for the subsequent network synthesis are difficult and do not guarantee a minimal solution.

Arising from the original concepts, however, were attempts to produce look-up tables for all possible functions $f(x)$, similar to the Chow-parameter look-up tables for linearly separable functions but listing in this case the minimum input weights and number and value of thresholds necessary for a multi-threshold realization. The linearly separable functions would of course be a special case in this general look-up table, being the functions realizable with one threshold value only.

However, as we have already seen in Chapter 2, the $n + 1$ Chow parameters of an n-variable function $f(x)$ are inadequate to define and classify $f(x)$ unless $f(x)$ is linearly separable. Therefore for the general case of any function $f(x)$ we require an augmented set of characterizing parameters, of which the Rademacher–Walsh set is the most appropriate. The multi-threshold work of Spann, Haring, and Ohori, and Mow and Fu in particular has been directed towards the derivation and publication of such look-up tables [50,52,56].

For $n \leq 4$ there is a total of 222 positive canonic NPN entries to classify all the 65536 different Boolean functions (see Table 2.3 of Chapter 2). Ignoring the all 0/all 1 case, there are therefore 221 non-trivial functions of $n \leq 4$. The Rademacher–Walsh coefficients for all these 221 representative functions have been listed both by Haring and Ohori and by Mow and Fu, together with the input weights, number of thresholds, and threshold values necessary in order to realize each entry by a multi-threshold realization. Unfortunately there is not entire agreement between the two tabulations and no guarantee that the latest one is yet minimal. However, it is agreed that not more than five threshold values per gate are required to realize any function of $n \leq 4$; indeed only two of the 221 canonic functions require this maximum number of thresholds for their multi-threshold realization.

However, there seems to be something rather inefficient in this procedure, or possibly in the application of single-output multi-threshold gates alone for network synthesis. It may be recalled from Chapter 2 that we can classify all $n \leq 4$ functions by just eight Rademacher–Walsh entries when we allow full invariance operations on the Rademacher–Walsh coefficients (see Table 2.12 of Chapter 2). Yet in this multi-threshold area we are using over 200 entries for the same set of functions. Admittedly the invariance operations to reduce the Rademacher–Walsh entries down to eight involve exclusive-OR considerations, but the question remains whether concentration on single-output multi-threshold gates by themselves is a viable exercise. Certainly more research around this area is wanted.

However, let us now continue and look at *multi-output gates,* which we shall argue are potentially more useful than the single-output gates we have just been considering. This continued consideration will in due course bring us back to the previous paragraph.

4.4.3. Multi-Output Threshold-Logic Gates

In the preceding section we did not consider at all how the single-output gate with multiple threshold values might be constructed. When one begins to consider this point, it becomes clear that a separate gate detection mechanism is necessary for each gate threshold $t_1, t_2, \ldots, t_k$. Thus the single-output of our previous multi-threshold case is the result of appropriately combining the discrimination produced by several threshold detectors, for example as illustrated in Fig. 4.25.

From these illustrations it is clear that the output must respond to

{(one threshold detection output t_j) AND
  (NOT the following detection output t_{j+1})}

in order to cause the final output to cycle between 1 and 0 as the input summation varies. Thus an exclusive-OR form of relationship is inherent in this requirement. Notice, however, that because the situation where output t_{j+1} is 1 can never occur when t_j is 0, $t_j < t_{j+1}$, the precise output requirement of combining two adjacent thresholds is as shown

Design of Threshold-Logic Networks

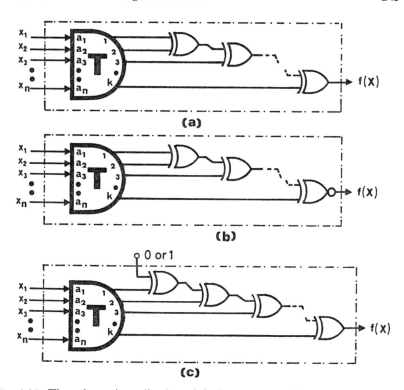

Fig. 4.25. The schematic realization of single-output multi-threshold gates.

(a) k-threshold gate, specification

$$\langle a_1, a_2, \ldots, a_n \rangle_{t_1, t_2, t_3, \ldots, t_k}$$

k odd, or

$$\langle a_1, a_2, \ldots, a_n \rangle_{t_1, t_2, t_3, \ldots, t_k}$$

k even

(b) k-threshold gate, specification

$$\langle a_1, a_2, \ldots, a_n \rangle_{t_1, t_2, t_2, \ldots, t_k}$$

k odd, or

$$\langle a_1, a_2, \ldots, a_n \rangle_{t_1, t_2, t_3, \ldots, t_k}$$

k even.

Note that the single exclusive-NOR gate can be included anywhere in the cascade of exclusive gates.

(c) The change from the above type (a) to type (b) gates, realized by the addition of one additional exclusive-OR (or NOR) gate at the beginning (or at the end) of the cascade of gates.

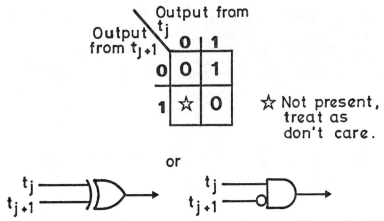

Fig. 4.26. The precise requirements for combining two adjacent threshold-detection outputs, corresponding to threshold values t_j and t_{j+1},

$1 \leq t_j < t_{j+1} \leq t_k$.

in Fig. 4.26, from which it may be seen that a "half exclusive-OR" function (see Fig. 1.3 of Chapter 1) is adequate for this particular duty. However, we shall normally continue to show a full exclusive-OR gate in these situations in subsequent diagrams, rather than a half exclusive-OR gate.

Having now introduced these two factors which are a feature of the previous single-output multi-threshold gates, namely (i) the fundamental necessity for separate threshold detection mechanisms and (ii) the necessity for exclusive-OR combining of the output from these several threshold detection mechanisms, the questions which may now be considered are as follows.

(*a*) If separate threshold-detection output signals are present, why not make them all available for multi-output use, connecting them together externally if/when required?

(*b*) If exclusive-OR gates have to be used, is it not more economical and useful to have them separate from the multi-threshold signals so that they may be used with gate *inputs or outputs or both* as required?

Whilst these arguments do not seem so far to have been exhaustively investigated and quantified, nevertheless simple considerations suggest that complex single-output gates such as shown in Fig. 4.25 are not the best concept for general-purpose use. Much more versatile would be the provision of multi-output threshold-logic gates together

Design of Threshold-Logic Networks

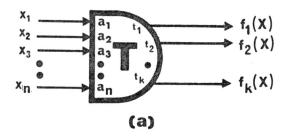

(a)

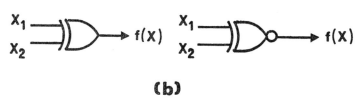

(b)

Fig. 4.27. The separation of the single-output multi-threshold gate concept into separate multi-output threshold-logic gates and exclusive-OR/NOR gates, to be used together however desired: (a) the multi-output gate; (b) conventional exclusive-OR/NOR gates, which may be used at the inputs or the outputs of (a).

with separate exclusive-OR gates, as illustrated in Fig. 4.27. This concept is supported by consideration of the canonic classification tabulations for all possible functions. As was mentioned in the preceding section, all possible $n \leq 4$ functions may be realized by an appropriate single-output multi-threshold gate with not more than five threshold-detection levels, assuming we allow the full NPN invariance operations on inputs and output, but there are 221 non-trivial such canonic functions for $n \leq 4$. However, if exclusive-OR operations are allowed at inputs and outputs as well as the NPN operations, then we may realize all possible functions from the Rademacher–Walsh canonic classification entries, which number only eight entries for $n \leq 4$. This possibility was originally suggested in Fig. 2.11 of Chapter 2, and will be extensively referred to again in the following chapter.

Therefore there seems to be a limited attraction in pursuing the possible application of the single-output multi-threshold gate for network synthesis; ordinary single-output single-threshold gates plus separate exclusive-OR/NOR gates may be far more useful allies. However, multi-threshold gates with more than one output per gate may be relevant for *multi-function* purposes, that is where two or more

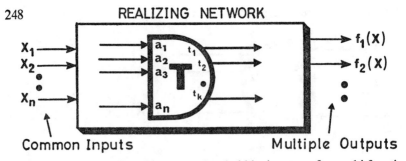

Fig. 4.28. The use of multi-output threshold-logic gates for multi-function realization.

functions $f_1(x)$, $f_2(x)$, etc. are simultaneously required from the same set of input variables x_i.[4] Let us spend the rest of this section considering possibilities in this particular area, which is as shown schematically in Fig. 4.28.

The change of logic coverage which results from different gate-threshold values but with a fixed set of input weights has been illustrated in passing in Fig. 4.24. However, if we look at this point more specifically we shall see that the change of logic coverage is in general a complex changing pattern.

Consider a gate with the input weightings 2,1,1,1 and with respective input signals x_1, $\bar{x}_2$, x_3, and $\bar{x}_4$. The total input summation $\sum_{i=1}^{n} a_i x_i$ at each minterm is illustrated in Fig. 4.29a. The output coverage available from threshold values 1, 2, 3, 4, and 5 is shown in the subsequent Karnaugh maps in Fig. 4.29b–f, respectively, from which the complexity of the changing coverage which results as t is increased is obvious. When we further begin to consider exclusive OR-ing of the various threshold outputs, a further complex family of coverages can be generated, for example as shown in Fig. 4.29g.

Clearly we here have a difficult design problem if we wish to utilize fully the considerable power and flexibility of the multi-output threshold-logic gate for multi-function realization. If the required functions happen to be isobaric, then by definition one such gate all by itself will fulfil our design requirements, but this is a rather artificial and special case and in general some more complex solution will be necessary.

---

[4] Once again we must be careful with terminology; what we are here calling multi-output threshold-logic gates are called multi-function threshold gates by Hampel[57]. However, we shall here continue to use our previous choice of terminology in preference to the latter and reserve the term "multi-function" to refer to *overall* network characteristics rather than single-gate characteristics.

Design of Threshold-Logic Networks

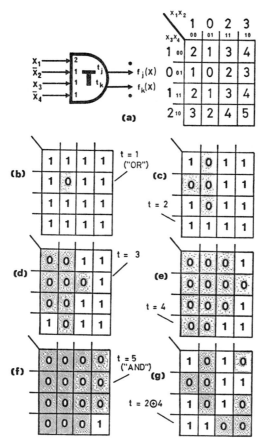

Fig. 4.29. Multi-function capability of a single multi-output threshold-logic gate: (*a*) example gate, inputs $x_1, \bar{x}_2, x_3, \bar{x}_4$, weights 2, 1, 1, 1, respectively, and associated Karnaugh map legend giving effective column, row, and minterm weights;

(*b*) coverage from $t = 1$, $f_1(x) = [x_1 + \bar{x}_2 + x_3 + \bar{x}_4]$;
(*c*) coverage from $t = 2$, $f_2(x) = [x_1 + \bar{x}_2 x_3 + \bar{x}_2 \bar{x}_4 + x_3 \bar{x}_4]$;
(*d*) coverage from $t = 3$, $f_3(x) = [x_1(\bar{x}_2 + x_3 + \bar{x}_4) + \bar{x}_2 x_3 \bar{x}_4]$;
(*e*) coverage from $t = 4$, $f_4(x) = [x_1(\bar{x}_2 x_3 + \bar{x}_2 \bar{x}_4 + x_3 \bar{x}_4)]$;
(*f*) coverage from $t = 5$, $f_5(x) = [x_1 \bar{x}_2 x_3 \bar{x}_4]$;
(*g*) coverage from a single-output gate,

$$f(x) = [f_2(x) \oplus f_4(x)]$$
$$= [x_1 x_2 (\bar{x}_3 + x_4) + x_1 \bar{x}_3 x_4 + \bar{x}_1 \bar{x}_2 (x_3 + \bar{x}_4) + \bar{x}_1 x_3 \bar{x}_4].$$

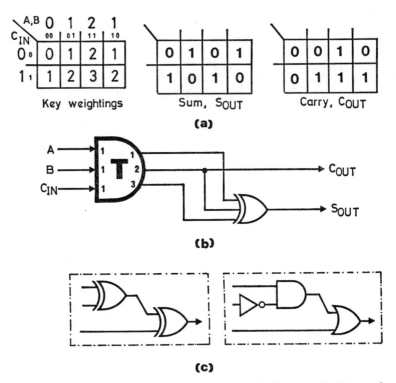

Fig. 4.30. Multi-threshold design of a two-input full-adder circuit: (*a*) mapping of the requirements; (*b*) the multi-threshold realization; (*c*) functional equivalents of the three-input exclusive-OR gate, remembering that "not possible" conditions arise at the outputs of the multi-threshold gate.

There are, however, some areas where multi-output threshold-logic gates may be intuitively applied, such as in voting and parity applications and in simple arithmetic requirements. In these areas the strong structure of the output functions and the *co-equal importance* of the input variables frequently allow heuristic methods to be applied. For illustration, take the case of a simple two-input full-adder, with inputs A and B together with a carry C_{IN}. By definition all these three inputs are of exactly equal importance in determining the sum and carry outputs S_{OUT} and C_{OUT}, respectively. Therefore if a threshold realization is being sought we can immediately allocate equal weights to all three inputs, which conventionally we would make 1.0. Figure 4.30 therefore illustrates the build-up of the design for this particular problem.

Design of Threshold-Logic Networks

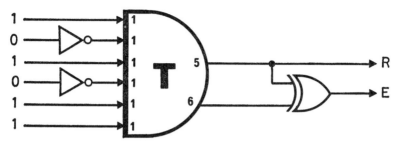

Fig. 4.31. Multi-threshold design for a six-bit word detector to recognize the input word 101011. Output R equals word recognized; output E equals one bit error present in the recognized word.

A similar technique may be applied to parity and code recognition problems. As a similar design exercise, consider the requirement to recognize a particular six-bit code word, say 101011, which is to be recognized when all six bits are correct, or where one but not more than one is in error. When one bit error is present, as well as the "word recognized" output, we also require a "one bit in error" output to be given. Figure 4.31 illustrates the simple solution for this requirement, which again can be produced without the need for any complex design algorithms.

These concepts for simple, rigorously structured problems are illustrated in published papers[57–59]. However, where the given problem exhibits a more random structure then more sophisticated design tools for multi-function synthesis are required. This is an area where considerably more research and development is still required, but in which it seems that Rademacher–Walsh spectral techniques such as we shall consider in the next chapter may be the answer.

However, a partially formal technique for the realization of multi-function problems using multi-output threshold-logic gates may be suggested[59]. The design steps in this technique are as follows.

(i) Determine the Chow parameters or the full Rademacher–Walsh spectrum for each of the required outputs $f(x)$. If any one, say $f_1(x)$, is found to be a listed linearly separable function with consequent single-output threshold realization, adopt this realization for $f_1(x)$.

(ii) Using the weights a_i determined for the threshold solution of $f_1(x)$, evaluate the input summation $\sum_{i=1}^{n} a_i x_i$ for each of the input minterms. Hence tabulate the summation values for each true (1) output of the further function(s) $f_2(x)$ etc. and for each false (0) output.

(iii) Examine this listing to see whether the true (or false) minterms for each function are given by a distinct range of input summation values; for example all true minterms have a $\sum_{i=1}^{n} a_i x_i$ value of 2 and 4 only, but not 3.

(iv) If such a feature can be found, realize $f_2(x)$ etc. by the appropriate outputs from the threshold-logic gate used for $f_1(x)$ with appropriate exclusive-OR output selection as necessary; for example $[f_j(x) \oplus f_k(x)]$, where $f_j(x)$, $f_k(x)$ are the appropriate outputs.

(v) If a coverage such as (iii) above can be found for $f_2(x)$ etc. except for a small number of minterms which do not fall into the desired grouping, such minority minterms may be added to or deleted from the threshold coverage by a following AND/OR/exclusive-OR level of gating; for example $[f_j(x) \oplus f_k(x)] \oplus x_1 x_2 x_3$, where the vertex AND function $x_1 x_2 x_3$ is used to modify the coverage from $[f_j(x) \oplus f_k(x)]$ alone.

(vi) If no exact threshold coverage for $f_1(x)$ can be found in test (i) above, choose a near-threshold cover for $f_1(x)$ and repeat the summation tabulation (ii) for all required functions $f_1(x)$, $f_2(x)$, etc. Then repeat procedure (iii) to (v) for all required functions.

As an illustration of this technique, consider the case of a certain arithmetic processor which was required to give two outputs, given by

$$f_1(x) = [x_1(x_3 + x_2 x_4) + x_2 x_3 x_4]$$

and

$$f_2(x) = [x_1(\bar{x}_2 \bar{x}_3 + \bar{x}_3 \bar{x}_4 + x_2 x_3 x_4) + \bar{x}_1(\bar{x}_2 x_3 + x_3 \bar{x}_4 + x_2 \bar{x}_3 x_4)].$$

Figures 4.32a and b give the mapping of the two required outputs.

Determination of the Chow parameters for the two functions, or the full Rademacher–Walsh (R.W.) spectral coefficients, gives the following.

Output $f_1(x)$
Chow parameters b_0 b_1 b_2 b_3 b_4
        4   8   4   8   4

R.W. spectral coefficients

r_0	r_1	r_2	r_3	r_4	r_{12}	r_{13}	r_{14}	r_{23}	r_{24}	r_{34}	r_{123}	r_{124}	r_{134}	r_{234}	r_{1234}
−4	8	4	8	4	0	−4	0	0	−4	0	−4	0	−4	0	−4

Output $f_2(x)$
Chow parameters: b_0 b_1 b_2 b_3 b_4
         0   0   0   0   0

Design of Threshold-Logic Networks

R.W. spectral coefficients

r_0	r_1	r_2	r_3	r_4	r_{12}	r_{13}	r_{14}	r_{23}	r_{24}	r_{34}	r_{123}	r_{124}	r_{134}	r_{234}	r_{1234}
0	0	0	0	0	8	0	0	0	0	8	0	8	0	−8

In descending magnitude order, which is the order in which Chow parameters are listed in look-up tables, these two functions have look-up values of

8, 8, 4, 4, 4

and

0, 0, 0, 0, 0

for $f_1(x)$ and $f_2(x)$, respectively. The first of these is a listed entry, and hence $f_1(x)$ is a linearly separable function with a single-threshold realization, but $f_2(x)$ is not listed and therefore cannot be realized with one threshold-logic gate. The look-up table lists that the realizing weights for $f_1(x)$ are 2, 1, 2, and 1 for x_1, x_2, x_3, and x_4, respectively, with a gate threshold value of 4. Thus $f_1(x)$ is realized by the gate $\langle 2x_1 + x_2 + 2x_3 + x_4 \rangle_4$.

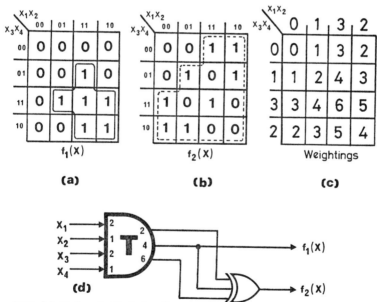

Fig. 4.32. Multi-threshold design for a two-output arithmetic processor problem: (a) Karnaugh map of output $f_1(x)$; (b) Karnaugh map of output $f_2(x)$; (c) column, row, and minterm weightings for the minimum-integer threshold realization for $f_1(x)$; (d) final realization for $f_1(x)$ and $f_2(x)$.

Before continuing with the remainder of the synthesis procedure to attempt to generate $f_2(x)$ using the input weightings appropriate for $f_1(x)$, it is instructive to look at the other information which we also have available to us in the values of the higher-order Rademacher–Walsh spectral coefficients. In $f_1(x)$ the spectral coefficients consist of a dominant grouping at the beginning of the list, with not so significant values associated with the higher-ordered coefficients r_{12}, r_{13}, etc. up to r_{1234}. This means that the output of the given function $f_1(x)$ has direct correlations with its input variables x_1, x_2, x_3, and x_4, but has very little correlation with any exclusive-OR combinations of its inputs, i.e. $[x_1 \oplus x_2]$, $[x_1 \oplus x_3]$, etc. A function with low exclusive-OR correlations but strong direct dependency is likely to be a linearly separable function, as indeed $f_1(x)$ proves to be.

On the other hand $f_2(x)$ exhibits a completely different structure. The low-order coefficients are all zero valued, indicating a wide scatter of the 0 and 1 output values of the function. On the other hand coefficients r_{13}, r_{123}, r_{134}, and r_{1234} are dominant, indicating strong exclusive-OR correlations between the inputs and output. Notice that 1 and 3 are represented in all these coefficients, indicating that inputs x_1 and x_3 are very important. In fact the synthesis which would be given from this spectral coefficient data, if we were not concerned with any other simultaneous output function $f(x)$, would be the exclusive-OR realization

$$f_2(x) = [x_1 \oplus x_3 \oplus x_2 x_4].$$

We are here anticipating somewhat the area we shall be covering in the following chapter, but as we have this Rademacher–Walsh data here before us a slight digression is possibly illuminating.

However, to return to our particular problem, that of synthesizing the two required functions $f_1(x)$ and $f_2(x)$, adoption of the input weights of 2, 1, 2, and 1 for x_1, x_2, x_3, and x_4 respectively gives us the overall input minterm summation values shown in Fig. 4.32c. It may now be seen by comparing Figs. 4.32b and c that $f_2(x)$ is selected by the threshold discrimination of $t = 2$, 3, and 6, leaving out $t = 4$ and 5, which in our previous multi-threshold terminology is

$$\langle 2x_1 + x_2 + 2x_3 + x_4 \rangle_{2,\bar{4},6}.$$

Hence we may construct the final realization shown in Fig. 4.32d.

It is significant in this example to note that although the original Boolean functions for $f_1(x)$ and $f_2(x)$ between them involved every input variable in both true and complemented form, the final realization shown in Fig. 4.32d does not require the complements of *any* inputs to be provided. This is a very significant feature of many of these design solutions, and of course the minimization of actual connections into a logic network often has great practical advantages in circuit layout.

This simple example is easily displayed and may be readily synthesized directly from the maps shown in Fig. 4.32. For functions with more than, say, five input variables computer listing and sorting of the data is necessary. The fast Walsh transform used to generate the full Rademacher–Walsh spectrum is a trivial computational procedure, and print-out of the coefficient values can be arranged in any grouping or order desired.

However, when none of the simultaneous functions to be realized is exactly realizable by one threshold-logic gate output, the choice of an appropriate threshold-logic gate around which to build up the final synthesis is still a matter of some intuition. Hence in such computer programs as have been developed to date to generate and sort the Chow parameter or spectral data, fully interactive CAD working has been maintained to enable the designer to choose or confirm a starting point from the available data. Further research is necessary in order to provide an efficient design procedure, which can optimize the synthesis with some form of convergence upon an optimum solution.

Certainly it would appear that the multi-output threshold-logic gate has a useful place in random-logic design where multiple outputs are involved, provided elegant synthesis techniques can be evolved to utilize fully its considerable power and flexibility.

4.5. Sequential Circuits

The design of asynchronous, that is non-clocked, sequential systems forms a large part of the total field of logic theory. The classic representation of a sequential system is shown in Fig. 4.33, from which analysis and synthesis techniques may be developed. The many standard texts available adequately cover these techniques[34-36,60,61]. The synthesis procedure for asynchronous systems basically is as follows.

(i) The compilation of a flow table, formally listing the system requirements.

(ii) Elimination of any redundancies in the first flow table, usually known as "merging" procedures, to produce a state-assignment table.

(iii) Synthesis of the state-assignment table, including the elimination of any critical races in the sequences.

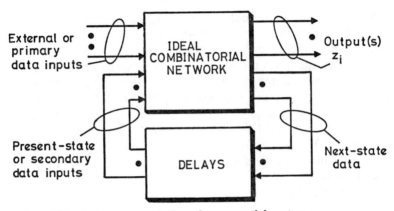

Fig. 4.33. The classic representation of a sequential system.

(iv) Derivation of the logic excitation equations of the system, which in the case where few variables only are involved may be presented on Karnaugh-type "excitation" or "output" maps.

The result of these procedures is to produce appropriate Boolean equations or Karnaugh maps which realize the required sequential action. The actual circuit realization, conventionally in vertex gates, can equally be carried on from here using threshold-logic gates, using the techniques illustrated in earlier sections of this chapter.

The principal published material in this area is given by Masters and Mattson[62] and others[63,64]. The former authors indicate that further work on more general state-assignment procedures specifically for threshold-logic realization may be profitable, but no such work seems to have been published so far in this area.

Turning from asynchronous systems to synchronous (or "clocked") systems, rather more published information is available. Again if classic design methods are followed to synthesize the logic gating requirements of a clocked system, as illustrated in Fig. 4.34, then substitution of threshold-logic gates rather than vertex gates can be directly considered in the final circuit realization.

There are, however, two further interesting topics in this general area of synchronously operating systems: firstly the possible design of clocked bistable circuit "building blocks" using threshold-logic rather than vertex gates, and secondly the (mainly intuitive) design of simple shift registers and the like using majority or threshold-logic gates. Let us first look at the possibilities of the bistable building blocks.

Design of Threshold-Logic Networks 257

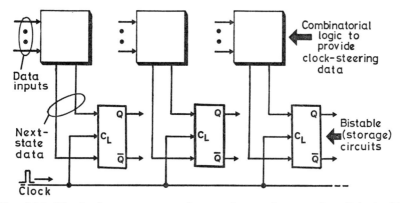

Fig. 4.34. The basic arrangement of a synchronously operating ("clocked") sequential system.

Type D, T, RS, and JK clocked bistable assemblies are well-known items of conventional sequential counter assemblies. The logic action of these assemblies, particularly the type D, T, and JK, have direct relevance for most sequential requirements. To build each of these assemblies with threshold-logic gates instead of the usual NAND gates is readily possible, as illustrated in Fig. 4.35.

Points of interest in these basic circuit configurations are as follows.

(i) The circuits are d.c. coupled throughout; in the simple type D and RS configurations, however, the clock-steering data inputs D or RS should not be changed whilst the clock is at logic 1, but the T and JK circuits are effectively master–slave configurations.

(ii) All four configurations require the use of two cross-coupled threshold-logic gates with complemented and uncomplemented outputs to give the required bistable action; this is in contrast to certain shift-register arrangements, which will be illustrated later, which utilize only one threshold-logic gate per stage.

(iii) The fundamental similarity is readily observed between the D and RS circuits, and between the T and JK circuits, this similarity not always being so obvious in the more usual vertex-gate configurations.

(iv) Clock C and complemented clock $\overline{C}$ signals are required in the T and JK circuits to achieve unambiguous bistable action.

These various threshold-logic assemblies may be employed as building blocks in sequential assemblies in exactly the same manner as conventional vertex assemblies. However, it remains a debatable point whether the adoption of threshold-logic gates is really justified in these assemblies, as relatively uncomplicated logic is involved. Thus the full potential power of the threshold-logic gate is not exploited to such an extent as to warrant the additional component complexity per gate, in comparison with conventional vertex-gate realizations.

However, if we turn away from considering standard "building blocks" with which any form of sequential system may be made, the more attractive use of threshold-logic gates would appear to be in simple ring counters, shift registers, and the like.

Majority gates, being a simple type of threshold-logic gate, have received individual attention as possible circuits for such sequential

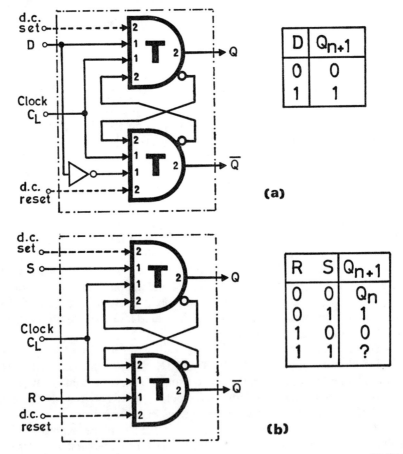

(a)

(b)

(continued)

Design of Threshold-Logic Networks

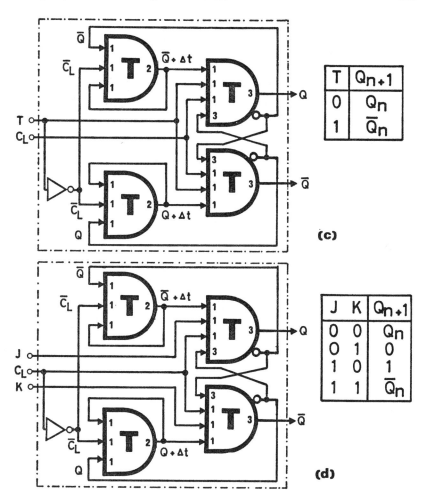

Fig. 4.35. A circuit realization for the principal types of clocked bistable circuits: (a) the type D circuit; (b) the type RS circuit; (c) the type T circuit; (d) the type JK circuit.

applications. Price has disclosed a number of basic possibilities using three-input majority gates, with complemented and uncomplemented outputs per gate[65], such as illustrated in Fig. 4.36. The circuit action of these assemblies is generally obvious by inspection. A further elaboration on Price's circuits may be found[66] using five-input majority gates.

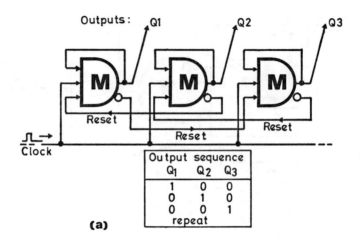

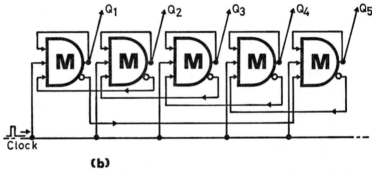

Fig. 4.36. Simple sequential circuits of Price, using three-input majority gates: (*a*) a three-stage ring counter; note, however, that this circuit also has a spurious stable all-0 and all-1 output sequence, which should be guarded against by some additional circuitry; (*b*) a similar five-stage circuit, but again spurious stable sequences can exist unless appropriately guarded against.

A circuit arrangement using five-input gates is shown in Fig. 4.37. Notice that a block of 1 outputs circulates in the latter circuits, rather than a single 1 output.

If input weightings other than the all-unity condition of majority gates are allowed, then with such more general threshold-logic gates other circuit arrangements may be proposed[67]. Figure 4.38 illustrates some of these possibilities. It will be noticed in all of these circuits from Price onwards that only one gate is being used per stage of the com-

Design of Threshold-Logic Networks

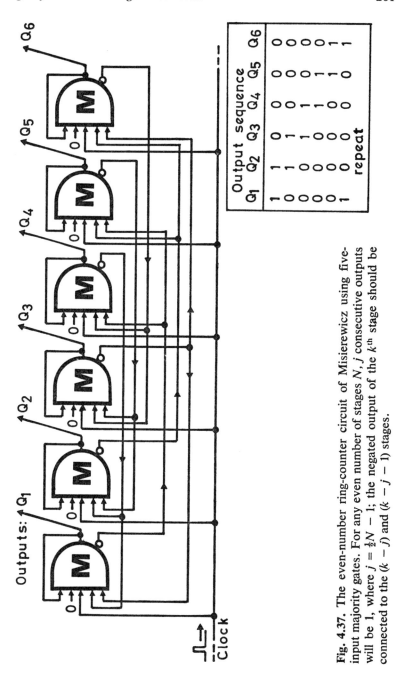

Fig. 4.37. The even-number ring-counter circuit of Misierewicz using five-input majority gates. For any even number of stages N, j consecutive outputs will be 1, where $j = \frac{1}{2}N - 1$; the negated output of the kth stage should be connected to the $(k - j)$ and $(k - j - 1)$ stages.

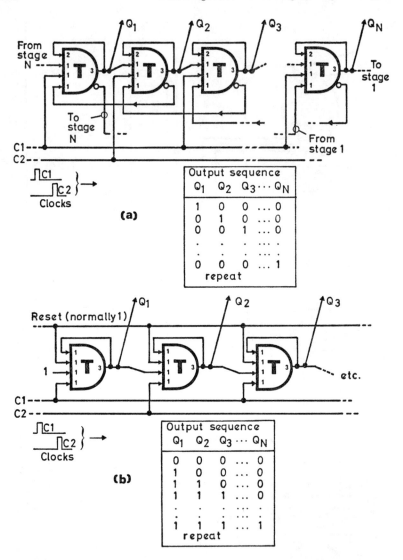

Fig. 4.38. Ring and chain circuits using threshold-logic gates: (*a*) an *N*-stage ring counter with reset using two interleaved clock-pulse lines; notice that more than one 1 may circulate in this type of circuit unless appropriately guarded against; (*b*) an *N*-stage circuit which "fills up" with 1 outputs and then requires resetting; notice that no spurious sequences can permanently exist in this configuration.

Design of Threshold-Logic Networks

plete assembly, a situation which is fundamentally impossible with conventional vertex gates. The circuit arrangements are also very simple, and the iterative nature of the designs leads to simple synthesis and analysis procedures.

To summarize this section, the potentially most useful application of threshold-logic gates to sequential problems would appear to be in the area of simple ring and chain counters and the like, rather than in binary coded decimal (BCD) or other coded sequences. The circuit of Fig. 4.38b, for example, is very simple for many industrial sequence controllers, and is very readily testable.

4.6. Chapter Summary

This chapter has ranged fairly extensively but not unduly mathematically over the possible application of various forms of threshold-logic gate to binary network design. Leaving aside for the moment the question of availability of such gates, we have seen that their use in network synthesis often is not a straightforward procedure. Indeed the number of papers which have been published on the theme of "the application of threshold-logic gates to network design" over the past decade has been very large, each often purporting to give the "best" method for efficient network synthesis. The fact that further papers on this topic continue to appear is proof that an easy and best method has not yet crystallized from all this work.

For problems with a small number of input variables, say $n \leq 5$, the Karnaugh mapping techniques offer a ready basis of design, allowing a visual build-up of the synthesis to be maintained. This limit can readily be extended to, say, $n \leq 8$ if visual display unit (VDU) facilities are available, with a CAD program rapidly to compute and display the problem with any chosen set of map axes and map identifications at will, that is to rotate the map identifications on the VDU display. However, these approaches are not entirely satisfactory for multi-function problems, where some efficient sharing of a single multi-output threshold-logic gate to generate several final network outputs is sought. Yet the multi-output threshold-logic gate theoretically is a very powerful tool, if it can be effectively used.

The standardization on threshold-logic gates only for network design would appear to be far too restrictive, even if a range of such gates with different input weightings and output thresholds is allowed. The use of simple vertex gates and exlusive-OR/NOR gates in association with threshold-logic gates would appear to be a much more balanced approach towards the goal of elegant network synthesis.

However, what precisely consitutes an "elegant" or "efficient" or "best" network realization, terms which we have frequently used without always defining them? In general this will ultimately depend upon the form of circuit realization. So far, however, we have tended to minimize the *number of gate symbols* in our network realizations, and thereby the *number of gate interconnections,* without concerning ourselves with the detailed circuit complexity or otherwise which may be involved within each gate. We have therefore ignored one side of the problem, since one type of gate may certainly be more complex in detailed design than another type of gate.

However, with present-day integrated circuits, circuit complexity by itself is not usually a limitation. Very often interconnections prove to be a far more restrictive parameter. For example, supposing we are considering a network realization using s.s.i. packages, available m.s.i. or l.s.i. packages not being relevant for the particular application. Under these conditions then gate count or pin (interconnection) count is an appropriate parameter to consider; the actual s.s.i. package cost is not usually dominant, as this tends to be swamped by assembly, testing, reliability, and other factors. If we apply this criterion, say, to the threshold-logic realization shown in Fig. 4.32, then assuming each gate has its independent input–output connections we have the following.

	Pin connection count	No. of inputs into the complete network
Circuit of Fig. 4.32*d*, assuming two two-input exclusive-OR gates used to generate the three-input exclusive-OR.	13	4
Comparable realization, with with the threshold gate replaced by an equivalent net of vertex gates.	about 36	4
Classic vertex gates only realizing $f_1(x)$ and $f_2(x)$ independently.	about 51	8

An efficiency criterion for subsystems within m.s.i./l.s.i. packages is the area of silicon necessary to produce the required function(s). In this respect the adoption of read-only memories or programmable logic

Design of Threshold-Logic Networks 265

arrays is not always efficient, although they may ease the work of the logic designer, and the adoption of non-vertex subsystems may show good improvements. In particular the reduction of interconnections per network which non-vertex designs can give is of great significance.

Two major areas therefore remain; firstly the need for yet further design techniques than those we have considered in the preceding pages, and secondly the question of the circuit design of the non-vertex gates themselves. In the next chapter we shall continue with the former area, whilst the consideration of gate designs will be the substance of the final chapter of this book.

References

1. Halligan, J. Using majority logic blocks. *Electron. Equip. News* **16**, 105, Nov. 1974.

2. Hurst, S.L. Improvements in general purpose s.s.i. logic packages and m.s.i./l.s.i. logic subsystems. *Electron. Lett.* **11**, 78–9, Feb. 1975.

3. Edwards, C.R. Some novel Exclusive-OR/NOR circuits, *Electron. Lett.* **11**, 3–4, Jan. 1975.

4. Wooley, B.A., and Baugh, C.R. An integrated m-out-of-n detection using threshold logic. *Trans. IEEE* **SC9**, 297–306, 1974.

5. Lewis, P.M., and Coates, C.L. *Threshold Logic.* John Wiley, New York, 1967.

6. McNaughton, R. Unate truth functions. *Trans. IEEE* **EC10**, 1–6, March 1961.

7. Muroga, S., Toda, I., and Takasu, S. Theory of majority decision elements. *J. Franklin Inst.* **271**, 376–418, May 1961.

8. Paul, M.C. and McClusky, E.J. Boolean functions realizable with single threshold devices. *Proc. IRE* **48**, 1335–7, July 1960.

9. Winder, R.O. *Fundamentals of Threshold Logic.* Air Force Cambridge Research Lab. Rep. No. 1, Contract AFCRL-68-0066, Jan. 1968.

10. Yajima, S., and Ibareki, T. Á theory of completely monotonic functions and its application to threshold logic. *Trans. IEEE* **C17**, 214–28, March 1968.

11. Yajima, S., and Ibareki, T. Realization of arbitrary logic functions by completely monotonic functions and its application to threshold logic. *Trans. IEEE,* **C17**, 338–51, March 1968.

12. Gabelman, I.J. The synthesis of Boolean functions using a single threshold element. *Trans IRE* **EC11**, 639–42, Oct. 1962.

13. Muroga, S., Tsuboi, T., and Baugh, C.R. Enumeration of threshold functions of eight variables. *Trans. IEEE* **C19**, 818–25, Sept. 1970.

14. Elgot, C.C. Truth functions realizable by single threshold organs. *Proc. 1960 Annual Symp. Switching Circuit Theory and Logical Design*, Sept. 1960, 225–45.

15. Cobham, A. *The Assumability Condition for Seven-Variable Functions*. IBM Res. Note NC. 483, March 1965.

16. Hu, S.T. *Threshold Logic*. University of California Press, Berkeley, Calif, 1965.

17. Minnick, R.C. Linear-input logic. *Trans IRE* **EC10**, 6–16, March 1961.

18. Winder, R.O. *An Evaluation of Heuristics for Threshold-Function Test-Synthesis*. Air Force Cambridge Research Lab. Rep. No. 4, Contract AF49/638-1184, May 1968.

19. Hawkins, J.K. Self-organising systems—a review and commentary. *Proc. IRE* **49**, 31–48, Jan. 1961.

20. Dertouzos, M.L. An approach to single-threshold-element synthesis. *Trans. IEEE* **EC13**, 519–28, Oct. 1964, **EC14**, 247, April 1965.

21. Winder, R.O. *Threshold Logic in Artificial Intelligence*, 107–28. IEEE Special Publ. No. S.142, Jan. 1963.

22. Winder, R.O. Chow parameters in threshold logic. *J. Ass. Computing Mach.* **18**, 265–89, April 1971.

23. Kaplan, K.R., and Winder, R.O. Chebyshev approximation and threshold functions. *Trans. IEEE* **EC14**, 250–2, April 1965.

24. Winder, R.O. *Threshold Functions through n = 7*. Air Force Cambridge Research Lab. Rep. No. 64–66, Contract AF19/604-8423, April 1963.

25. Winder, R.O. Enumeration of seven-argument threshold functions. *Trans. IEEE* **EC4**, 315–25, June 1965.

26. Muroga, A., Tsuboi, T., and Baugh, C.R. *Threshold Functions of Eight Variables*. Rep. No. 245, Dept. of Computer Science, University of Illinois, Aug. 1967.

27. Muroga, A., Tsuboi, T., and Baugh, C.R. Enumeration of threshold functions of eight variables. *Trans IEEE,* **C19**, 818–25, Sept. 1970.

28. Dertouzos, M.L. *Threshold Logic: A Synthesis Approach*. MIT Press, Cambridge, Mass., 1965.

29. Muroga, S. *Threshold Logic and its Applicatoin*. John Wiley Interscience, New York, 1971.

30. Hurst, S.L. *Threshold Logic: An Engineering Approach to the Synthesis of Multi-Level Threshold Logic Networks*. Ph.D. Thesis, University of London, 1971.

31. Hurst, S.L. Synthesis of threshold-logic networks using Karnaugh-mapping techniques. *Proc. IEE* **119**, 1119–28, Aug. 1972.

32. Hurst, S.L. Threshold logic network synthesis with specific threshold-gate sensitivities. *Radio Electron. Eng.* **42**, 295–300, June 1972.

33. Sheng, C.L. *Threshold Logic*. Academic Press, New York, 1969.

34. Hill, F.J., and Peterson, G.R. *Introduction to Switching Theory and Logical Design*. John Wiley, New York, 1968.

35. Lewin, D. *Logical Design of Switching Circuits*. Elsevier/North Holland, New York; Nelson, London, 1974.

36. Friedman, A.R., and Menon, P.R. *Theory and Design of Switching Circuits*. Computer Science Press, California, 1975.

37. Amodei, J.J., Hampel, D., Mayhew, T.R., and Winder, R.O. An integrated threshold gate. *Dig. Int. Solid-State Circuits Conf.* New York, 1967, 114–5.

38. Hurst, S.L. Specification of threshold-logic gates for optimum s.s.i. logic packaging. *Electron. Lett.* **8**, 514–5, Oct. 1972.

39. Cohen, S. and Winder, R.O. Threshold-gate building blocks. *Trans. IEEE* **C18**, 816–23, Sept. 1969.

40. Brown, R.J. *Adaptive Multiple-Output Threshold Systems and their Storage Capacities*. Ph.D. Thesis, Stanford University, Calif. 1964.

41. Meisel, W.S. Variable-threshold threshold elements. *Trans. IEEE* **C17**, 656–66, July 1968.

42. Meisel, W.S. Nets of variable-threshold threshold elements. *Trans. IEEE* **C17**, 667–76, July 1968.

43. Yajima, S., and Ibaraki, T. A lower bound on the number of threshold functions. *Trans. IEEE* **EC14**, 926–9, Dec. 1965.

44. Mays, C. *Adaptive Threshold Logic*. Ph.D. Thesis, Stanford University, Calif. 1963.

45. Nilsson, N.J. *Learning Machines*. McGraw-Hill, New York, 1968.

46. Aleksander, I. *Microcircuit Learning Computers*. Mills and Boon, London, 1971.

47. Amari, S. Learning patterns and pattern sequences by self-organising nets of threshold elements. *Trans. IEEE* **C21,** 1197–206, Nov. 1972.

48. Watanabe, S. *Frontiers of Pattern Recognition*. Academic Press, New York, 1972.

49. Haring, D.H. Multi-threshold elements. *Trans. IEEE* **EC15,** 45–65, Feb. 1966.

50. Haring, D.H., and Ohori, D. A tabular method for the synthesis of multi-threshold threshold elements. *Trans. IEEE* **EC16,** 216–20, April 1967.

51. Haring, D., and Diephuis, R.J. A realization procedure for multi-threshold threshold elements. *Trans. IEEE* **EC16,** 828–35, Dec. 1967.

52. Mow, C.-W., and Fu, K.-S. An approach for the realization of multi-threshold threshold elements. *Trans. IEEE* **C17,** 32–46, Jan. 1968.

53. Mow, C.-W., and Fu, K.-S. Input tolerance considerations for multi-threshold threshold elements. *Trans. IEEE* **C17,** 46–54, Jan. 1968.

54. Yen, Y.T. Some theoretical properties of multithreshold realizable functions. *Trans. IEEE* **C17,** 1081–8, Nov. 1968.

55. Necula, N.N. An algorithm for multithreshold synthesis. *Trans. IEEE* **C17,** 978–85, Oct. 1968.

56. Spann, R.N. *Generalized Threshold Functions*. Ph.D. Thesis, MIT, Cambridge, Mass., Feb. 1966.

57. Hampel, D. Multifunction threshold gates. *Trans. IEEE* **C22,** 197–203, Feb. 1973.

58. Hurst, S.L. Logic network synthesis using digital-summation threshold-logic gates *Proc. 12th Annual Allerton Circuits and Systems Conf., University of Illinois, Oct. 1974,* 525–34.

59. Hurst, S.L. The application of multi-output threshold-logic gates to digital network design. *Proc. IEE* **123,** 128–34, Feb. 1976.

60. Dietmeyer, D.L. *Logic Design of Digital Systems*. Allyn and Bacon, Boston, Mass, 1971.

61. Marcovitz, A.B., and Pugsley, J.H. *An Introduction to Switching System Design*. John Wiley, New York, 1971.

62. Masters, G.M., and Mattson, R.L. The application of threshold logic to the design of sequential machines. *Proc. 1967 Annual Symp. on Switching and Automation Theory, Oct. 1966,* 184–94.

63. Gustafson, C.H., Haring, D.R., Susskind, A.K., and Wills-Sandford, T.G. Synthesis of counters with threshold elements. *Proc. 1965 Annual Symp. Switching Circuit Theory and Logical Design, Oct. 1965,* 25–35.

64. Hadlock, F.O., and Coates, C.L. Realization of sequential machines with threshold elements. *Trans IEEE* **C18,** 428–39, May 1969.
65. Price, J.E. Counting with majority-logic networks. *Trans IEEE* **EC14,** 256–60, April 1965.
66. Misiurewicz, P. Comment on "Counting with majority logic networks *Trans. IEEE* **EC15,** 262, April 1966.
67. Hurst, S.L. Sequential circuits using threshold-logic gates. *Int. J. Electron.* **29,** 495–9, Nov. 1970.

Chapter 5

The Design of Binary Logic Networks Using Spectral and Other Techniques

Introduction

In this chapter we shall look at some of the possible ways of synthesizing binary logic networks using in the main conventional binary gates, other than the familiar methods which are in widespread present-day use. The latter are well known and very adequately referenced in many standard texts[1-7].

It is realistic to state at the outset that some of the concepts chosen to be reviewed in this chapter have not yet reached maturity and general application. Indeed some may prove to be more useful than others, with the result that the others may represent a line of research which was not ultimately profitable. However, what may be confidently stated is that more powerful design concepts will be finalized and adopted to replace in part, if not in whole, the present logic-design methods with their limited capabilities.

5.1. Boolean Matrices

The matrices which we shall review in this section provide a means of manipulating binary data which may be used in the analysis or synthesis of combinatorial and sequential networks. This approach has not received very wide attention, although it has interesting possibilities

Design Using Spectral and Other Techniques

particularly for multiple-output combinatorial networks. In the sequential area the concepts have been applied particularly to clocked (synchronous) systems, but not as far as is known to asynchronous operating systems. Results so far seem more attractive for network analysis than for network synthesis.

For practical reasons we shall continue to confine ourselves to fairly simple examples only, which can be readily illustrated, although it may be appreciated that computer handling of the matrix information for larger problems is entirely straightforward. The matrices considered in this section must not be confused in any way with the orthogonal matrices involved in the Rademacher–Walsh spectral area, as considered in the sections following, but instead should be regarded as essentially truth-table matrices of the binary data.

The first published work in this area was that of Campeau, although further subsequent publications may be found[8-10,12]. The matrix requirement for a combinatorial problem may be expressed as

$$\begin{bmatrix} \text{COEFFICIENTS} \\ \text{defining the} \\ \text{function or} \\ \text{functions} \end{bmatrix} \begin{bmatrix} \text{INPUT} \\ \text{VARIABLES} \\ \text{of} \\ \text{the function(s)} \end{bmatrix} = \begin{bmatrix} \text{The} \\ \text{REQUIRED} \\ \text{FUNCTION or} \\ \text{FUNCTIONS} \end{bmatrix}$$

where the input variables and the required function(s) are expressed in normal 0, 1 binary data.

This may be directly compared with normal matrix algebra, for example

$$\begin{bmatrix} 7 & 5 \\ -3 & 9 \end{bmatrix} \begin{bmatrix} x \\ y \end{bmatrix} = \begin{bmatrix} R \\ S \end{bmatrix}$$

whence

$$R = 7x + 5y$$
$$S = -3x + 9y.$$

Notice that in conventional matrix algebra the defining coefficients must be in a particular order such that under the normal rules of matrix multiplication the correct coefficients are associated with the particular functions. This is equally true and necessary in the Boolean matrix case.

The order which is used for the Boolean matrices is based upon the usual canonic sum-of-products minterm expansion of any function

$f(x)$. Consider any two-variable function with inputs x_1 and x_2, then this may be expanded into the canonical form.

$$f(x) = [c_1 \bar{x}_1 \bar{x}_2 + c_2 \bar{x}_1 x_2 + c_3 x_1 \bar{x}_2 + c_4 x_1 x_2].$$

At any particular input minterm, say $x_1 = 0$, $x_2 = 1$, we have

$$f(x) = [c_1 \cdot 1 \cdot 0 + c_2 \cdot 1 \cdot 1 + c_3 \cdot 0 \cdot 0 + c_4 \cdot 0 \cdot 1]$$
$$= c_2 \cdot 1 \cdot 1 = c_2.$$

Hence the coefficients c_i, $i = 1 \ldots, 2^n$, $c_i \in \{0, 1\}$, fully define the function $f(x)$. In matrix form this expansion may be written as

$$[c_1 \ c_2 \ c_3 \ c_4] \begin{bmatrix} x_1 \\ x_2 \end{bmatrix} = f(x)],$$

where $\begin{bmatrix} x_1 \\ x_2 \end{bmatrix}$ implies $\begin{bmatrix} \bar{x}_1 \bar{x}_2 \\ \bar{x}_1 x_2 \\ x_1 \bar{x}_2 \\ x_1 x_2 \end{bmatrix}.$

Notice that the c_i terms are precisely the same as the vertical output values of a normal input–output truth table for the function $f(x)$.

Equally there is no reason why we may not write any number of output functions simultaneously in this manner, for example three two-variable functions $f_1(x)$, $f_2(x)$, and $f_3(x)$:

$$\begin{bmatrix} c_{11} & c_{21} & c_{31} & c_{41} \\ c_{12} & c_{22} & c_{32} & c_{42} \\ c_{13} & c_{23} & c_{33} & c_{43} \end{bmatrix} \begin{bmatrix} x_1 \\ x_2 \end{bmatrix} = \begin{bmatrix} f_1(x) \\ f_2(x) \\ f_3(x) \end{bmatrix}.$$

As an example suppose we have

$$\begin{bmatrix} 0 & 1 & 1 & 0 \\ 1 & 0 & 0 & 0 \\ 1 & 0 & 0 & 1 \end{bmatrix} \begin{bmatrix} x_1 \\ x_2 \end{bmatrix} = \begin{bmatrix} f_1(x) \\ f_2(x) \\ f_3(x) \end{bmatrix}.$$

It is left as a trivial exercise for the reader to show that this matrix defines the three functions

$$f_1(x) = [x_1 \oplus x_2]$$
$$f_2(x) = [\bar{x}_1 \bar{x}_2] = [\overline{x_1 + x_2}]$$

and

$$f_3(x) = [x_1 \overline{\oplus} x_2].$$

The order in which we have defined the c_i matrix means that when we are evaluating the output value of any function $f(x)$ we effectively

Design Using Spectral and Other Techniques

look at the extreme left-hand coefficient for the output value of $f(x)$ when the x_i inputs are all logic 0, progressing along the individual c_i values to the c_{2^n} value when all x_i inputs are logic 1. For example, in the previous matrix we have that the three function outputs when $x_1 = 1$, $x_2 = 0$ are given by

$$\begin{bmatrix} 0 & 1 & 1 & 0 \\ 1 & 0 & 0 & 0 \\ 1 & 0 & 0 & 1 \end{bmatrix} \begin{bmatrix} 1 \\ 0 \end{bmatrix} = \begin{bmatrix} 1 \\ 0 \\ 0 \end{bmatrix}.$$

More formally, we can write below the coefficient matrix [C] an *identity* or *unit* matrix [A], which represents the binary progression from 00... through to 11... reading from left to right. This is of course in the order chosen for the c_i values in the coefficient matrix. Thus for a three-variable case with three outputs $f_1(x)$, $f_2(x)$, and $f_3(x)$, we may have, for example,

$$\begin{bmatrix} 0 & 1 & 0 & 1 & 1 & 0 & 1 & 1 \\ 1 & 1 & 0 & 1 & 1 & 1 & 0 & 0 \\ 1 & 0 & 1 & 1 & 0 & 0 & 0 & 0 \end{bmatrix} \begin{bmatrix} x_1 \\ x_2 \\ x_3 \end{bmatrix} = \begin{bmatrix} f_1(x) \\ f_2(x) \\ f_3(x) \end{bmatrix}$$

$$\begin{bmatrix} 0 & 0 & 0 & 0 & 1 & 1 & 1 & 1 \\ 0 & 0 & 1 & 1 & 0 & 0 & 1 & 1 \\ 0 & 1 & 0 & 1 & 0 & 1 & 0 & 1 \end{bmatrix} \longleftarrow \text{identity or unit matrix}$$

which enables us to read off directly the three output values on any input combination of x_1, x_2, x_3 by merely reading above this input combination the resulting output states.

A further way of writing such matrices is to use the equivalent decimal number for each binary column in the coefficient and identity matrices. The previous example may be rewritten as follows:

[3 6 1 7 6 2 4 4] X] = f(X)]

[0 1 2 3 4 5 6 7]

where X] is the column vector x_1, x_2, x_3 and $f(X)$] is the column vector $f_1(x)$, $f_2(x)$, and $f_3(x)$.

Whilst it is entirely straightforward to read off directly the outputs from these matrix formats, they do nevertheless obey formal mathematical rules. Briefly, if one identifies the matrix row and column entries as follows:

$$j\downarrow [C] \xrightarrow{p} \quad X]\downarrow_n^1 = f(X)]\downarrow_j^1$$
$$n\downarrow [A]$$

it may be shown[8] for any particular function output $f_j(x)$ that

$$f_j(x) = \left\{ \sum_{p=1}^{2^n} c_{jp} \prod_{n=1}^{n} (a_{np}x_n + \bar{a}_{np}\bar{x}_n) \right\}.$$

For a simple two-input function $f_1(x)$ this expression expands into

$$\begin{aligned}f_1(x) = \{&c_{11}(a_{11}x_1 + \bar{a}_{11}\bar{x}_1)(a_{21}x_2 + \bar{a}_{21}\bar{x}_2) \\+ &c_{12}(a_{12}x_1 + \bar{a}_{12}\bar{x}_1)(a_{22}x_2 + \bar{a}_{22}\bar{x}_2) \\+ &c_{13}(a_{13}x_1 + \bar{a}_{13}\bar{x}_1)(a_{23}x_2 + \bar{a}_{23}\bar{x}_2) \\+ &c_{14}(a_{14}x_1 + \bar{a}_{14}\bar{x}_1)(a_{24}x_2 + \bar{a}_{24}\bar{x}_2)\}.\end{aligned}$$

All except one of the terms in brackets is zero valued at any particular time, the non-zero term selecting the appropriate c_{jp} value for $f(x)$.

So far our matrices have been merely a revised form of truth table, to which rules of matrix algebra apply, each matrix representing the behavior of some known combinatorial logic network. However, let us continue this approach to consider an overall network composed of networks in series. As we shall see this will correspond to appropriate multiplication of the defining matrices.

Figure 5.1a illustrates our previous considerations. Suppose now we have two networks, as shown in Fig. 5.1b. The overall system response may now be expressed as

[E] [D] X] = f(X)]

where [D] and [E] are the defining coefficients for the two individual networks, with X] and f(X)] as the input column vector and the output column vector, respectively, as before. What we now require is a procedure for multiplying the two coefficient matrices in order to produce one logically equivalent coefficient matrix, say [F], as illustrated in Fig. 5.1c, that is

[F] X] = f(X)]

where

[E] [D] = [F].

The hand procedure for performing this matrix multiplication is straightforward. The identity matrix [A] is written under the matrix which is being multiplied by the input matrix [D]. Suppose now the first column of [D] matches the fourth column of [A], then the first column

Design Using Spectral and Other Techniques 275

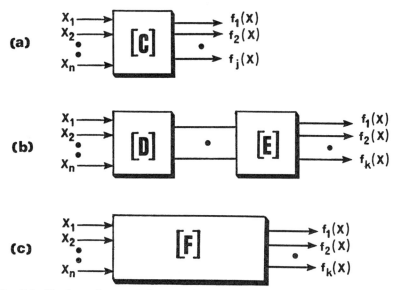

Fig. 5.1. Single and cascaded networks: (a) a single network, input column matrix X], output column matrix $f(X)$], where $f(X)$] = [C] X]; (b) two networks in cascade, the inputs to the second network being [D] X], the final outputs being $f(X)$] = [E] [D] X]; (c) the equivalent of (b), where [F] = [E] [D].

of the resultant matrix [F] is equal to the fourth column of [E]. Again, suppose the second column of [D] matches the second column of [A], then the second column of [F] is equal to the second column of [E]. This procedure continues until all the columns of [D] have been considered.

For example consider the following two-input, two-output network which fits the previous paragraph. The first two of the four operations to obtain the final matrix have been indicated by the arrows. The subsequent operations are readily followed.

[E] [D] = [F]

$$\begin{bmatrix} 1 & 0 & 1 & 0 \\ 0 & 0 & 1 & 1 \end{bmatrix} \begin{bmatrix} 1 & 0 & 0 & 0 \\ 1 & 1 & 0 & 1 \end{bmatrix} = \begin{bmatrix} 0 & 0 & 1 & 0 \\ 1 & 0 & 0 & 0 \end{bmatrix}$$

$$\begin{bmatrix} 0 & 0 & 1 & 1 \\ 0 & 1 & 0 & 1 \end{bmatrix}$$

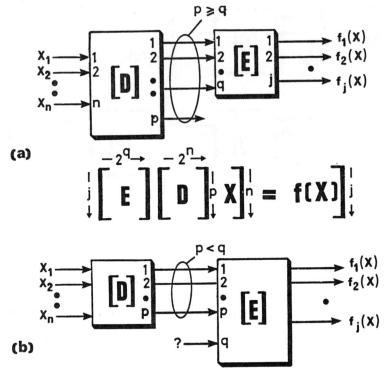

Fig. 5.2. Necessary dimensions of cascaded matrices: (*a*) satisfactory, $p \geq q$; (*b*) not satisfactory, as the outputs available from [D] are less than the number of inputs required by [E].

We may carry out this same operation as readily using the decimal notation, for example

$$[2\ 0\ 3\ 1][3\ 1\ 0\ 1] = [1\ 0\ 2\ 0]$$
$$[0\ 1\ 2\ 3]$$

Note that in these operations in general [E] [D] $\neq$ [D] [E]. One exception, however, is when one of the two matrices being multiplied is mathematically identical to the matrix operator [A], as then

[A] [D] = [D] [A] = [D].

It may be noticed that these matrix multiplications are not limited to matrices of precisely the same dimensions, the essential requirement being only that there are not more inputs required by a particular network than there are available signals to the network. This is illustrated in Fig. 5.2.

Design Using Spectral and Other Techniques

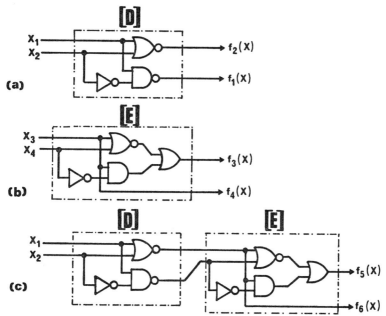

Fig. 5.3. The networks of the worked example: (*a*) network with the matrix [D]; (*b*) network with the matrix [E]; (*c*) cascade of the two networks, with an overall function given by the matrix multiplication [E] [D].

The full mathematical treatment and justification of this matrix multiplication may be found published[11,12]. However, let us look back and see what we have accomplished in the simple example we have just multiplied.

The matrix [D] of this example is the network shown in Fig. 5.3*a* where

$$f_1(x) = [\bar{x}_1 \bar{x}_2] = \overline{[x_1 + x_2]}$$

and

$$f_2(x) = \overline{[x_1 \bar{x}_2]} = [\bar{x}_1 + x_2].$$

The second network, defined by matrix [E], is shown in Fig. 5.3*b*, and is

$$f_3(x) = [\bar{x}_3 \bar{x}_4 + x_3 \bar{x}_4]$$

and

$$f_4(x) = [x_3].$$

Putting these two networks in series such that outputs $f_1(x)$ and $f_2(x)$ of the first become the x_3 and x_4 inputs of the second, then according to our matrix multiplication the new overall function is

$$f_5(x) = [x_1 \bar{x}_2]$$

and

$$f_6(x) = [\bar{x}_1 \bar{x}_2] = \overline{[x_1 + x_2]}$$

as shown in Fig. 5.3c. It is clearly more tedious to arrive at the last result by conventional means than by this matrix-multiplication approach, particularly when the functions in cascade are more complex than this simple example.

Still considering for the moment two networks in cascade as shown in Fig. 5.3c, suppose we know one of the two functions and the overall function, can we determine by matrix methods the second (unknown) of the two functions in cascade? This involves the same matrices as previously, namely

[E] [D] = [F]

but now say [D] and [F] are known but not [E]. In conventional matrix algebra we could express this problem as

[D] = [E]$^{-1}$[F]

but in this case does [E]$^{-1}$ exist and if so what is its relationship to [E]?

It has been shown[10] that the inverse of a Boolean matrix exists only if the matrix is "non-singular," that is if all columns of the matrix are different. For example

$$\begin{bmatrix} 0 & 1 & 0 & 1 \\ 1 & 1 & 0 & 0 \end{bmatrix} \text{ is non-singular}$$

but

$$\begin{bmatrix} 0 & 1 & 1 & 1 \\ 1 & 1 & 0 & 0 \end{bmatrix} \text{ is singular.}$$

This is a severe restriction, and hence in only a minority of problems can the inverse of a Boolean matrix be found.

However, when the matrix is non-singular then we have that the matrix times its inverse is equal to the unit or identity matrix [A]. The inverse matrix for any given Boolean matrix, therefore, may be rapidly evaluated by inspection, for example as follows:

Design Using Spectral and Other Techniques

given $[E] = \begin{bmatrix} 1 & 0 & 1 & 0 \\ 1 & 0 & 0 & 1 \end{bmatrix} = [3\ 0\ 2\ 1]$

then applying the multiplication

$[E]^{-1}[E] = [A]$

$[A]$

we have

$\begin{bmatrix} & ? & \end{bmatrix} \begin{bmatrix} 1 & 0 & 1 & 0 \\ 1 & 0 & 0 & 1 \end{bmatrix} = \begin{bmatrix} 0 & 0 & 1 & 1 \\ 0 & 1 & 0 & 1 \end{bmatrix}$

$\begin{bmatrix} 0 & 0 & 1 & 1 \\ 0 & 1 & 0 & 1 \end{bmatrix}$

or in decimal form

$[\ ?\ ][3\ 0\ 2\ 1] = [0\ 1\ 2\ 3]$

$[0\ 1\ 2\ 3]$

whence by inspection of the result we construct the unknown inverse matrix

$[E]^{-1} = \begin{bmatrix} 0 & 1 & 1 & 0 \\ 1 & 1 & 0 & 0 \end{bmatrix} = [1\ 3\ 2\ 0].$

If we continue an example using the above matrix, supposing our overall function [F] in a two-network cascade (see Fig. 5.4a) was

$[F] = \begin{bmatrix} 1 & 0 & 1 & 1 \\ 0 & 1 & 1 & 0 \end{bmatrix} = [2\ 1\ 3\ 2]$

then the unknown function [D] is given by

$[D] = [E]^{-1}[F]$
$= \begin{bmatrix} 0 & 1 & 1 & 0 \\ 1 & 1 & 0 & 0 \end{bmatrix} \begin{bmatrix} 1 & 0 & 1 & 1 \\ 0 & 1 & 1 & 0 \end{bmatrix}$

or

$[1\ 3\ 2\ 0]\ [2\ 1\ 3\ 2]$

whence by the normal procedure we obtain

$[D] = \begin{bmatrix} 1 & 1 & 0 & 1 \\ 0 & 1 & 0 & 0 \end{bmatrix} = [2\ 3\ 0\ 2].$

Thus the second network requires to be as illustrated in Fig. 5.4b.

It is also possible to consider the cross-over of interconnections between functions, and other simple logical operations on the interconnection lines.

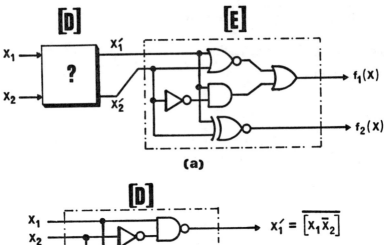

Fig. 5.4. Determination of an unknown matrix, given:
$f_1(x) = \overline{[\bar{x}_1 x_2]}$,
$f_2(x) = [x_1 \oplus x_2]$,
[E] = [3 0 2 1].

(a) The problem; (b) the matrix realization for [D], where [D] = [2 3 0 2].

Consider the cross-overs shown in Fig. 5.5a. Without such crossovers we have the same situation as previously evaluated, namely [E] [D] X] = f(X)], but with the cross-overs we effectively have an additional function between [D] and [E], say [P], giving us

[E] [P] [D] X] = f(X)].

However, what is this additional matrix?

Cross-over matrices have been shown to be extremely simple. If we take the unit or identity matrix [A], all that is necessary is to interchange rows of this matrix in accordance with the cross-overs of the circuit, and the result is the required cross-over matrix [P]. For the example shown in Fig. 5.5a we have

$$[A] = \begin{bmatrix} 0 & 0 & 0 & 0 & 1 & 1 & 1 & 1 \\ 0 & 0 & 1 & 1 & 0 & 0 & 1 & 1 \\ 0 & 1 & 0 & 1 & 0 & 1 & 0 & 1 \end{bmatrix}$$

whence the required cross-over matrix is

$$[P] = \begin{bmatrix} 0 & 1 & 0 & 1 & 0 & 1 & 0 & 1 \\ 0 & 0 & 1 & 1 & 0 & 0 & 1 & 1 \\ 0 & 0 & 0 & 0 & 1 & 1 & 1 & 1 \end{bmatrix}.$$

Notice that with no cross-overs we could (unnecessarily!) include the identity matrix [A] between terms, but

$$[E][A][D] \equiv [E][D]$$

as can be readily verified.

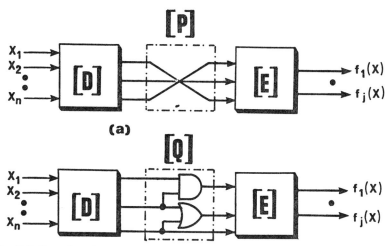

Fig. 5.5. Interfunction operations: (*a*) signal-line interchanges; (*b*) signal-line logic operations.

More complex operators on interconnections can also be accommodated. For example the interfunction logic shown in Fig. 5.5*b* requires an interfunction matrix operator, say [Q], which may be quickly obtained by logically AND-ing and OR-ing the rows of the identity matrix operator as follows:

$$[A] = \begin{bmatrix} 0 & 0 & 0 & 0 & 1 & 1 & 1 & 1 \\ 0 & 0 & 1 & 1 & 0 & 0 & 1 & 1 \\ 0 & 1 & 0 & 1 & 0 & 1 & 0 & 1 \end{bmatrix}$$

$$[Q] = \begin{bmatrix} 0 & 0 & 0 & 0 & 0 & 0 & 1 & 1 \\ 0 & 1 & 1 & 1 & 0 & 1 & 1 & 1 \\ 0 & 1 & 0 & 1 & 0 & 1 & 0 & 1 \end{bmatrix}$$

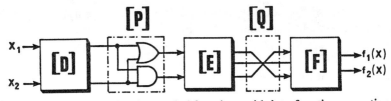

Fig. 5.6. A final example of cascaded functions with interfunction operations.

As a final example of applying this approach to the analysis of cascaded functions, consider the network shown in Fig. 5.6, where

$$[D] = \begin{bmatrix} 1 & 0 & 1 & 1 \\ 0 & 0 & 0 & 1 \end{bmatrix} = [2\ 0\ 2\ 3]$$

$$[E] = \begin{bmatrix} 1 & 1 & 0 & 1 \\ 1 & 0 & 1 & 0 \\ 1 & 0 & 1 & 1 \end{bmatrix} = [7\ 4\ 3\ 5]$$

$$[F] = \begin{bmatrix} 0 & 1 & 1 & 0 & 1 & 1 & 0 & 1 \\ 0 & 0 & 1 & 1 & 1 & 0 & 0 & 1 \end{bmatrix} = [0\ 2\ 3\ 1\ 3\ 2\ 0\ 3].$$

The two interfunction operators [P] and [Q] may be evaluated from the [A] matrices, giving us

$$[P] = \begin{bmatrix} 0 & 1 & 1 & 1 \\ 0 & 0 & 0 & 1 \end{bmatrix} = [0\ 2\ 2\ 3]$$

and

$$[Q] = \begin{bmatrix} 0 & 1 & 0 & 1 & 0 & 1 & 0 & 1 \\ 0 & 0 & 1 & 1 & 0 & 0 & 1 & 1 \\ 0 & 0 & 0 & 0 & 1 & 1 & 1 & 1 \end{bmatrix} = [0\ 4\ 2\ 6\ 1\ 5\ 3\ 7].$$

The final matrix multiplication is therefore as follows, with the identity matrices added below to show the order of the matrix-multiplication terms. The route of the first multiplication is shown, yielding the final value 0 for the first term:

Design Using Spectral and Other Techniques

$$[0\ 2\ 3\ 1\ 3\ 2\ \overset{0}{\overset{\uparrow}{0}}\ 3][0\ 4\ 2\ 6\ 1\ 5\ 3\ 7][7\ 4\ 3\ 5][0\ 2\ 2\ 3][2\ 0\ 2\ 3]\mathbf{X}] = f(X)]$$
$$[0\ 1\ 2\ 3\ 4\ 5\ 6\ 7][0\ 1\ 2\ 3\ 4\ 5\ 6\ 7][0\ 1\ 2\ 3][0\ 1\ 2\ 3]$$

The final complete result is the matrix [0 3 0 2], giving us

$$f_1(x) = 0\ 1\ 0\ 1$$
$$f_2(x) = 0\ 1\ 0\ 0$$

that is

$$f_1(x) = [x_2]$$

and

$$f_2(x) = [\bar{x}_1 x_2].$$

These matrix methods therefore are very powerful for the analysis of the overall performance of cascaded binary logic networks. Computer implementation of the procedures is straightforward, with fast implementation using store interchanging and word comparison techniques and without the need for large machine capacity. The procedures have been extended to the analysis of two-dimensional cellular-array structures without difficulty[11].

Unfortunately what is not directly possible with these techniques is cascaded-network or cellular-array *synthesis,* where one is given the required overall function(s) and wishes to determine what functions, ideally identical, can be cascaded to realize the overall requirements. If the cascaded functions are identical, this effectively poses the problem of *what is the nth root of the overall function matrix,* where n is the number of cells in cascade. Some iterative procedure may be possible to solve this problem, but so far no promising work along these lines is known.

Turning to sequential networks, again analysis rather than synthesis is more direct[1,8,9,12]. Also all published work to date refers to clocked (synchronous) systems, with no obvious extension to asynchronous operation.

Matrices identical to those we have been considering for multiple-output combinatorial functions may be used to express the successive outputs from the individual stages of a sequential network. Our previous p columns wide, j rows high coefficient matrices, which detailed the j individual combinatorial outputs, remain p wide by j high,

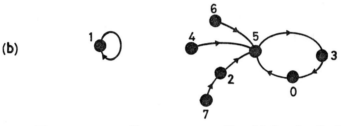

Fig. 5.7. Three-stage type D counter assembly: (*a*) the circuit; (*b*) the sequence diagram; note that state 1 ($x_1 = 0$, $x_2 = 0$, $x_3 = 1$) is a locked state.

but where j is now the number of binary stages in the assembly and $p = 2^j$, that is all possible output combinations of the j stages.

For example, consider the simple network composed of three type D (delay) elements, shown in Fig. 5.7a. If $x_{1(n)}$, $x_{2(n)}$, and $x_{3(n)}$ represent the present outputs of the three circuits, and $x_{1(n+1)}$, $x_{2(n+1)}$, and $x_{3(n+1)}$ represent the next-state outputs, that is after the receipt of one clock pulse only, we may write the next-state outputs for the type D assembly as

$$x_{1(n+1)} = [\bar{x}_3]$$
$$x_{2(n+1)} = [x_1 x_3]$$

and

$$x_{3(n+1)} = [x_2 x_3]$$

which in our usual matrix form becomes

$$[C] = \begin{bmatrix} 1 & 0 & 1 & 0 & 1 & 0 & 1 & 0 \\ 0 & 0 & 0 & 0 & 0 & 1 & 0 & 1 \\ 1 & 1 & 1 & 0 & 1 & 1 & 1 & 0 \end{bmatrix}$$
$$= [5\ 1\ 5\ 0\ 5\ 3\ 5\ 2].$$

Design Using Spectral and Other Techniques

To find the next state of this assembly after any given present state, locate the unit or identity matrix [A] below this coefficient matrix. The next state is directly found by locating the present state in the [A] matrix and reading off the next state directly above it in the [C] matrix. For example (see below) from the present state 0, i.e. $x_1 = 0$, $x_2 = 0$, $x_3 = 0$, we have that the next-state output is 5, that is $x_1 = 1$, $x_2 = 0$, $x_3 = 1$, and so on. The full sequence diagram of this assembly is therefore as shown in Fig. 5.7b.

[5 1 5 0 5 3 5 2]   coefficient matrix

[0 1 2 3 4 5 6 7]   unit matrix

For type JK circuits rather than type D the next-state equations require to accommodate the next-state characteristic of a type JK circuit, which is given by the expression

$$x_{(n+1)} = [J\bar{x}_n + \bar{K}x_n].$$

For the three-stage assembly shown in Fig. 5.8a the three next-state equations therefore are

$$x_{1(n+1)} = [\bar{x}_1 + \bar{x}_3 x_1]$$

$$x_{2(n+1)} = [\bar{x}_1\bar{x}_2 + 0]$$

and

$$x_{3(n+1)} = [x_1\bar{x}_3 + \bar{x}_2 x_3].$$

It is left as a simple exercise for the reader to construct the [C] matrix and prove the state-sequence diagram shown in Fig. 5.8b.

Hence it may be appreciated that this matrix method provides a very attractive way of presenting and interpreting the data to analyze clocked sequential systems. For synthesis, however, the entries in the [C] matrix are unknown, but provided the structure of the network is a simple counter structure as above then the excitation matrix may be obtained by considering the same matrix multiplication process, namely

[C] X] = [D]
[A]

where [D] is the *known* next-state output matrix. Having determined the [C] matrix there still remains the problem of producing minimal logic configuration for the clock-steering inputs to generate this matrix.

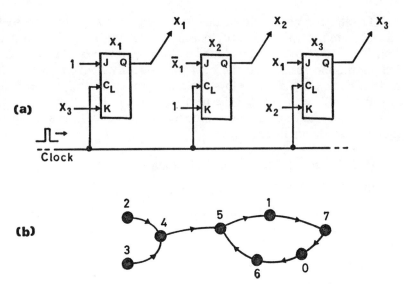

Fig. 5.8. Three-stage type JK counter assembly: (*a*) the circuit; (*b*) the sequence diagram.

More difficult, however, is the initial problem of specifying the structure for a sequential machine to perform a given duty with several controlling inputs, rather than the simple counter-type structures we have here considered. Further work in this area may prove useful.

5.2. Rademacher–Walsh Spectral Translation Techniques.

The mathematical background of the Rademacher–Walsh spectra has been covered in Chapter 2 of this book, and applied particularly to the classification of Boolean functions by means of invariance operations. No detailed attempt, however, was made in our earlier work to apply the spectral data to the synthesis of given functions, although some pointers to possible methods were given.

The Rademacher–Walsh coefficient values themselves can be directly employed to synthesize any binary function by using an arithmetic means as shown in Fig. 5.9. This is based upon the inverse transform from the spectral domain back to the Boolean domain, namely

$$[T]^{-1} S] = F]$$

where $[T]$ is the Rademacher–Walsh transform matrix,

$$= \frac{1}{2^n}[T]^t S] = F]$$

Design Using Spectral and Other Techniques

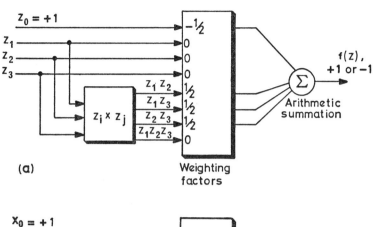

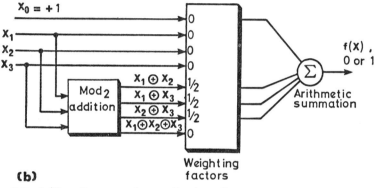

Fig. 5.9. Arithmetic-summation realization of

$$f(x) = [x_1\bar{x}_2 + x_2\bar{x}_3 + \bar{x}_1 x_3]$$

equal to the un-normalized Rademacher–Walsh spectrum of

r_0	r_1	r_2	r_3	r_{12}	r_{13}	r_{23}	r_{123}
−4	0	0	0	4	4	4	0

(a) The $\{+1,-1\}$ realization $f(z)$. (b) The $\{0,1\}$ realization $f(x)$, where the minterm value at $p = 0$ replaces the r_0 spectrum value. (Note: in both these realizations the normalizing factor $1/2^n = \frac{1}{8}$ could be incorporated after the output summation instead of in each individual weighting factor.)

the $1/2^n$ being the normalizing factor required in the inverse transform. Note, however, that the transform [T] must be the orthogonal $\{+1, -1\}$ matrix and not the non-orthogonal $\{0, 1\}$ matrix, and therefore the resulting function column matrix F] will be the $\{+1, -1\}$ realization $f(z)$ rather than the $\{0, 1\}$ realization $f(x)$. This is as shown in

Fig. 5.9a, where the output function may be expressed by the arithmetic summation:

$$f(z) = \frac{1}{2^n}\{r_0 z_0 + r_1 z_1 + \ldots + r_n z_n + r_{12}(z_1 \cdot z_2) + \ldots$$
$$+ r_{12\ldots n}(z_1 \cdots z_n)\}.$$

However, it may be shown[13] that, if the first Rademacher–Walsh coefficient r_0 is replaced by the value of $f(x)$ at minterm $p = 0$, a $\{0, 1\}$ realization for $f(x)$ is possible, as shown in Fig. 5.9b. The output $f(x)$ is now given by arithmetic summation:

$$f(x) = \left(f_{\underline{00\ldots 0}}(x) + \frac{1}{2^n}\{r_1 x_1 + \ldots + r_n x_n + \right.$$
$$\left. + r_{12}(x_1 \oplus x_2) + \ldots + r_{12\ldots n}(x_1 \oplus \ldots \oplus x_n)\}\right).$$

Such forms of realization, however, are not really practical, as we have no convenient means of reliably implementing the weighting and summing operations. Therefore, rather than attempting to use the actual values of the spectral coefficients in some arithmetic-type circuit configuration, we shall instead use the coefficient values to guide our synthesis of required functions.

5.2.1. The Invariance Operations Applied to a Given Function $f(x)$

As covered in Chapter 2, the Rademacher–Walsh spectral coefficients for any n-variable function $f(x)$ are always 2^n in number. We have so far seen how these coefficients may be rearranged under rigorous rules to provide a canonic classification for any binary function.

We shall now consider in more detail the five invariance operations introduced for these classification purposes in Chapter 2, and their significance for possible network synthesis. First let us define the result of the Rademacher–Walsh transform which generates the 2^n coefficient values, in terms of the $\{0, 1\}$ valued matrices. This is

$$r_i = 2^n - 2\left\{\sum_{k=0}^{2^n-1}\{t_{i,k} \oplus f_k(x)\}\right\}$$

where $f_k(x)$ is the value 0 or 1 of $f(x)$ at minterm k, and where the $\{0, 1\}$ matrices, see §2.4.3, have been identified row- and column-wise as follows:

Design Using Spectral and Other Techniques

$$[T] \times F] = S]$$

$$\underbrace{\begin{matrix} i = 0 \\ \text{to } 2^n - 1 \end{matrix} \downarrow \begin{bmatrix} 0 & 0 & \cdots \\ 0 & \cdot & \\ \vdots & & \end{bmatrix}}_{k = 0 \ldots 2^n - 1} \underbrace{\begin{bmatrix} f_0(x) \\ f_1(x) \\ \vdots \end{bmatrix}}_{\begin{matrix} k = 0 \\ \text{to } 2^n - 1 \end{matrix} \downarrow} = \underbrace{\begin{bmatrix} r_0 \\ r_1 \\ \vdots \end{bmatrix}}_{\begin{matrix} i = 0 \\ \text{to } 2^n - 1 \end{matrix} \downarrow}$$

Recall that the first (primary) rows of the $\{0, 1\}$ transform $[T]$ are identical to the x_i inputs, $i = 0 \ldots n$, whilst the subsequent (secondary) rows are identical to all exclusive-OR combinations of the x_i inputs, $i = 1 \ldots n$. Hence for the primary coefficients we may rewrite the previous expression as

$$r_i, \; i = 0 \ldots n, \; = 2^n - 2 \left\{ \sum_{k=0}^{2^n - 1} \{ x_{i,k} \oplus f_k(x) \} \right\}$$

and for the secondary coefficients we may write

$$r_{ij\ldots}, \; ij\ldots = 12, 13, \cdots, 12..n,$$

$$= 2^n - 2 \left\{ \sum_{k=0}^{2^n - 1} \{ (x_{i,k} \oplus x_{j,k} \oplus \ldots) \oplus f_k(x) \} \right\}.$$

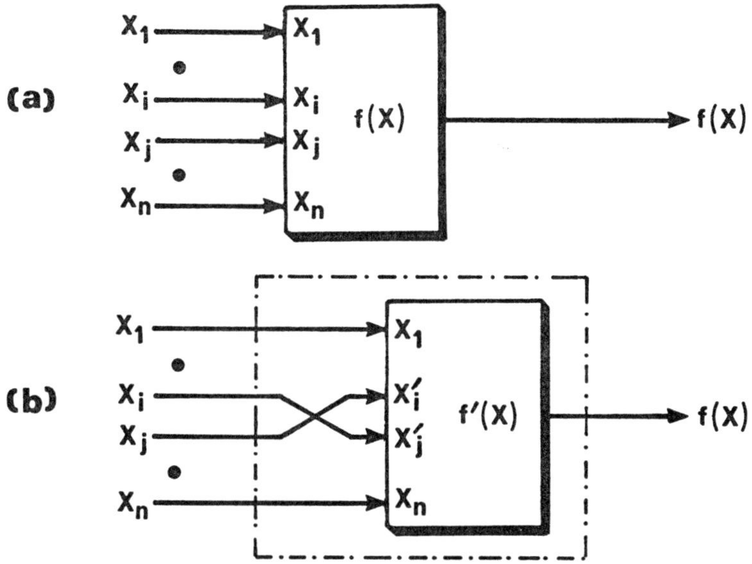

Fig. 5.10. Equivalent networks under spectral translation: (*a*) a given function $f(x)$ of n input variables; (*b*) the same overall function $f(x)$, but with an internal function $f'(x)$ and input cross-overs.

Suppose now we interchange two network inputs, say x_i and x_j, $i \neq j \neq 0$, of any given function $f(x_1, \ldots, x_i, x_j, \ldots, x_n)$. Let us also define a new function $f'(x_1, \ldots, x_i', x_j', \ldots, x_n)$ whose output is identical to the given function and where $x_i' = x_j$ and $x_j' = x_i$. This is illustrated in Fig. 5.10. The question which can now be posed is: What are the spectral coefficients of the new function $f'(x)$ such that output $f'(x) \equiv$ output $f(x)$?

Consider first the primary spectral coefficients. The spectral coefficient value of $f'(x)$ associated with input x_i' is

$$r_i' = 2^n - 2\left\{ \sum_{k=0}^{2^n-1} \{x_{i,k}' \oplus f_{\underline{k}}'(x)\} \right\}.$$

Similarly the spectral coefficient of $f(x)$ associated with input x_j is

$$r_j = 2^n - 2\left\{ \sum_{k=0}^{2^n-1} \{x_{j,k} \oplus f_{\underline{k}}(x)\} \right\}.$$

On the right-hand side of these two expressions we have $x_i' \triangleq x_j$, and $f_{\underline{k}}'(x) \triangleq f_{\underline{k}}(x)$. Thus these two expressions must be numerically equal, and hence $r_i' = r_j$. Similarly we may go through all other spectral coefficient values, and prove in total that

$$r_j' = r_i$$

$$r_{ik}' = r_{jk}$$

etc.

that is the spectral coefficients of $f'(x)$ may be obtained from the spectral coefficients of $f(x)$ by replacing i by j and *vice versa* in all the coefficient subscript identifications of $f(x)$. This is the result we have already used in Chapter 2, but was not formally proved at that stage.

Although, as we shall see later, this particular spectral conversion has no useful significance for synthesis purposes, as distinct from later ones, what we have found by doing such invariance operations on the spectrum of $f(x)$ is the specification for the new function $f'(x)$. Thus we could realize the original function $f(x)$ by synthesizing the function $f'(x)$ and feeding into it the revised order of the original input signals.

In a precisely similar manner the next two invariance operations on the spectral coefficients, that is negation of any one or more input variable and negation of the whole function, may be proved to modify the spectral coefficient values, as outlined in Chapter 2. It may be recalled that (i) negation of any input variable x_i results in the change of

Design Using Spectral and Other Techniques

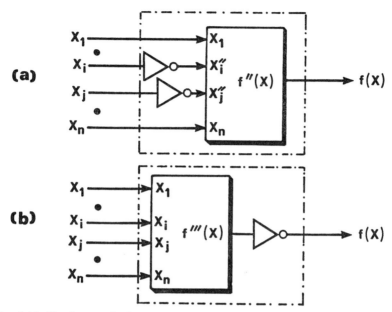

Fig. 5.11. Further equivalent networks: (*a*) same overall function $f(x)$, but with input negations; (*b*) same overall function $f(x)$, but with output negation.

sign of all spectral coefficient values containing i in their subscript identification and (ii) negation of the whole function results in the change of sign of all 2^n coefficient values. Proof follows readily from the previous example. Again it would be possible to synthesize any given function $f(x)$ by applying either (or both) of these operations and realizing the resultant functions as shown in Fig. 5.11, but still no significant attraction or advantages would accrue.

However, now let us consider the last two of the five possible invariance operations. These will be seen to be potentially useful.

First consider the replacement of any input variable x_i by the exclusive-OR signal $[x_i \oplus x_j]$, $i \neq j \neq 0$, as illustrated in Fig. 5.12*a*. This operation results in the interchange of the coefficient values of $f(x)$ as follows:

$r_i \leftrightarrow r_{ij}$

$r_{ik} \leftrightarrow r_{ijk}$

etc.

that is in all coefficients which contain i delete j if it also appears and append it if it does not appear, leaving unchanged all coefficients not containing i in their subscripts. Finally, if a feed-forward path as shown in Fig. 5.12b is made, then the spectral coefficients for $f''(x)$ are given by the interchange of the coefficient values of $f(x)$ as follows:

$r_i \leftrightarrow r_0$

$r_{ij} \leftrightarrow r_j$

$r_{ijk} \leftrightarrow r_{jk}$

etc.

that is in all 2^n coefficients append i if it does not appear and delete i if it does appear. Formal proof of these final two invariance operations may be found in Edwards[11,14].

What we must now distinguish between is for what purpose we are going to use the spectral invariance operations. In Chapter 2 we were principally concerned with the *classification* of functions, whereby

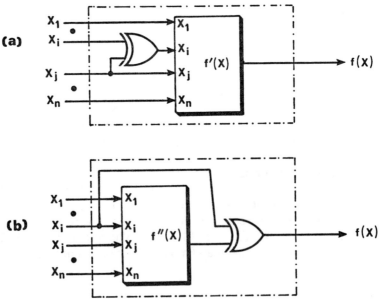

Fig. 5.12. Further equivalent networks: (a) same overall function $f(x)$, but with input exclusive-OR; (b) same overall function $f(x)$, but with output exclusive-OR.

Design Using Spectral and Other Techniques 293

many dissimilar functions $f(x)$ were transformed into a standard canonic representative function. What we are now interested in is maintenance of a given overall function $f(x)$, and seeing what form an intermediate "core" function $f'(x)$ may take. If this core function can be made a simple one in comparison with the overall function $f(x)$, then we have a possibly useful means of synthesizing $f(x)$. This concept was illustrated in outline form in Fig. 2.11, Chapter 2. Notice also from the preceding Fig. 2.10 how an initially complex function was transformed into a simple function by repeated invariance operations; our proposed synthesis technique is now to try to use the simple function, for example Fig. 2.10g, in the synthesis of the more complex required function $f(x)$.

The first requirement in this possible synthesis approach is some guide to which invariance operations are useful in deriving the core function. Consider the two functions shown in Fig. 5.13. It will be observed that the first is fairly trivial to realize by conventional means, but the second is more inconvenient as it involves a fairly wide scatter of 0- and 1-valued minterms. These characteristics are reflected in the distribution of the spectral coefficient values for each function.

From this illustration we may make the following general observations.

(i) If the spectral coefficient values for any given function $f(x)$ are predominately first order, that is they lie in the primary range r_1 to r_n, then the function is straightforward and may be realized without very great circuit complexity.

(ii) If the spectral coefficient values are predominantly higher ordered, that is they lie in the secondary ranges r_{12} to $r_{12..n}$, then the function is inconvenient to synthesize by classical means, particularly using vertex gates.

A possible synthesis procedure therefore is as follows. Given a function $f(x)$ which does not have a simple first-order spectrum pattern:

(*a*) Apply appropriate spectral translation operations to the given spectrum, for example as illustrated in Table 2.11 of Chapter 2, in order to maximize the first-order coefficient values.

(*b*) Record each spectral translation operation performed and its equivalent operation in the Boolean domain, e.g. exclusive-OR of inputs x_i, x_j, etc.

(*c*) Realize the final simplified core function which results from (*a*) by any appropriate type(s) of gates, and pre- and post-connect this core as required by (*b*).

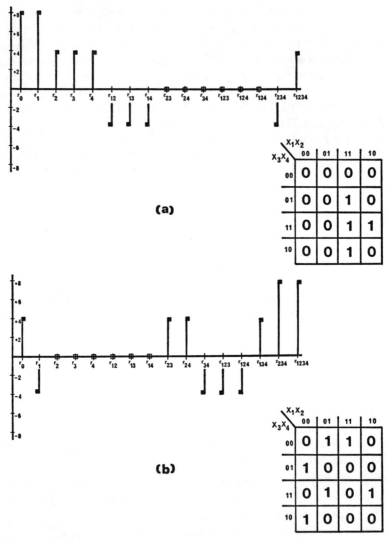

Fig. 5.13. Example functions and their spectral coefficients:
(a) function $f(x) = [x_1x_2(x_3 + x_4) + x_1x_3x_4]$
spectrum 8, 8, 4, 4, 4, −4, −4, −4, 0, 0, 0, 0, 0, 0, −4, 4
(b) function $f(x) = [x_2\bar{x}_3\bar{x}_4 + \bar{x}_1\bar{x}_2\bar{x}_3x_4 + x_3x_4(\bar{x}_1x_2 + x_1\bar{x}_2) + \bar{x}_1\bar{x}_2x_3\bar{x}_4]$
spectrum 4, −4, 0, 0, 0, 0, 0, 0, 4, 4, −4, −4, −4, 4, 8, 8.

Design Using Spectral and Other Techniques

This approach is summarized in Table 5.1.

Table 5.1. The method of synthesis of a given function $f(x)$ by spectral translation techniques

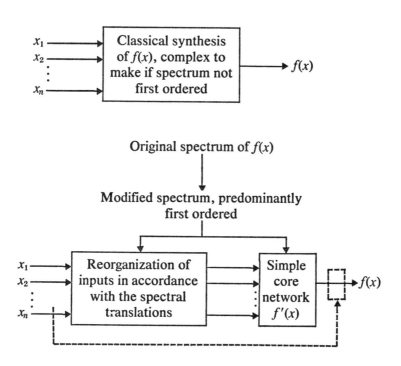

If we now consider the five possible spectral operations, which for reference have been summarized in Table 5.2, it will now be clear that operations (1) to (3), the NPN classification operations, have no immediate use as none will reorder a spectrum from predominantly second ordered to predominantly first ordered, but the final two operations which involve spectral translations between the two groups are the operations of importance. Notice also that both of these translations involve the exclusive-OR operation, and hence we now have a synthesis technique which can profitably utilize the exclusive-OR function which classic Boolean techniques are unable to do.

Table 5.2. Summary of the five spectral translation operations and their corresponding logical meaning

Spectral translation	Corresponding operation in the binary domain
(1) Interchange of subscript identifications i, j, $i \neq j \neq 0$, in all coefficients containing i, j	Interchange of the input variables x_i, x_j into the network
(2) Change of sign of all coefficients containing i in their subscript identification, $i \neq 0$	Negation of the input variable x_i into the network
(3) Change of sign of all 2^n spectral coefficients	Negation of the output of the network
(4) Interchange of the subscript identifications $r_i \leftrightarrow r_{ij}$, $r_{ik} \leftrightarrow r_{ijk}$, etc., $i \neq j \neq 0$, in all coefficients containing i.	Replacement of the original input x_i into the network by $[x_i \oplus x_j]$
(5) Interchange of the subscript identifications $r_i \leftrightarrow r_0$, $r_{ij} \leftrightarrow r_j$, etc., $i \neq 0$, in all 2^n coefficients.	Replacement of the network output $f(x)$ by $[f(x) \oplus x_i]$

Let us look at some simple syntheses using this spectral translation approach.

5.2.2. Some Example Syntheses Using Spectral Translations

The first example we shall show illustrates how several translation operations may be required to maximize the primary spectral coefficient values, and hence maximally simplify the core function. However, it will also indicate that with some functions maximum simplification of the core does not necessarily guarantee maximum simplification of the whole function realization, as the surrounding exclusive-OR gates may become excessive.

Design Using Spectral and Other Techniques

The given function $f(x)$ is shown in Fig. 5.14a, together with its spectrum. Three spectral translation operations are shown in the subsequent Figs. 5.14b–d, these in order being as follows.

(i) Replacement of input x_1 by $[x_1 \oplus x_4]$, which moves down the prominent r_{14} spectral coefficient value into the primary set.

(ii) Replacement of input x_3 by $[x_3 \oplus x_4]$, which moves down the r_{34} coefficient value.

(iii) Replacement of input x_2 by $[x_2 \oplus x_3']$, which moves down the r_{23} coefficient value of (c).

The final result of these spectral translations is the core function of Fig. 5.14d. The complete realization of $f(x)$ using this result is illustrated in Fig. 5.15a. In terms of the number of gates this realization using spectral translation techniques shows some savings over conventional AND/OR realizations, but notice two very interesting points which this particular example shows well. Firstly the spectral solution does not require the complements of any of the x_i inputs to be provided, and secondly the number of signal lines at any level of the realization never exceeds the number of input variables n, that is four in this case.

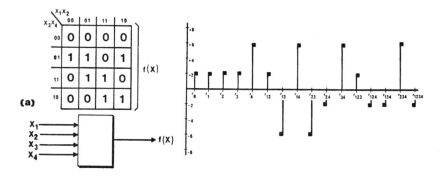

Fig. 15.4. *(continued)*

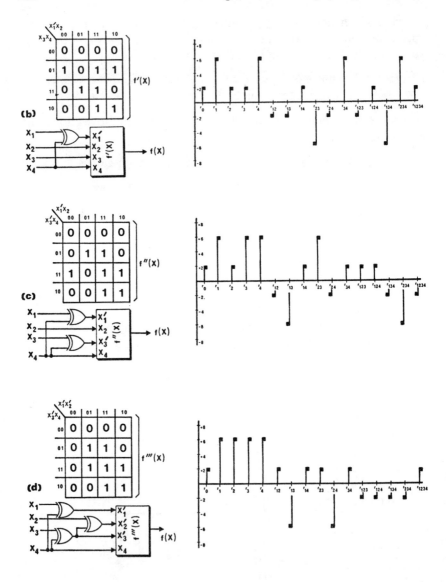

Fig. 5.14. Spectral translations on a given function
$$f(x) = [x_1x_3\bar{x}_4 + x_1x_2x_3 + \bar{x}_1x_2x_4 + \bar{x}_2\bar{x}_3x_4]:$$
(a) the given function and its spectrum; (b) translation of x_1 and $[x_1 \oplus x_4]$; (c) translation of x_3 and $[x_3 \oplus x_4]$; (d) translation of x_2 and $[x_2 \oplus x_3']$.

Design Using Spectral and Other Techniques

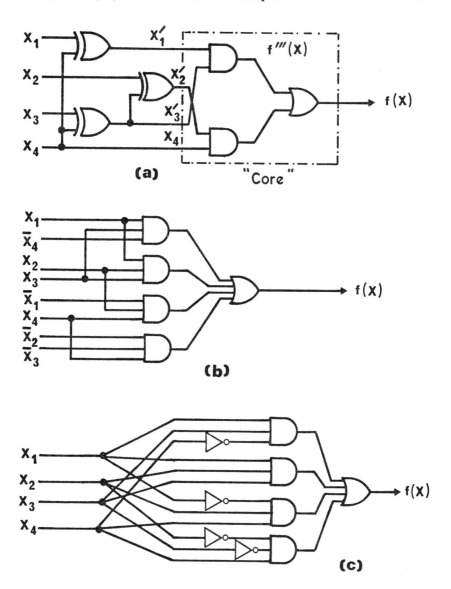

Fig. 5.15. Realizations for the function of Fig. 5.14: (a) realization based upon the final spectral translation of Fig. 5.14d; (b) classic AND/OR realization, assuming the complements of all inputs are available; (c) as (b) but with complements not available.

Thus a simple count of the number of gates alone is not necessarily a measure of the "goodness" of a realization; the number of interconnections, the fan-out of interconnections within the network, and the number of cross-overs of interconnections may all be very important and indeed critical factors.

As a second example, consider the function shown in Fig. 5.16. Note that we are still limiting our examples to $n \leq 4$ functions purely for ease of illustration. It is left as an exercise for the reader to confirm the realization shown, and to try other spectral translations to see if preferable alternative solutions can be found.

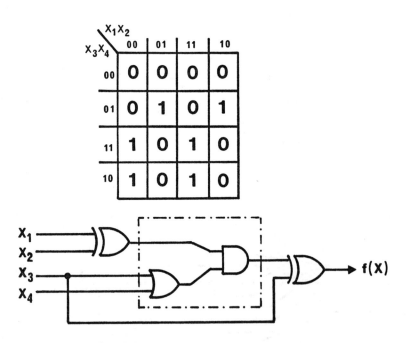

Fig. 5.16. A possible realization for the function
$f(x) = \Sigma\ 2,3,5,9,14,15$
using spectral translation techniques.

Design Using Spectral and Other Techniques

5.2.3. The Incorporation of Threshold-Logic Gates

It will be recalled from Chapter 2 and from Appendix B that in the classification of functions by spectral invariance operations the final canonic entries frequently are linearly separable functions, since the primary spectral coefficient magnitudes r_0 to r_n become identical to the listed Chow-parameter magnitudes b_0 to b_n. From Table 2.12 it was noted that seven of the eight $n \leq 4$ spectral classification entries are threshold-logic functions, whilst in Appendix B 21 of the 47 $n \leq 5$ entries are threshold-logic functions.

What this means for network synthesis is that spectral translation techniques may enable the core of the final realization to be a single threshold-logic gate, even though the function being realized is not linearly separable. The final realization therefore may be an appropriate network of exclusive-OR and threshold-logic gates, the translation techniques giving us a unique method of handling both of these non-vertex types of gates in our synthesis procedure.

Consider again the function previously synthesized in Fig. 5.15. The spectral coefficients of $f(x)$ are given on line (a) of Table 5.3. Examination of the $n \leq 4$ Chow-parameter values will show that the only possible threshold-logic function which may be embedded in this given function is the Chow-parameter entry 6, 6, 6, 6, 6. Therefore it is necessary to see if we can perform appropriate spectral translations in order to move down all 6's into the primary set of spectral coefficients. This indeed can be done. The following is one order of so doing, but again is not necessarily the only order which can be followed.

(i) Translate $r_0 \leftrightarrow r_4$, which involves in the final realization exclusive-OR of the core output with x_4; this translation moves all the coefficient values to the order shown in line (b) of Table 5.3.

(ii) Translate the new $r_2 \leftrightarrow r_{23}$, which involves the exclusive-OR of x_2 and x_3; this gives line (c) of Table 5.3.

(iii) Translate the new $r_4 \leftrightarrow r_{24}$, which involves the exclusive-OR of the x_4 and the new x_2; this gives line (d) of Table 5.3.

(iv) As we have a minus sign for coefficient r_4 in this last spectrum, let us make this positive so as not to have negative weightings in our threshold-logic core; this changes all relevant signs, as shown in line (e) of Table 5.3, and involves an inverter function in the final x_4 core input. This final spectrum may be checked with the canonic function 7 of Table 2.12 for correctness.

Table 5.3. Spectral translation operations on the given function $f(x) = \sum 1, 5, 7, 9, 10, 14, 15$ in order to generate a threshold logic core function

	r_0	r_1	r_2	r_3	r_4	r_{12}	r_{13}	r_{14}	r_{23}	r_{24}	r_{34}	r_{123}	r_{124}	r_{134}	r_{234}	r_{1234}
(a)	2	2	2	2	6	2	−6	6	−6	−2	6	2	−2	−2	6	−2
(b)	6	6	−2	6	2	−2	−2	2	6	2	2	−2	2	−6	−6	2
(c)	6	6	6	6	2	−2	−2	2	−2	−6	2	−2	2	−6	2	2
(d)	6	6	6	6	−6	−2	−2	2	−2	2	2	−2	2	2	2	−6
(e)	6	6	6	6	6	−2	−2	−2	−2	−2	−2	−2	−2	−2	−2	6

The threshold/exclusive-OR realization for this given function therefore is as shown in Fig. 5.17. The specific threshold-logic gate required as the core function can of course be looked up from the 6, 6, 6, 6, 6 entry on the $n \leq 4$ Chow-parameter tables (see Appendix A).

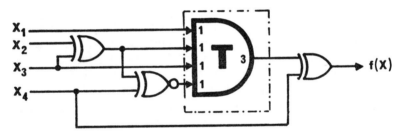

Fig. 5.17. An alternative realization for the function of Fig. 5.14 using a threshold-logic core.

Whilst such single-output syntheses using threshold-logic/exclusive-OR gates are academically very fascinating, practical viability hinges entirely on the design and availability of non-vertex gates. We shall be looking at gate design in Chapter 7 in some detail, and it will be seen that very compact exclusive circuits can be proposed, but threshold-logic gates still represent a problem. Thus the viability of at least vertex/exclusive-OR realizations seems assured, but realizations employing threshold-logic gates are more controversial.

However, let us look at one further but related feature. This is the problem of multi-function synthesis, that is several output functions required from the same set of x_i inputs. This problem was raised in Chapter 4, and the use of multi-threshold gates plus exclusive-OR gates was suggested in Fig. 4.27 as being potentially useful. So far we do not have a synthesis technique to handle this problem efficiently, but spectral translation techniques may be applicable. In outline such a synthesis procedure might involve spectral translations to be made on each required output in order to converge on some desirable spectrum or spectra, or alternatively to use common translations between the different outputs. The coverage (spectrum) provided by the different thresholds on a multi-threshold gate may, for example, be the target of the spectral translations. If this is possible then we have the facility to employ the very powerful multi-threshold gate capability together with exclusive-OR gate power. Research along these lines may prove to be useful.

To summarize this area, the examples shown have been fairly simple functions for convenience. Where the spectral translation techniques score even more heavily is in the computer-aided design of networks with $n > 4$, where mapping and geometric techniques cannot be applied. Interactive CAD programs can readily be prepared which enable the designer to interchange coefficient values and see how the function he is synthesising may be realized around a hopefully simplifying core function. The magnitude of the spectral coefficients is of course the key to which translations to choose in such interactive design procedures. For a more rigorous and formal mathematical background to this work reference may be made particularly to the published works of Dertouzos and others[14-16]. A minimum standard set of core functions is a possible further advantage of this method of function synthesis[14,17,18].

5.3. Synthesis based upon Rademacher–Walsh spectral statistics

It will be appreciated from previous use of the Rademacher–Walsh spectra that (*a*) the numerical values of the individual spectral components are not mutually independent but collectively bear certain (complex) relationships, and (*b*) the higher the magnitude of any particular component the "more like" the overall function $f(x)$ is to the input(s) which define such a component. Both of these attributes are of course unique to the spectral domain and are completely absent in normal Boolean working.

Therefore, from examination of the spectrum of any function $f(x)$, an accurate guide as to what may constitute a realization for $f(x)$ can be made by noting the largest-magnitude components. The trivial extreme of this approach is when one r_i spectral component is maximally valued, that is $\pm 2^n$, with all other components zero; in this extreme case the function $f(x)$ is merely composed of the one input combination which defines this $\pm 2^n$ valued component.

However, if we discount this latter trivial extreme and concentrate more on the case where there are two or more large-magnitude components, then we shall be faced with a spread of information in several prominent components, each of which may provide a viable complex factor for a final realization of $f(x)$. For example, suppose the spectral coefficients r_{13} and r_{123} are prominent, then this indicates some strong relationship in the required function with $[x_1 \oplus x_3]$ and $[x_1 \oplus x_2 \oplus x_3]$, and hence these two complex factors may be significant for the realization of $f(x)$. An iterative CAD program to synthesize $f(x)$ based upon this approach therefore can be proposed, as illustrated in Table 5.4.

As an example of this possible approach consider the function illustrated in Fig. 5.18a. Examination of the spectrum suggests that because the second-order spectral components r_{14} and r_{34} are both equally large, then possibly $[(x_1 \oplus x_4) \text{ AND } (x_3 \oplus x_4)]$ is a factor of the given function $f(x)$. The check is illustrated in Fig. 5.18b, which will be seen to cover four of the seven 1-valued minterms of $f(x)$.

Table 5.4. Possible CAD flow diagram for the extraction of complex factors predicted from the Rademacher–Walsh spectral coefficient values

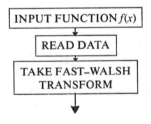

(continued)

Design Using Spectral and Other Techniques

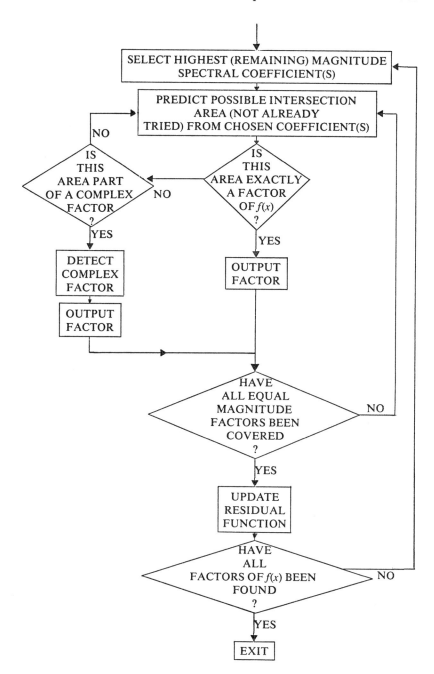

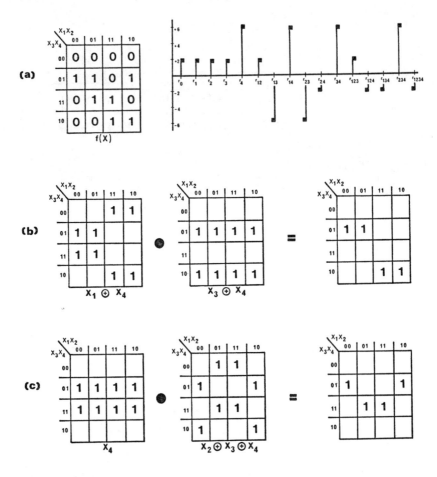

Fig. 5.18. Synthesis of the function $f(x) = \sum 1,5,7,9,10,14,15$: (a) the map and spectrum of the function;
(b) cover given by $[(x_1 \oplus x_4)(x_3 \oplus x_4)]$, which is a factor of $f(x)$;
(c) cover given by $[x_4(x_2 \oplus x_3 \oplus x_4)]$, which is also a factor of $f(x)$.

Similarly $[x_4 \text{ AND } (x_2 \oplus x_3 \oplus x_4)]$ may possibly be a factor of $f(x)$, and the check shown in Fig. 5.18c confirms that this is so. These two complex factors of Figs. 5.18b and c together cover the complete function $f(x)$, and hence the realization

$$f(x) = [(x_1 \oplus x_4)(x_3 \oplus x_4) + x_4(x_2 \oplus x_3 \oplus x_4)]$$

is produced. It is left as an exercise for the reader to try other possible combinations of complex factors for this problem which will realize the same overall function $f(x)$.

However, what is so far lacking in this approach is a good mathematical background, based on statistics and probabilities, upon which to make the first choice of spectral coefficient values. If a detailed statistical analysis of the possible distributions of spectral coefficient values were undertaken, then it may be possible to state that if, say, r_{ij} and r_{jk} have magnitudes greater than or equal to some fraction of 2^n, then it is statistically very probable (say $\geq 95\%$) that $[(x_i \oplus x_j)(x_j \oplus x_k)]$ is a factor of $f(x)$, irrespective of all other spectral coefficient values. This is clearly a significant pointer towards the synthesis of $f(x)$, particularly when n is large and therefore cannot be geographically or algebraically handled. A detailed statistical analysis of spectral coefficient values would be very valuable.

5.4. Synthesis Based Upon the Symmetrical Properties of Functions

5.4.1. Preliminary Considerations

The concept of simple symmetries which may exist in combinatorial functions was introduced in Chapter 1, and elaborated in more detail in using geometric Karnaugh map layouts in Chapter 3. It will be recalled that several types of symmetries may be present in a given function $f(x)$, which may be identified by similar patterns of 0- and 1-valued minterms in parts of a Karnaugh map (see Figs. 3.9 to 3.11 for example).

Before we look at the main problem in this area, which is the *detection* of what symmetries are present in any given function $f(x)$ assuming that Karnaugh mapping is inconvenient or impractical, let us illustrate how symmetric properties may advantageously be used in syntheses. Because symmetries in the input variables x_1 and x_2 are in general easier to see on a Karnaugh map layout than those which involve say x_1 and x_3, or x_2 and x_4, etc., we shall initially illustrate the basic design concepts using these x_1 and x_2 symmetries only; clearly the principles are applicable to any other symmetry pairs.

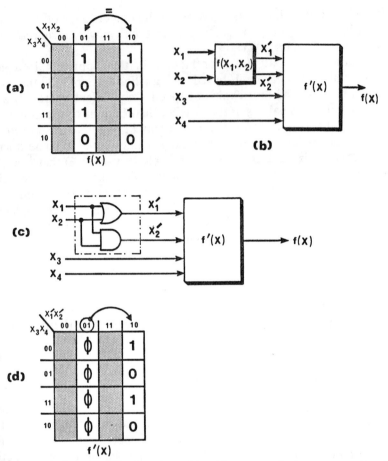

Fig. 5.19. A four-variable function $f(x)$ with non-equivalence symmetry in x_1, x_2: (a) plot of $f(x)$, not detailing the $\bar{x}_1 \bar{x}_2$ and $x_1 x_2$ minterm values; (b) conversion of the original input variables of symmetry x_1, x_2 into second-level inputs x'_1, x'_2; (c) the required first-level remapping network; (d) the plot of the required second-level function $f'(x)$, where ϕ are the don't-care conditions generated by the first-level network, all other minterms being the same as in the given function $f(x)$.

Consider the function shown partially plotted in Fig. 5.19a. The entries in the $\bar{x}_1 \bar{x}_2$ and the $x_1 x_2$ columns are irrelevant for our immediate purpose, but notice that the $\bar{x}_1 x_2$ and the $x_1 \bar{x}_2$ column entries are identical. This could be shown algebraically by the Boolean expansion of $f(x)$ into

Design Using Spectral and Other Techniques

$$f(x) = [\bar{x}_1\bar{x}_2 f_0(x_3,x_4) + \bar{x}_1 x_2 f_1(x_3,x_4) + x_1\bar{x}_2 f_2(x_3,x_4) + x_1 x_2 f_3(x_3,x_4)]$$

where $f_1(x_3,x_4) \equiv f_2(x_3,x_4)$. This function is therefore "non-equivalence symmetric" in x_1,x_2 (see Fig. 3.9) which may be written as NES$\{x_1,x_2\}$.

If we now consider a realization as outlined in Fig. 5.19b, should the original input combination of $\bar{x}_1 x_2$ never be addressed to the second-level network but instead be converted into $x_1\bar{x}_2$, all other input combinations unchanged, then the second-level network may now have "don't-cares" in all the input conditions which it never receives. The required truth table for this first-level "remapping" function therefore is as follows.

x_1	x_2	x_1'	x_2'
0	0	0	0
0	1	1	0
1	0	1	0
1	1	1	1

Hence for this particular symmetry condition the first-level network requires to be as shown in Fig. 5.19c—a possible circuit realization for the AND/OR network will be covered in Fig. 7.24, Chapter 7.

Having now one complete column of don't-cares in the function $f'(x)$, its realization can be made in the most economic manner possible, fully utilizing these don't-care conditions in the optimum manner. This may of course involve further symmetry operations similar to that employed in the first level of the realization. Notice also that if the original function $f(x)$ had any don't-cares in its specification, then we may be able to choose symmetries by optimally allocating 0 or 1 values to these input minterms. In such cases we are most effectively incorporating the original don't-care minterms into bigger don't-care groupings.

In a similar manner, suppose the $\bar{x}_1\bar{x}_2$ and $x_1 x_2$ minterm columns of a given function $f(x)$ are identical, as illustrated in Fig. 5.20a. This corresponds to "equivalence symmetry" in x_1,x_2 (seeFig. 3.11) which may be written as ES$\{x_1,x_2\}$. The Boolean proof of this symmetry would be that $f_0(x_3,x_4) \equiv f_3(x_3,x_4)$ in the previous algebraic decomposition.

By the same concepts as previously, we may eliminate one of these two identical columns, say $\bar{x}_1\bar{x}_2$, by an appropriate first-level remapping network which obeys the following relationship.

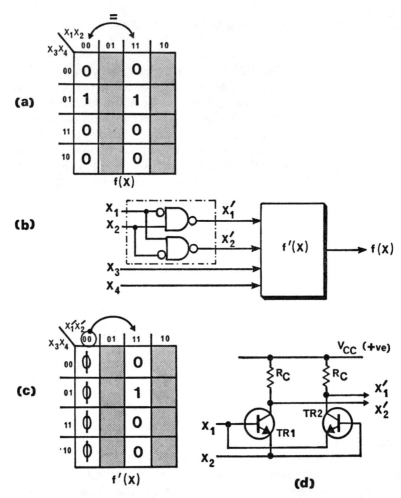

Fig. 5.20. A four-variable function $f(x)$ with equivalence symmetry in x_1, x_2: (a) plot of $f(x)$, not detailing the $\bar{x}_1 x_2$ and $x_1 \bar{x}_2$ input minterms; (b) the required first-level remapping network; (c) the plot of the required second-level function $f'(x)$; (d) one possible circuit realization for the first-level network.

x_1	x_2	x_1'	x_2'
0	0	1	1
0	1	0	1
1	0	1	0
1	1	1	1

Design Using Spectral and Other Techniques 311

Hence a simpler second-level realization fully utilizing all available don't-care conditions may be constructed. This is illustrated in Figs. 5.20b and c. Notice that non-vertex gates, as originally shown in Fig. 1.3 of Chapter 1, are now being introduced into our syntheses. The particular gates required here are negated half-exclusive-OR's, which, anticipating the ideas which will be covered in Chapter 7, may be realized by the basic circuit configuration shown in Fig. 5.20d.

More than one symmetry may be simultaneously present in $f(x)$. The function illustrated in Fig. 5.21a possesses "multiform symmetry" in x_1, x_2 (see Fig. 3.12 of Chapter 3) which may be written as

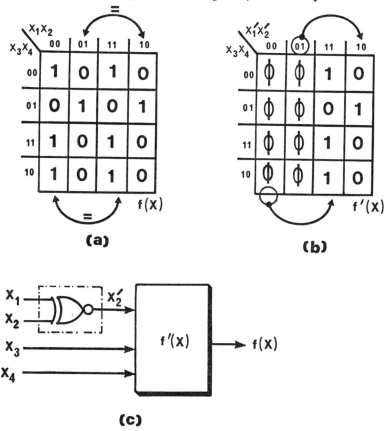

Fig. 5.21. A four-variable function with multiform symmetry in x_1, x_2: (a) plot of $f(x)$; (b) resultant don't-care columns after appropriate first-level remapping; (c) the first-level network realization.

MS $\{x_1,x_2\}$. The Boolean algebraic proof of this multiform symmetry would be that in the previous decomposition of $f(x)$ we have

$$f_0(x_3,x_4) \equiv f_3(x_3,x_4)$$

and

$$f_1(x_3,x_4) \equiv f_2(x_3,x_4).$$

The appropriate truth table for a first-level remapping network which does not allow, say, the $\bar{x}_1\bar{x}_2$ and $\bar{x}_1 x_2$ minterm columns of Fig. 5.21a to be addressed to the second level is as follows.

x_1	x_2	x_1'	x_2'
0	0	1	1
0	1	1	0
1	0	1	0
1	1	1	1

This illustrates that with multiform symmetry only one variable signal is required from the first-level remapping network to the second-level network, this being an exclusive signal generated from the two variables of symmetry. This is illustrated in Fig. 5.21c.

A further very powerful attraction of this symmetry approach to network synthesis is where multiple-output problems are involved. Here the obvious technique is to search for symmetries which are common to several outputs, which means that common first-level remapping functions may be used in the overall realization. However, notice that *this does not require that the several outputs be identical to each other* in their symmetry groups of minterms, as is the case with other methods of multi-output synthesis.

As a simple illustration of this technique consider the case of a two-bit two-word comparator circuit, with two input words A and B, $A = A_1, A_2$, $B = B_1, B_2$, and three outputs $A > B$, $A = B$, and $A < B$. The maps for these three outputs are shown in Fig. 5.22a. Examination will show that each is equivalence symmetric in A_1, B_1 and also equivalence symmetric in A_2, B_2. Suppression of the $\bar{A}_1\bar{B}_1$ minterm column and also the $\bar{A}_2\bar{B}_2$ minterm row therefore will give the second-level maps shown in Fig. 5.22b, from which the three outputs may be minimized to

$$A > B = [A_1'(\bar{B}_1' + \bar{B}_2')] = [A_1'(\overline{B_1'B_2'})]$$
$$A = B = [A_1'A_2'B_1'B_2']$$
$$A > B = [B_1'(\bar{A}_1' + \bar{A}_2')] = [B_1'(\overline{A_1'A_2'})].$$

Design Using Spectral and Other Techniques 313

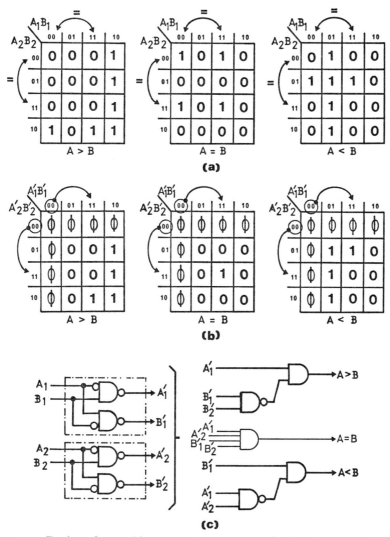

Fig. 5.22. Design of a two-bit two-word comparator circuit using symmetries: (a) the three outputs required, showing equivalence symmetries in A_1, B_1 and in A_2, B_2; (b) the reduced second-level functions after appropriate first-level remapping networks; (c) a final realization.

The first-level remapping networks which are shared between these three outputs again employ the negated half-exclusive-OR gate previously encountered in Fig. 5.20, giving a final realization for the complete comparator shown in Fig. 5.22c.

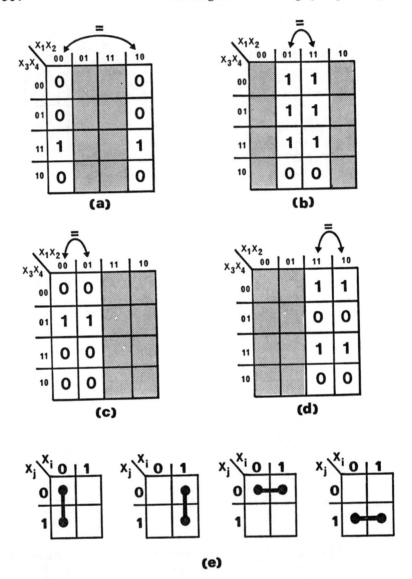

Fig. 5.23. Single-variable symmetries in a four-variable function $f(x)$: (a) SV symmetry $(SVSx_1)\bar{x}_2$; (b) SV symmetry $(SVSx_1)x_2$; (c) SV symmetry $(SVSx_2)\bar{x}_1$; (d) SV symmetry $(SVSx_2)x_1$ (note a total of 24 such SV symmetries exist in $n = 4$); (e) summary details $(SVSx_j)\bar{x}_i$, $(SVSx_j)x_i$, $(SVSx_i)\bar{x}_j$, and $(SVSx_i)x_j$, respectively.

Design Using Spectral and Other Techniques

The symmetries so far considered are symmetries involving the invariance of a function $f(x)$ when two inputs x_i, x_j are permutated in some manner. Notice that we are deliberately not considering the case of *completely symmetric* functions, as defined in §1.5.2 of Chapter 1; such functions are extreme cases, possessing multiple $\text{ES}\{x_i, x_j\}$ or $\text{NES}\{x_i, x_j\}$ symmetries, where x_i, x_j are all possible different pairs of the input variables $x_1, x_2, x_1, x_3, \ldots, x_{n-1}, x_n$.

However, we may go still further and define additional individual symmetries in $f(x)$ involving two input variables x_i, x_j which may be significant for network synthesis. Looking at the previous equivalence and non-equivalence symmetries it will be seen that they are concerned with like-valued minterms spaced apart by a Hamming distance of 2. Hence these are cases where normal Boolean minimization techniques do not apply. However, we may also consider the symmetries of like-valued minterms which lie entirely within any given x_i or x_j area, which are now separated by Hamming distances of 1 and are thus amenable to normal Boolean minimization. Nevertheless we may certainly consider the symmetry implications of such cases.

Consider any function $f(x) = f(x_1, \ldots, x_i, x_j, \ldots, x_n)$. Suppose now that the $\bar{x}_i \bar{x}_j$ and $x_i \bar{x}_j$ minterm areas are identical, that is

$$f(x_1, \ldots, 0, 0, \ldots, x_n) = f(x_1, \ldots, 1, 0, \ldots, x_n)$$

then we may say that $f(x)$ exhibits "single-variable symmetry" in x_i in the n-space $\bar{x}_j$. We may express this as $\{\text{SVS } x_i\}\ \bar{x}_j$. Similarly we may possibly have

$$f(x_1, \ldots, 0, 1, \ldots, x_n) = f(x_1, \ldots, 1, 1, \ldots, x_n)$$

in which case $f(x)$ exhibits single-variable symmetry in x_i in the n-space x_j, written as $\{\text{SVS } x_i\}\ x_j$. Such cases are illustrated for $n = 4$ in Fig. 5.23. Notice that if any function exhibits *both* these symmetries in the $x_j, \bar{x}_j$ n-space, then such a function is independent of x_i, this variable being an entirely redundant input variable. Combinations of such single-variable symmetries in x_i, x_j may also exist. Figure 5.24 illustrates certain possibilities, omitting those which become completely independent of x_i and/or x_j.

In a similar manner to that employed in the equivalence-symmetric, non-equivalence-symmetric, and multiform-symmetric cases we may remap single-variable-symmetric functions such that only one of the two identical minterm areas is addressed to the next level of realization. For example, suppose we have $\{\text{SVS } x_i\}\ x_j$ present

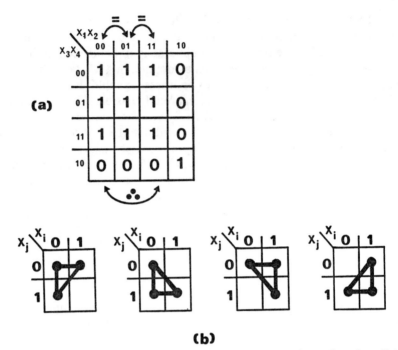

Fig. 5.24. Multiple single-variable symmetries in a given function $f(x)$: (a) function $f(x)$, SV symmetric in $(SVSx_2)\bar{x}_1$ and $(SVSx_1)x_2$, and hence also equivalence symmetric in x_1,x_2; (b) summary details of multiple single-variable symmetries in x_i,x_j (cf. Figs. 3.9 and 3.11 of Chapter 3 for the diagonal symmetries).

and we wish to eliminate say the $\bar{x}_i x_j$ minterm entries from the second level of realization, then we require the first-level remapping

x_i	x_j	x_i'	x_j'
0	0	0	0
0	1	1	1
1	0	1	0
1	1	1	1

that is the input signal x_j is passed on unchanged to the second level of realization, but x_i is replaced by the OR function $[x_i + x_j]$.

What we are now building up is a library of functions which are appropriate for first-level remappings. In fact we may in principle define even more complex symmetry pattern relationships with their required first-level remapping functions if we so desire, although clearly if we choose too unrealistic symmetries then we shall be involved in undesirable first-level functions. Nevertheless it may eventually be

Design Using Spectral and Other Techniques

useful to consider more forms of symmetry than the simple ones we have looked at here.

If we refer to Appendix E we shall see in detail the first-level remapping functions which are appropriate for the various types of symmetry we have so far considered. It will be noticed that if we wish to realize a given function using a specific type of gate, then with this symmetry method of synthesis we should look for particular types of symmetries in the given function. Hence we now have a synthesis technique which may be constrained to using specific types of gate to best advantage, unlike classic Boolean-based methods which are basically confined to AND-OR syntheses.

5.4.2. Detection of Symmetries from Spectral Coefficients

The classical methods of detecting simple symmetries in a given binary function $f(x)$ are based upon algebraically expanding the function and checking whether appropriate pairs of reduced ("residue") functions are identical. For example, expansion of the simple function

$$f(x) = [\bar{x}_1\bar{x}_2\bar{x}_3 + x_1x_2 + x_1x_3 + x_2x_3]$$

about the variables x_1, x_2 gives

$$f(x) = [\bar{x}_1\bar{x}_2(\bar{x}_3) + \bar{x}_1x_2(x_3) + x_1\bar{x}_2(x_3) + x_1x_2(\bar{x}_3 + x_3)]$$

which shows that $f(x)$ is symmetric in $\bar{x}_1x_2$ and $x_1\bar{x}_2$, that is NES $\{x_1,x_2\}$, as the residue functions, merely x_3 in this case, are identical. Details of such methods will be found in published literature[1], including binary number tests for identifying identical residue functions which are more applicable than algebraic tests when n is large[19-23].

Nevertheless such methods still tend to be cumbersome, with no immediate guide available towards which symmetries may be present in any given function. Even in the very simple three-variable example above, where it was algebraically shown that a symmetry exists between x_1 and x_2, it is not immediately apparent that non-equivalence symmetry also exists between x_1,x_3 and between x_2,x_3, which the reader may confirm by appropriately expanding or mapping $f(x)$. More direct indication of possible symmetry relationships, therefore, is very desirable.

[1] The reader must be careful to distinguish between the terminology used by various authors for the different types of symmetry. For example Das and Sheng[23] and others use the single term "partially symmetric" to cover what we have here termed non-equivalence symmetric or equivalence symmetric, whilst Edwards[24] uses the term "partial symmetric" for what we have here called single-variable symmetric. Some general agreement on terminology would be advantageous.

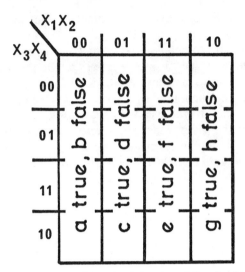

Fig. 5.25. The column count of 1-valued (true) and 0-valued (false) minterms for any four-variable function $f(x)$.

However, let us look at whether the Rademacher–Walsh spectral coefficients of $f(x)$ may help us in this area. For convenience we shall examine functions of four variables $x_1, \ldots, x_4$, which we can readily illustrate on Karnaugh maps and which do not contain an excessive number of spectral coefficients for our development purposes. If we can establish appropriate rules for $n = 4$, then hopefully their extension to $n > 4$ will be clear.

Suppose we wish to prove the existence of equivalence symmetry in x_1, x_2; that is columns $\bar{x}_1 \bar{x}_2$ and $x_1 x_2$ of a Karnaugh map plot are identical. Let us identify the number of 1-valued minterms in each of the x_1, x_2 columns by $a, c, e,$ and g, and the number of 0-valued minterms in the same columns by $b, d, f,$ and h, respectively, as shown in Fig. 5.25. Thus the (true − false) minterm count in each column is $(a - b)$, $(c - d)$, $(e - f)$, and $(g - h)$, respectively. If we now refer back to Fig. 3.14 of Chapter 3, we shall quickly see that the r_1 and r_2 spectral coefficient values are defined by

$$r_1 \triangleq \{-(a - b) - (c - d) + (e - f) + (g - h)\}$$

and

$$r_2 \triangleq \{-(a - b) + (c - d) + (e - f) - (g - h)\}.$$

Design Using Spectral and Other Techniques

Therefore adding we have

$$r_1 + r_2 = 2\{-(a - b) + (e - f)\}$$

which now involves information relating to the $\bar{x}_1\bar{x}_2$ and x_1x_2 areas only, that is the symmetry areas of our present interest. Clearly, if these two minterm areas of the given function are identical, then $(a - b) = (e - f)$, and hence a necessary (but not yet sufficient) condition for equivalence symmetry in x_1,x_2 is that $r_1 + r_2 = 0$.

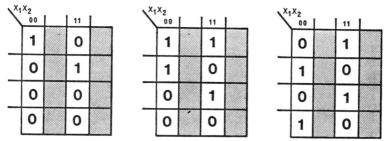

Fig. 5.26, Three example functions where $(a - b) = (e - f)$, and hence where $r_1 + r_2 = 0$, but where clearly equivalence symmetry between the two minterm columns does not hold.

However, sufficiency is not yet proved, as is illustrated by the counter-examples shown in Fig. 5.26. Further spectral coefficient data are necessary in order to ensure that the two patterns are indeed identical, and not merely the same total number of (true − false) minterms.

Let us bring x_3 information into the data which we are evaluating. By the same reasoning as previously, if we add the r_{13} and r_{23} coefficient values the $\bar{x}_1x_2$ and $x_1\bar{x}_2$ column data will be eliminated as before—see Fig. 3.14 for the r_{13} and r_{23} covers. However, what we are now left with is a true/false minterm agreement count in the $\bar{x}_1\bar{x}_2$ and x_1x_2 areas but enumerated in a dissimilar order to that involved in the $r_1 + r_2$ coefficient addition.

Similarly we may bring in x_4 and $[x_3 \oplus x_4]$ into our evaluation by considering the further two additions of

$$r_{14} + r_{24}$$

and

$$r_{134} = r_{234}$$

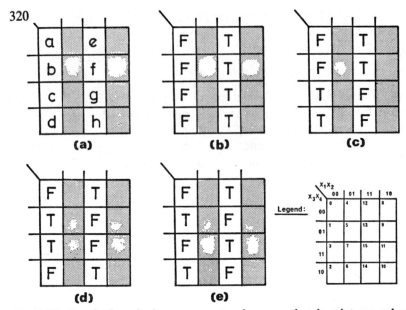

Fig. 5.27. Proof of equivalence symmetry in x_1,x_2, that is minterm column $\bar{x}_1\bar{x}_2 \equiv$ minterm column x_1x_2: (a) allocation of arbitrary symbols to each minterm square in $\bar{x}_1\bar{x}_2$ and x_1x_2 columns; (b) effective minterms in the Rademacher–Walsh addition $r_1 + r_2$, where F = 0-valued in both components, and T = 1-valued in both components; (c) as (b) for $r_{13} + r_{23}$; (d) as (b) for $r_{14} + r_{24}$; (e) as (b) for $r_{134} + r_{234}$. (Note: The $\bar{x}_1x_2$ and $x_1\bar{x}_2$ columns always completely cancel each other in the additions (b), (c), (d), and (e), and hence are not detailed.)

respectively. In all we now have four pairs of spectral coefficient values which should each sum to zero if the equivalence symmetry is present in the four-variable function $f(x)$. We may easily show that these four tests collectively are a necessary and sufficient test for equivalence symmetry in x_1,x_2 as follows.

Let $a, b, c, \ldots, g, h$ be values representing {the number of true minterms − the number of false minterms} in each of the squares shown in Fig. 5.27a. If equivalence symmetry exists in x_1,x_2 then we require to prove that

$a = e$
$b = f$
$c = g$
$d = h$.[2]

---

[2] For a four-variable function $f(x_1, x_2, x_3, x_4)$ as plotted in Fig. 5.27a the numbers a, b, c, etc. must of course be merely +1 or −1 for each minterm area.

Design Using Spectral and Other Techniques

The effective true and false minterm areas for the four pairs of Rademacher–Walsh additions are obtainable from Fig. 3.14, and are as detailed in Figs. 5.27b–e inclusive. Hence the spectral coefficient value for each pair, obtained by the usual {agreement − disagreement} count, is as follows.

From Figs. 5.27a and b

$$(r_1 + r_2) = 2\{-a - b - c - d + e + f + g + h\}$$

or if $(r_1 + r_2) = 0$, then

$$\{(a + b) + (c + d)\} = \{(e + f) + (g + h)\} \tag{5.1}$$

From Figs. 5.27a and c

$$(r_{13} + r_{23}) = 2\{-a - b + c + d + e + f - g - h\}$$

or if $(r_{13} + r_{23}) = 0$, then

$$\{(a + b) - (c + d)\} = \{(e + f) - (g + h)\}. \tag{5.2}$$

From Figs. 5.27a and d

$$(r_{14} + r_{24}) = 2\{-a + b + c - d + e - f - g + h\}$$

or if $(r_{14} + r_{24}) = 0$, then

$$\{(a - b) - (c - d)\} = \{(e - f) - (g - h)\}. \tag{5.3}$$

Finally from Figs. 5.27a and e:

$$(r_{134} + r_{234}) = 2\{-a + b - c + d + e - f + g - h\}$$

or if $(r_{134} + r_{234}) = 0$, then

$$\{(a - b) + (c - d)\} = \{(e - f) + (g - h)\}. \tag{5.4}$$

From (5.1) and (5.2) we now have by adding

$$(a + b) = (e + f)$$

and from (5.3) and (5.4) we have

$$(a - b) = (e - f)$$

whence

$$a = e$$
$$b = f.$$

In a similar manner the remaining two equalities

$$c = g,$$

and

$$d = h$$

may be proved. Hence if these four pairs of spectral coefficient values each sum to zero, then full symmetry between the $\bar{x}_1\bar{x}_2$ and x_1x_2 minterm areas is proven.

Thus for the equivalence symmetry $\mathrm{ES}\{x_1,x_2\}$ we have the conditions

$$r_1 + r_2 = 0$$
$$r_{13} + r_{23} = 0$$
$$r_{14} + r_{24} = 0$$

and

$$r_{134} + r_{124} = 0$$

which may be summarized in the necessary single collective check

$$\{r_1 + r_{13} + r_{14} + r_{134}\} + \{r_2 + r_{23} + r_{24} + r_{234}\} = 0$$

or

$$\sum r_1 \ldots + \sum r_2 \ldots = 0$$

where $r_1 \ldots$ are all spectral coefficients containing 1 but not 2 in their subscript identification and $r_2 \ldots$ are all spectral coefficients containing 2 but not 1 in their subscript identification.

By an identical train of development, non-equivalence symmetry $\mathrm{NES}\{x_1,x_2\}$ in any four-variable function, that is minterm area $\bar{x}_1 x_2 \equiv$ minterm area $x_1\bar{x}_2$, is present when

$$r_1 - r_2 = 0$$
$$r_{13} - r_{23} = 0$$
$$r_{14} - r_{24} = 0$$
$$r_{134} - r_{234} = 0$$

which may be rearranged and summarized in the necessary single collective check

Design Using Spectral and Other Techniques

$$\sum r_1\ldots - \sum r_2\ldots = 0$$

where $r_1\ldots, r_2\ldots$ are as before.

If multiform symmetry MS $\{x_1,x_2\}$ is present, then clearly both the previous sets of coefficient relationships must be present. This can only occur, however, if each appropriate coefficient is zero valued, and hence for multi-form symmetry in x_1,x_2 we have

$r_1 = r_2 = 0$

$r_{13} = r_{23} = 0$

$r_{14} = r_{24} = 0$

$r_{134} = r_{234} = 0.$

Turning now to single-variable symmetry, we may develop similar tests using pairs of spectral coefficient values. Consider the single-variable symmetry $\{\text{SVS } x_1\}\ \bar{x}_2$, that is the $\bar{x}_1\bar{x}_2$ minterm area $\equiv$ the $x_1\bar{x}_2$ minterm area, as for example shown in Fig. 5.23a.

The four pairs of spectral coefficients which taken together will eliminate contributions from the unwanted $\bar{x}_1 x_2$ and $x_1 x_2$ minterm areas are

$r_1 + r_{12}$

$r_{13} + r_{123}$

$r_{14} + r_{124}$

$r_{134} + r_{1234}.$

In a similar manner to that previously detailed, we may show that for the remaining $\bar{x}_1\bar{x}_2$ and $x_1\bar{x}_2$ minterm areas to be identical (symmetrical), then each of these pairs of coefficient values must sum to zero. This may be rearranged and summarized in the necessary single collective check

$$\sum r_1\ldots + \sum r_{12}\ldots = 0$$

where $r_1\ldots$ are as before and $r_{12}\ldots$ are all spectral coefficients containing 1 and 2 in their subscript identification.

In like manner, if the three remaining possible single-variable symmetries between x_1,x_2 are considered, then the appropriate identifying spectral coefficient pairs are as follows.

Table 5.5. Spectral coefficient value relationships which prove function symmetry

Type of symmetry between input variables x_i, x_j in the function $f(x_1,\ldots,x_i,x_j,\ldots,x_n)$ $i \neq j$, $1 \leq i,j \leq n$	First check involving the primary spectral coefficient values r_1 to r_n, (necessary but not sufficient)	Further checks necessary to satisfy function symmetry (necessary and sufficient)	Necessary single collective check (see note (iii) below)
Equivalence symmetry $ES\{x_i,x_j\}$ $f(x_1,\ldots,0,0,\ldots,x_n)$ $= f(x_1,\ldots,1,1,\ldots,x_n)$	$r_i + r_j = 0$	$r_{ik} + r_{jk} = 0$ $r_{il} + r_{jl} = 0$ $r_{ikl} + r_{jkl} = 0$ $\vdots$ where $k, l, \ldots$ are all different combinations of subscripts not containing i, j.	$\Sigma r_\alpha + \Sigma r_\beta = 0$ where α are all spectral coefficients containing i but not j in their subscripts, and β are all spectral coefficients containing j but not i in their subscripts
Non-equivalence symmetry, $NES\{x_i,x_j\}$ $f(x_1,\ldots,0,1,\ldots,x_n)$ $= f(x_1,\ldots,1,0,\ldots,x_n)$	$r_i - r_j = 0$	$r_{ik} - r_{jk} = 0$ $r_{il} - r_{jl} = 0$ $r_{ikl} - r_{jkl} = 0$ $\vdots$ where $k, l, \ldots$ are as above	$\Sigma r_\alpha - \Sigma r_\beta = 0$ where α, β are as above
Multiform symmetry $MS\{x_i,x_j\}$ $f(x_1\ldots,0,0,\ldots,x_n)$ $= f(x_1,\ldots,1,1,\ldots,x_n)$ and $f(x_1,\ldots,0,1,\ldots,x_n)$ $= f(x_1,\ldots,1,0,\ldots,x_n)$	$r_i = r_j = 0$	$r_{ik} = r_{jk} = 0$ $r_{il} = r_{jl} = 0$ $r_{ikl} = r_{jkl} = 0$ $\vdots$ where $k, l, \ldots$ are as above	(All coefficients containing i, j in their subscripts are zero valued)
Single-variable symmetry $\{SVSx_i\}\bar{x}_j$ $f(x_1,\ldots,0,0,\ldots,x_n)$ $= f(x_1,\ldots,1,0,\ldots,x_n)$	$r_i + r_{ij} = 0$	$r_{ik} + r_{ijk} = 0$ $r_{il} + r_{ijl} = 0$ $r_{ikl} + r_{ijkl} = 0$ $\vdots$ where $k, l, \ldots$ are as above	$\Sigma r_\alpha + \Sigma r_{\alpha j} = 0$ where α are as above, and αj are all spectral coefficients containing i and j in their subscripts
Single-variable symmetry $\{SVSx_i\}x_j$ $f(x_1,\ldots,0,1,\ldots,x_n)$ $= f(x_1,\ldots,1,1,\ldots,x_n)$	$r_i - r_{ij} = 0$	$r_{ik} - r_{ijk} = 0$ $r_{il} - r_{ijl} = 0$ $r_{ikl} - r_{ijkl} = 0$ $\vdots$ where $k, l, \ldots$ are as above	$\Sigma r_\alpha - \Sigma r_{\alpha j} = 0$ where α, αj are as above

Design Using Spectral and Other Techniques

Note: (i) Several types of symmetry may be simultaneously present in any given function $f(x)$. (ii) For any n-valued function a maximum of 2^{n-2} coefficient value checks must be made to prove a particular symmetry over the whole n-space. (iii) Mathematically the single collective summary equation for each type of symmetry does not prove that each individual equation $r_i \pm r_j$ etc. of the set sums to zero, as is required. However, as the spectral coefficient values are not independent value parameters, no counter-examples to taking the single summary equation as a necessary *and sufficient* test for the appropriate symmetry have yet been found. However, until it is formally proved that a single summary equation is sufficient, it remains prudent to check all 2^{n-2} individual equations when testing for a particular symmetry.

$\{SVS\ x_1\}x_2$ (e.g. Fig. 5.23b)

$r_1 - r_{12}$

$r_{13} - r_{123}$

$r_{14} - r_{124}$

$r_{134} - r_{1234}$

$\{SVS\ x_2\}\bar{x}_1$ (e.g. Fig. 5.23c)

$r_2 + r_{12}$

$r_{23} + r_{123}$

$r_{24} + r_{124}$

$r_{234} + r_{1234}$

$\{SVS\ x_2\}x_1$ (e.g. Fig. 5.23d)

$r_2 - r_{12}$

$r_{23} - r_{123}$

$r_{24} - r_{124}$

$r_{234} - r_{1234}.$

A necessary and sufficient test for the presence of each of these symmetries is for the set of four equations per symmetry condition to be all zero valued, as detailed for the $\{SVS\ x_1\}\bar{x}_2$ case.

To summarize all we have now developed for the individual symmetries between the input variables x_1, x_2 and to extend it to the more general case of such symmetries between any two variables x_i, x_j of $f(x)$, we have the details tabulated in Table 5.5. Notice that an immediate guide as to whether equivalence or non-equivalence symmetry is present in any given function is given by an inspection of the *magnitude of the first-ordered coefficient values* r_1 to r_n only, which must pair with similar magnitude values if such symmetries are present. This is a very powerful necessary (but not of course sufficient) first test, which has no counterpart whatsoever in the Boolean domain.

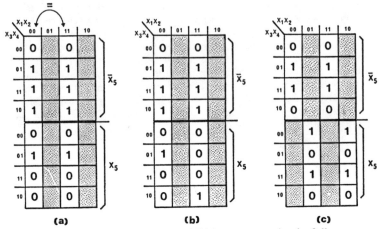

Fig. 5.28. Some example functions exhibiting symmetries in full n-space and reduced n-space: (a) function equivalence symmetric in x_1,x_2 in the full $n = 5$ space; (b) function equivalence symmetric in x_1,x_2 in the reduced n-space of $\bar{x}_5$ only; (c) function equivalence symmetric in x_1,x_2 in the reduced n-space $\bar{x}_5$ only, and non-equivalence symmetric in the reduced n-space of x_5 only.

Before we continue with some illustrations of the spectral coefficient values of symmetric functions, one further train of development is worth considering. So far we have been considering symmetries which may exist between x_i,x_j over the whole n-space of the given function $f(x)$. For example, in a five-variable function equivalence symmetry between, say, x_1,x_2 extends over the complete $\bar{x}_1\bar{x}_2$ and x_1x_2 minterm columns shown in Fig. 5.28a. The eight pairs of terms involved in the $\Sigma r_\alpha + \Sigma r_\beta = 0$ check would confirm this. However, we could possibly have equivalence symmetry (or any other symmetry) in, say, the $\bar{x}_5$ minterm area only, but not in the x_5 minterm area, or *vice versa*, or different symmetries in both areas, all of which constitute appropriate symmetries in (n − 1) space. Such symmetries may be termed "reduced n-space symmetries", giving us, for example

equivalence symmetry in x_i,x_j in n-space $\bar{x}_k$
equivalence symmetry in x_i,x_j in n-space x_k
non-equivalence symmetry in x_i,x_j in n-space $\bar{x}_k$

and so on. Examples are shown in Figs. 5.28b and c.

Clearly if we continue to reduce the n-space over which symmetries may be found, we shall ultimately be comparing two single minterms, and hence in the extreme all functions $f(x)$ can be shown to exhibit multiple reduced n-space symmetries. However, for practical purposes it is most desirable to find the symmetries which exist over

Design Using Spectral and Other Techniques 327

the *largest* area, ideally the whole n-space such as we have been considering so far.

The checks for reduced n-space symmetry conditions are clearly more complex than for complete n-space symmetries. We shall not detail these further developments in this book, except to indicate briefly typical checks which are necessary.

For example, suppose we wish to confirm the x_1, x_2 equivalence symmetry in n-space $\bar{x}_5$, as shown in Fig. 5.28b. We still wish to eliminate the complete $\bar{x}_1 x_2$ and $x_1 \bar{x}_2$ minterm columns as before, but now also the $\bar{x}_1 \bar{x}_2$ and $x_1 x_2$ columns in the x_5 minterm area only. It is left as an exercise for the reader to confirm that the following will give a necessary and sufficient test for this symmetry in $\bar{x}_5$, which may be compared with the checks for the full n-space $\bar{x}_1 \bar{x}_2$, $x_1 x_2$ symmetry in the $n = 4$ case:

$$r'_1 + r'_2 = 0$$
$$r'_{13} + r'_{23} = 0$$
$$r'_{14} + r'_{24} = 0$$

and

$$r'_{134} + r'_{234} = 0$$

where

$$r'_1 = r_1 + r_{15}$$
$$r'_2 = r_2 + r_{25}$$
$$r'_{13} = r_{13} + r_{135}$$
$$r'_{23} = r_{23} + r_{235}$$
$$r'_{14} = r_{14} + r_{145}$$
$$r'_{24} = r_{24} + r_{245}$$
$$r'_{134} = r_{134} + r_{1345}$$
$$r'_{234} = r_{234} + r_{2345}.$$

Finally if the same type of symmetry, say ES$\{x_1, x_2\}$, exists in the $\bar{x}_5$ and the x_5 reduced n-spaces, then a further interesting exercise is to show that the tests which *separately* prove the symmetry in the $\bar{x}_5$ and the x_5 areas collectively are *precisely the same* as the tests for the same symmetry over the whole n-space of the given function.

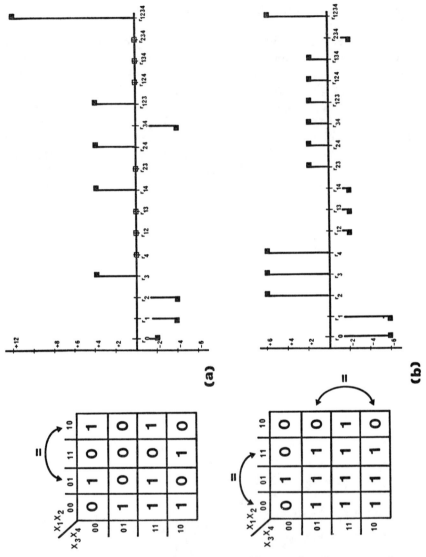

Fig. 5.29. Two examples and their spectra to illustrate function symmetries:
(a) function $f(x) = [\bar{x}_1\bar{x}_2(x_3 + x_4) + x_3x_4(\bar{x}_1 + \bar{x}_2)$
$\qquad + \bar{x}_3\bar{x}_4(\bar{x}_1x_2 + x_1\bar{x}_2) + x_1x_2x_3\bar{x}_4]$
(b) function $f(x) = [\bar{x}_1x_2 + x_3(\bar{x}_1 + x_2 + x_4) + x_4(\bar{x}_1 + x_2)]$.

5.4.3. Some Examples

As numerical illustrations of the developments we have now covered, consider the two examples illustrated in Figs. 5.29a and b. In Fig. 5.29a it is apparent from the Karnaugh map plot that non-equivalence symmetry exists in x_1,x_2, but no other x_1,x_2 symmetry is present. In Fig. 5.29b, however, equivalence symmetry exists in x_1,x_2, together with the obvious non-equivalence symmetry in x_3,x_4 as well. The minimal Boolean equations for each function are not particularly instructive without detailed processing.

The full Rademacher–Walsh coefficient values of these two functions are as follows, these values being readily evaluated by hand for such small functions by {agreement − disagreement} counts between the given functions and the 16 Rademacher–Walsh functions detailed in Fig. 3.14.

	r_0	r_1	r_2	r_3	r_4	r_{12}	r_{13}	r_{14}	r_{23}	r_{24}	r_{34}	r_{123}	r_{124}	r_{134}	r_{234}	r_{1234}
Fig. 5.29a:	−2	−4	−4	4	0	0	0	4	0	4	−4	4	0	0	0	12
Fig. 5.29b:	−6	−6	6	6	6	−2	−2	−2	2	2	2	2	2	2	−2	6

We consider first Fig. 5.29a and its spectrum. If we look first for possible equivalence and non-equivalence symmetries, then consideration of the magnitude of primary spectral coefficients r_1 to r_4 is our initial guide. The possible symmetries indicated by these magnitudes which may be present are

(a) NES$\{x_1,x_2\}$, prompted by $r_1 - r_2 = 0$

(b) ES$\{x_1,x_3\}$, prompted by $r_1 + r_3 = 0$

(c) ES$\{x_2,x_3\}$, prompted by $r_2 + r_3 = 0$.

The full tests which we must make to confirm whether these symmetries are indeed present are, in addition to the above,

(a) $r_{13} - r_{23} = 0$, $r_{14} - r_{24} = 0$, and $r_{134} - r_{234} = 0$

or in total $\{(r_1 + r_{13} + r_{14} + r_{134}) - (r_2 + r_{23} + r_{24} + r_{234})\} = 0$

(b) $r_{12} + r_{23} = 0$, $r_{14} + r_{34} = 0$, and $r_{124} + r_{234} = 0$

or in total $\{(r_1 + r_{12} + r_{14} + r_{124}) + (r_3 + r_{23} + r_{34} + r_{234})\} = 0$

(c) $r_{12} + r_{13} = 0$, $r_{24} + r_{34} = 0$, and $r_{124} + r_{134} = 0$

or in total $\{(r_2 + r_{12} + r_{24} + r_{124}) + (r_3 + r_{13} + r_{34} + r_{134})\} = 0$.

Checking now with the spectral coefficient values for the function, we shall see that all three of the above conditions are met. Hence the function has been proved to be

(a) non-equivalence symmetric in x_1, x_2

(b) equivalence symmetric in x_1, x_3

(c) equivalence symmetric in x_2, x_3.

If we now check back to our equivalence-symmetry Karnaugh map plots of Figs. 3.9 and 3.11, it will be clear that all these three symmetries are indeed present. Notice, however, that even in this simple four-variable problem the x_1, x_3 and the x_2, x_3 symmetries in the Karnaugh map are not immediately apparent unless one has a hint that they are present.

Let us check to see whether any single-variable symmetries are also present in this function. The initial clues which indicate whether such symmetries may be present are

(d) $\{SVS\ x_1\}\bar{x}_4$, prompted by $r_1 + r_{14} = 0$

(e) $\{SVS\ x_2\}\bar{x}_4$, prompted by $r_2 + r_{24} = 0$

(f) $\{SVS\ x_3\}\bar{x}_4$, prompted by $r_3 + r_{34} = 0$.

No other $(r_i \pm r_{ij}) = 0$, $1 \leq i \leq 4$, $i \neq j$, pair of spectral coefficient values can be found apart from those three.

The further spectral coefficient values which must sum to zero to prove that these possible single-variable symmetries are present are

(d) $r_{12} + r_{124} = 0$, $r_{13} + r_{134} = 0$, $r_{123} + r_{1234} = 0$
or in total $\{(r_1 + r_{12} + r_{13} + r_{123}) + (r_{14} + r_{124} + r_{134} + r_{1234})\} = 0$

(e) $r_{12} + r_{124} = 0$, $r_{23} + r_{234} = 0$, $r_{123} + r_{1234} = 0$
or in total $\{(r_2 + r_{12} + r_{23} + r_{123}) + (r_{24} + r_{124} + r_{234} + r_{1234})\} = 0$

(f) $r_{13} + r_{134} = 0$, $r_{23} + r_{234} = 0$, $r_{123} + r_{1234} = 0$
or in total $\{(r_3 + r_{13} + r_{23} + r_{123}) + (r_{34} + r_{134} + r_{234} + r_{1234})\} = 0$.

Checking with our spectral coefficient magnitudes, we shall quickly find that none of these three sets of equalities are satisfied, and hence the given function does not possess any single-variable symmetry relationship.

An interesting point is brought out by the latter development. It will be appreciated that in the numerical check for any single-variable symmetry the highest-possible spectral coefficient value $r_{12...n}$, in this case r_{1234}, will appear once in the total check. Hence if this spectral coefficient takes a value *different from all other spectral coefficient values*, this is an immediate and sufficient check for the absence of any single-variable symmetry. The reason why this is so is also consistent; a large-magnitude $r_{12...n}$ coefficient indicates that the function under consideration is strongly dependent upon the exclusive-OR (or NOR) of its input variables, but, if this is so, then like-valued minterms are separated by a Hamming distance of 2 and not 1 as is required by definition of single-variable symmetry.

Turning now to the second function given in Fig. 5.29, it is immediately apparent by a first inspection of the r_1 to r_4 coefficient values that x_1,x_2, x_1,x_3, x_1,x_4, x_2,x_3, x_2,x_4, and x_3,x_4 equivalence or non-equivalence symmetries may be present. However, no possible single-variable symmetry is present, as all first-order spectral coefficient values r_1 to r_4 are dissimilar in magnitude to the second-order coefficients r_{12}, r_{13}, etc.

Let us therefore comprehensively check for the possible equivalence and non-equivalence symmetries in the given function as follows:

(a) $\mathrm{ES}\{x_1,x_2\}$ requires the collective and individual pair checks
$\{(r_1 + r_{13} + r_{14} + r_{134}) + (r_2 + r_{23} + r_{24} + r_{234})\} = 0$

(b) $\mathrm{ES}\{x_1,x_3\}$ requires
$\{(r_1 + r_{12} + r_{14} + r_{124}) + (r_3 + r_{23} + r_{34} + r_{234})\} = 0$

(c) $\mathrm{ES}\{x_1,x_4\}$ requires
$\{(r_1 + r_{12} + r_{13} + r_{123}) + (r_4 + r_{24} + r_{34} + r_{234})\} = 0$

(d) $\mathrm{NES}\{x_2,x_3\}$ requires
$\{(r_2 + r_{12} + r_{24} + r_{124}) - (r_3 + r_{13} + r_{34} + r_{134})\} = 0$

(e) $\mathrm{NES}\{x_2,x_4\}$ requires
$\{(r_2 + r_{12} + r_{23} + r_{123}) - (r_4 + r_{14} + r_{34} + r_{134})\} = 0$

(f) $\mathrm{NES}\{x_3,x_4\}$ requires
$\{(r_3 + r_{13} + r_{23} + r_{123}) - (r_4 + r_{14} + r_{24} + r_{124})\} = 0.$

If we check the above, it will be found that the function satisfies all six criteria and therefore possess all six types of symmetry. Again this may be checked from the symmetry maps of Figs. 3.9 and 3.11, although again the diagonal map checks involving symmetries between horizontal and vertical map variables are not easy to see without some initial guidance.

However, let us look a little more closely at the function we chose to use in Fig. 5.29b. Examination of the first $n + 1$ spectral coefficients $r_0, \ldots, r_4$ will show that the function we have deliberately chosen is a linearly separable function, as reference to the Chow-parameter tabulations of Appendix A will confirm. It is indeed the simple threshold-logic function

$$f(x) = \langle \bar{x}_1 + x_2 + x_3 + x_4 \rangle_2.$$

However, from the threshold realization we may now immediately see the coequal importance of the four input variables x_1, x_2, x_3, and x_4—the function is a completely symmetric function in its four x_i inputs.

One further point may be made; if the function $f(x)$ we are analysing turns out to be a linearly separable function, then the first $n + 1$ Chow parameters *fully and unambiguously define the function*. The remaining $2^n - (n - 1)$ Rademacher–Walsh spectral coefficients are in such cases unnecessary to define $f(x)$. From this we may then state that, for linearly separable functions, (i) equivalence symmetry in x_i, x_j is present when $r_i + r_j = 0$, which is identical to $b_i + b_j = 0$, and (ii) non-equivalence symmetry in x_i, x_j is present when $r_i - r_j = 0$, which is identical to $b_i - b_j = 0$. However, tests for single-variable symmetry $\{\text{SVS } x_i\}\bar{x}_j$ or $\{\text{SVS } x_i\}x_j$ still require the higher-order Rademacher–Walsh spectral coefficient values to be inspected. This final tie-up of our symmetry considerations with the input weighting values of threshold-logic functions is an interesting feature of all these various but related aspects.

Having now introduced the concepts of symmetries and their detection based upon the Rademacher–Walsh spectral coefficient values, we now have a range of possible new techniques for binary synthesis, namely (i) the spectral translation techniques of §5.2, (ii) the statistical techniques of §5.3, and now (iii) the symmetry techniques of this latter section. Depending upon the particular function to be synthesized and any restrictions on the type of realization required so the use of one or other or combinations of these techniques may generate the optimum realization for a given problem. The greatest potential of all these approaches is the possibility of being able to apply the techniques to the CAD of large functions without difficulty, based upon the principles which we have illustrated with our fairly small problems. Appropriate constraints may be built into CAD programs in order to optimize the syntheses for any given range of gates, or fan-in, or any other technological constraint which may be required. Finally, the potential ability of being able optimally to allocate don't-care conditions to input minterms is a very powerful attraction.

Design Using Spectral and Other Techniques

Work to bring this whole area of logic synthesis to full fruition is still continuing, and further developments and rationalization of techniques may be expected. Publications so far available must be considered as preliminary, to be followed by others which will hopefully extend and consolidate these techniques[14,24-26].

For a final illustration of these principles, let us look at a possible approach to synthesizing the function illustrated in Fig. 5.30a. Let us assume that we are free to use any types of gate. From the spectrum of the function (or from examination of the Karnaugh map for this mappable problem) we may identify the following symmetries present in $f(x)$: (a) non-equivalence symmetry $\text{NES}\{x_3,x_4\}$, (b) multiple single-variable symmetries $\{\text{SVS } x_3\}\bar{x}_4$ and $\{\text{SVS } x_4\}\bar{x}_3$, and (c) single-variable symmetries $\{\text{SVS } x_1\}\bar{x}_3$ and $\{\text{SVS } x_1\}\bar{x}_4$.

Using the collective symmetries (a) and (b), we may reduce the size of the remaining function to be synthesised by combining these three identical rows into one row by a first-level AND remapping function (see Appendix E, §5). In effect we are making two adjacent minterm rows of the map into don't cares, and then discarding them entirely. This gives us the part-realization shown in Fig. 5.30b.

Now if we apply a *spectral translation* operation to replace input x_1 by $[x_1 \oplus x_2]$, this will interchange the $\bar{x}_1x_2$ and x_1x_2 minterms as shown in Fig. 5.30c—see Fig. 3.18. The reason for this operation is that we now have in the residual function of Fig. 5.30c multiple symmetries in x_1',x_3' which the spectrum of this residual function will show. We may more clearly see this if we replot Fig. 5.30c with alternative axes, as shown in Fig. 5.30d.

A remapping function to utilize these symmetries and reduce still further the size of the residual function is another AND remapping function as shown in Fig. 5.30e. The final residual function is now merely the exclusive-NOR of these two inputs, which if the two two-input AND gates are combined into one three-input AND gate gives us the final realization shown in Fig. 5.30f. Notice particularly the minimization of (i) cross-overs of connections and (ii) the number of inverters on input lines which this synthesis technique frequently provides.

Should we wish to put constraints on the synthesis, for example not to use exclusive functions, or to use gates with restricted fan-in, etc., then alternative routes to a final realization for $f(x)$ may be followed. This gives these methods of synthesis much greater flexibility than conventional Boolean-based techniques, one which becomes increasingly valuable as function size increases beyond that which can be readily mapped or algebraically manipulated. For further details refer-

334 The Logical Processing of Digital Signals

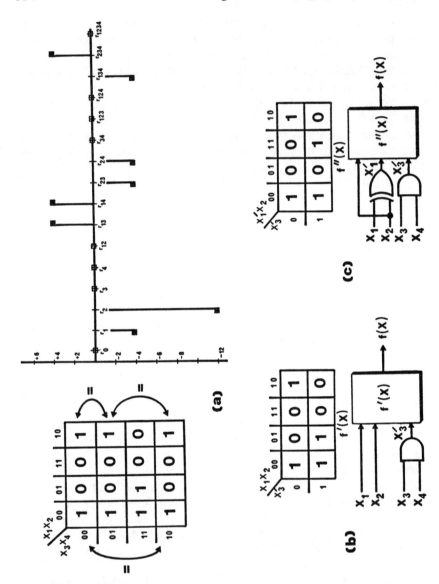

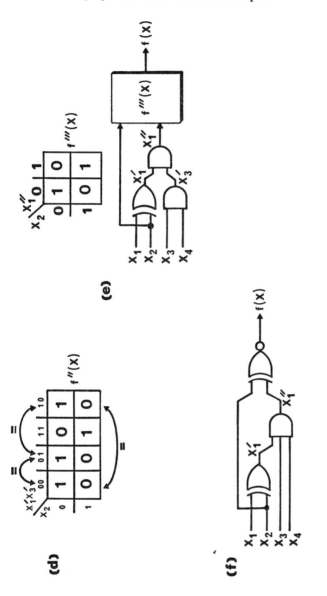

Fig. 5.30. A final example: (a) the function
$$f(x) = [\bar{x}_1\bar{x}_2 + \bar{x}_2(\bar{x}_3 + \bar{x}_4) + \bar{x}_1x_3x_4]$$
and its spectrum; (b) first symmetry reduction; (c) spectral translation between x_1 and $[x_1 \oplus x_2]$; (d) remapping of (c) to show x_1', x_3' symmetries; (e) second symmetry reduction; (f) final realization.

ence may be made to the preliminary published work of Edwards[24]. This publication also introduces the possibility of complex remapping functions, so that use may be made of symmetries which are, for example, dissimilar in the x_i and $\bar{x}_i$ n-space of a given function.

5.4.4. Some Final Considerations

In the preceding pages we have shown how the Rademacher–Walsh spectral coefficient values may be used to show symmetries in combinatorial functions. Such methods appear preferable to those based upon algebraic decompositions of Boolean expressions, as ready guidance is available towards which symmetries may possibly be present.

Another possible mathematical approach to the detection of symmetries, however, is to use Boolean differentials. A Boolean differential of a function $f(x)$ with respect to any input variable x_j is given by

$$\frac{d f(x)}{d x_j} \triangleq [f(x_1,\ldots,x_i,\bar{x}_j,\ldots,x_n) \oplus f(x_1,\ldots,x_i,x_j,\ldots,x_n)]$$

that is $d f(x)/d x_j$ is 0 when the two functions are identical (either both 0 or both 1) and 1 when they are dissimilar. Thus in a simple case such as shown in Fig. 5.31, we have

$$\frac{d f(x)}{d x_3} = 0 \text{ on } \bar{x}_1 \bar{x}_2$$

$$= 0 \text{ on } \bar{x}_1 x_2$$

$$= 1 \text{ on } x_1 x_2$$

$$= 1 \text{ on } x_1 \bar{x}_2.$$

x_3 \ $x_1 x_2$	00	01	11	10
0	0	1	1	0
1	0	1	0	1

Fig. 5.31. A simple function $f(x)$ with $d.f(x)/dx_3 = 0$ on $\bar{x}_1$, $= 1$ on x_1.

Design Using Spectral and Other Techniques

Notice that this operator is not a true differential operator in the full mathematical sense, as it does not distinguish between a change of $f(x)$ from 1 to 0 and a change from 0 to 1 as x_j changes in a given direction. Hence it is more usual to refer to this operator as the "Boolean difference" operator, since it only distinguishes *differences* in function value $f(x)$ and not the sign of the differences.

Consider now the AND function

$$f'(x) = \left[x_i \frac{d.f(x)}{dx_j} \right].$$

When $x_i = 1$, the value of $f'(x)$ will take the value of the Boolean difference function. However, if $f'(x) = 0$ when $x_i = 1$, then we must have

$$f(x_1, \ldots, 1, \bar{x}_j, \ldots, x_n) = f(x_1, \ldots, 1, x_j, \ldots, x_n)$$

which is precisely the definition of the single-variable symmetry condition in x_j in the n-space $x_i = 1$, that is $\{\text{SVS } x_j\} x_i$. Hence

$$\left[x_i \frac{d.f(x)}{dx_j} \right] = 0$$

defines the presence of the single-variable symmetry $\{\text{SVS } x_j\} x_i$. In a precisely similar manner we may show that

$$\left[\bar{x}_i \frac{d.f(x)}{dx_j} \right] = 0$$

defines the presence of the single-variable symmetry $\{\text{SVS } x_j\} \bar{x}_i$.

Continuing further, we may define the second Boolean difference of any function $f(x)$ as

$$\left[\frac{d.f(x)}{d(x_i, x_j)} \right] \triangleq [f(x_1, \ldots, \bar{x}_i, \bar{x}_j, \ldots, x_n) \oplus f(x_1, \ldots, x_i, x_j, \ldots, x_n)]$$

Consider now the AND function

$$f''(x) = \left[(x_i \oplus x_j) \frac{d.f(x)}{d(x_i, x_j)} \right].$$

If now $f''(x) = 0$, either $(x_i \oplus x_j)$ or the difference function, or both, must equal 0, but when $x_i \neq x_j$, then the difference function itself must be 0, that is

$$f(x_1, \ldots, 0, 1, \ldots, x_n) = f(x_1, \ldots, 1, 0, \ldots, x_n)$$

which is the definition of non-equivalence symmetry in x_i, x_j. Thus we may define non-equivalence symmetry $\text{NES}\{x_i, x_j\}$ by

$$\left[(x_i \oplus x_j) \frac{d.f(x)}{d(x_i, x_j)} \right] = 0.$$

In a similar manner it will be clear that equivalence symmetry $\text{ES}\{x_i, x_j\}$, that is

$$f(x_1, \ldots, 0, 0, \ldots, x_n) = f(x_1, \ldots, 1, 1, \ldots, x_n),$$

may be defined by

$$\left[(x_i \overline{\oplus} x_j) \frac{d.f(x)}{d(x_i, x_j)} \right] = 0.$$

Finally multiform symmetry $\text{MS}\{x_i, x_j\}$, that is the presence of both equivalence and non-equivalence symmetry in x_i, x_j, is present when both of the latter conditions are simultaneously present, which can only arise when

$$\left[\frac{d.f(x)}{d(x_i, x_j)} \right] = 0.$$

Thus all the complete n-space symmetries which we previously considered may be identified in terms of these Boolean difference operators.

Now these difference operators basically involve Boolean algebraic manipulations for their execution and are not directly related to the Rademacher–Walsh spectral domain. However, if we recall Chapter 2, the underlying mathematics of the Rademacher–Walsh transform involved the binary-valued set +1, −1 for the Walsh functions instead of the set 0, 1, and this suggests that we might define a differentiation operator in the +1, −1 domain. Such differentiation has indeed been proposed by Gibbs[27,28], but unlike the Boolean difference operator this Gibbs differential operator maintains sign as well as magnitude information in its differential. It is thus a true differentiation operator, unlike the previous Boolean difference operator $d.f(x)/dx_j$. Continuing, it has been shown that (a) the Gibbs differential operator may be performed under the normal Walsh transform[28], and (b) the differential operator is simply related to the Boolean difference operator[24]. The overall result is that it is possible to extract symmetry information from a given function $f(x)$ by appropriate transform operations, rather than undertaking the previous method of comparing the spectral coefficient values after completion of the normal Rademacher–Walsh transform

Design Using Spectral and Other Techniques

from the Boolean to the spectral domain. The full possibilities of this approach are not yet fully investigated, and considerable further work is required in order to clarify our understanding of this mathematical area. However, irrespective of which technique finally proves to be preferable in given circumstances for identifying available symmetries, the use of symmetries in network synthesis seems assured.

A further related possibility in this general area may also be raised, this being fault diagnosis and diagnostability of combinatorial networks. Currently fault diagnosis is rather a specialist topic in the digital engineering area, one which has not permeated far into industrial usage and which we have not the space to deal with in this book. However, existing information will be found published in many excellent sources[29-38] with a comprehensive bibliography being provided by Bennetts et al [39,40].

Now the Boolean difference operator and exclusive-OR/NOR relationships feature prominently in this area, as may be gathered from the cited references. However, all this work is concerned with the *analysis* of existing networks and the generation of optimal test sequences, whereas we have been viewing these same features from the point of view of their usage in network synthesis. Thus it is not difficult to postulate that future developments may combine these two areas, such that diagnostability becomes a parameter which can be specified at the original network design stage.

Finally, mention may be made of the possibility of using symmetry conditions in the design of sequential finite-state machines. Again existing design concepts for finite-state machines have not been covered in this book, as current theory is readily available from many standard texts[1-4], but it may be possible to apply tests for symmetries to the state coding assignments for any given machine. If symmetric state assignments can be recognized, then it may be possible to reduce the size of the final finite-state machine by a sharing process, in much the same way as that by which we combined two (or more) symmetric minterm areas in combinatorial problems into one common minterm area realization. This concept has been suggested[24,25], but much work remains to be done in this potential area also.

5.5. Logic Design Using LSI Programmable Devices

The majority of information which we have been considering in this chapter has been dealing with basic design concepts, without considering in detail practical realizations using standard commercial packages,

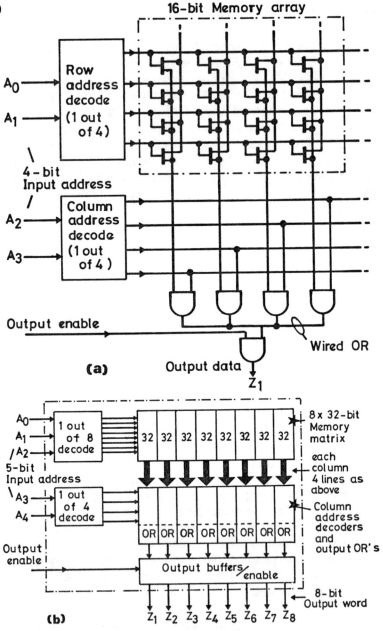

Fig. 5.32. Schematic arrangements of PROM's: (*a*) a simple single-output configuration using FET's as the memory array switches; (*b*) the arrangement of an 8-bit-output PROM.

Design Using Spectral and Other Techniques

the availability or economics of which may have over-riding implications. However, all this information is still relevant, as, in spite of the high technology of software-oriented systems etc. using, say, microprocessor and memory sophistication, there still remains an appreciable area of random-logic design to "glue" together a complete complex system[41].

One of the fundamental problems of the random-logic area is how best to use the resources of l.s.i. technology, but at the same time maintain the flexibility of the single logic gate approach, custom design of special logic requirements and realization in unique monolithic form being in general a completely uneconomic approach for small-quantity requirements. A more viable approach, however, is to consider programmable devices, that is l.s.i. circuits which are general-purpose networks and which finally may be arranged so as to perform specific duties. In this area we shall principally look at two main contenders, the programmable read-only memory ("PROM") and the programmable logic array ("PLA").

5.5.1. Programmable Read-Only Memories ("PROM's")

Probably the current most often suggested l.s.i. circuit for random-logic applications is the read-only-memory circuit assembly, which may be programmed by the manufacturer or in particular cases by the user to realize any logic requirement within the capacity of its array size.

The basic concept of the read-only-memory is an array of undirectional switches, usually a transistor configuration, which may be addressed by a binary input code (the *"input address"*), the outputs from the addressed switches forming the 0, 1 binary output or outputs (the *"output word"*). The memory of the array consists in making each switch permanently operative ("closed") or permanently inoperative ("open"), and does not consist of any form of conventional bistable elements. A simple 16-bit single-output ROM is illustrated in Fig. 5.32a.

It will be noticed from this figure that in order to economize in input connections to the complete package binary decoding of the inputs into individual decode lines is made. Only one output line per decode circuit is therefore at logic 1 on each input address. The switches shown are FET switches in this particular illustration; in order to program the device for any specific application each FET switch is either made (or left) fully operational, or omitted (or made open-circuit) when not required. Multiple memory arrays are also usually provided on commercial packages; for example Fig. 5.32b illustrates an eight-output (8-bit output word) ROM with a 256-bit (8 × 32) memory matrix.

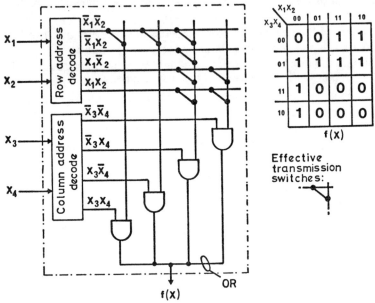

Fig. 5.33. Outline PROM realization of $f(x) = \sum 1,2,3,5,8,9,12,13$.

The use of such arrays to realize any combinatorial function is straightforward. However, unlike realizations using separately recognizable logic gates, where minimization of the function is of prime importance in order to minimize the realization, with ROM realizations the function must be expressed in minterm form. For example, take the simple function $f(x) = [\bar{x}_1\bar{x}_2x_3 + x_1\bar{x}_3 + \bar{x}_3x_4]$. Expanding this into minterm form we will find that $f(x) = \sum 1,2,3,5,8,9,12,13$. Therefore to realize this function using the 16-bit ROM shown in Fig. 5.32a, we require the memory array to be manufactured or programmed as shown in Fig. 5.33, assuming that no signal inversion occurs at the data output. It may be noted that for an n-variable combinatorial function $f(x)$ with 2^n input minterms, it is never necessary to make effective more than $\frac{1}{2}(2^n)$ paths in the array, provided a choice of inversion or non-inversion is available at the data output. For example, suppose the required function output was 1 on more than $\frac{1}{2}(2^n)$ input minterms, then the minority minterms which correspond to $f(x) = 0$ can be processed with output inversion to realize finally the correct $f(x) = 1$ minterms.

Where the requirement of the system exceeds the capacity of a single ROM package, then multiple package assemblies can be adopted. For more individual outputs than available per package,

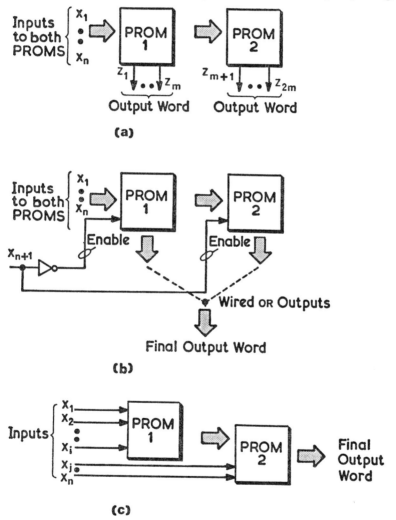

Fig. 5.34. Expansion of PROM capability: (a) output word expansion to double the output functions; (b) input address expansion to cater for an additional input variable; (c) cascaded multi-rail PROM configuration.

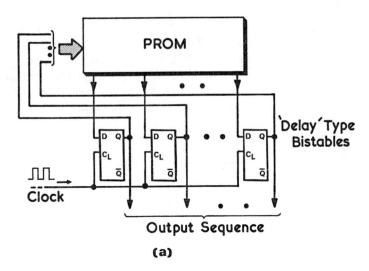

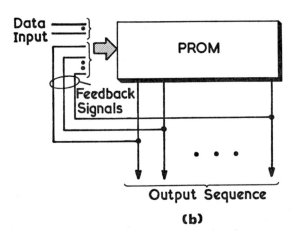

Fig. 5.35. Use of PROM's in sequential applications: (*a*) a clocked system using type D bistables (note, as an alternative to this arrangement, the clocked bistables may be placed in the feedback paths and the PROM outputs used directly); (*b*) an unclocked system, with direct feedback from outputs to inputs to control the next state.

packages in parallel as shown in Fig. 5.34*a* are an obvious solution. Where more input minterms are involved than the capacity of the memory matrix will handle, then expansion as shown in Fig. 5.34*b* is possible. Notice that in such cases the additional x_i inputs which cannot be accommodated on each ROM package effectively multiplex the package outputs such that each package is selected in turn for its appropriate minterm coverage. Further practical details of package duplication may be found[1,42,43].

An alternative and potentially more powerful method of increasing the logic capability is to adopt cascaded ROM configurations, as illustrated in Fig. 5.34c. However, we now have the problem of partitioning the function in the most efficient way, and although a great deal of intuitive partitioning may be done for strongly structured functions, no formal design methods are yet available for optimum partitioning. Some simple examples of this approach have been published[1,44].

ROM packages are also directly applicable for sequential applications, both clocked (synchronous) and unclocked (asynchronous), as Fig. 5.35 illustrates. The basic design techniques for either are entirely conventional, the ROM merely serving to realize the combinatorial requirements which would conventionally be realized with appropriate logic gates. Here of course the logic would be executed within the ROM in minterm form, exactly as we have already demonstrated. The problem of hazards in state assignments of asynchronous realizations, however, is still present, to which must be added the problem that the propagation delays for all outputs of the ROM are not always the same, and hence transitory wrong output words are generated during changes of input data. Examples of simple sequential designs using ROM's may be found in published literature, including consideration of hazard conditions[44,45].

The master monolithic layout of a ROM provides for every one of the elements in the memory array to be available. For a specific logic application some must be operative and some inoperative, as we have for example shown in Fig. 5.33. The principle differences in programmable ROM's are (i) the technology, that is bipolar or unipolar (MOS), and (ii) the techniques for programming the particular paths required in the memory array for specific applications.

An obvious method of providing a particular memory array is merely to connect the requisite paths in the array at the final metalization stage of production by including or excluding conducting jumpers. This means that a unique mask has to be made for each customer requirement, which tends to be expensive and hence only viable for fairly large quantities of the same circuit. The final ROM is then a "mask-programmed" ROM and generally is not considered to be a programmable ROM (PROM) in the sense of the following alternative techniques.

Truly programmable ROM's on the other hand are fabricated with their whole array fully interconnected, and then the appropriate paths are selected or destroyed by electrical or other means. By the use of such techniques we have the truly device-programmable PROM, as the device can be programmed to individual customer requirements after manufacture and packaging. In bipolar devices the memory-array programming is performed by passing pulses of current through selected paths, of sufficient energy to blow a fuse link or to short an emitter–base junction, whilst in MOS devices an excess voltage may be applied to the selected paths to break down a floating gate path and make each field-effect transistor switch operative. With the latter technique restoration of the memory array to its original state may be possible by ultraviolet radiation, thus giving us the possibility of *erasable programmable* ROM's, sometimes known as EPROM's. The current state of the art may be found in detail in several publications[42,43,46,47], fitting into a general hierarchy as shown in Table 5.6. If the customer can himself program the PROM to his own requirements, as is generally the case with well-established techniques, then such devices may be additionally defined as *field programmable*.

Table 5.6. The read-only-memory families

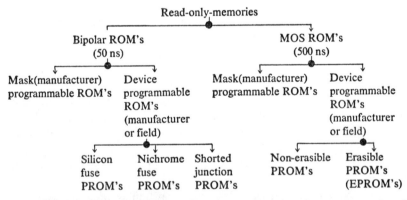

Before we leave this section the distinction between read-only memories and random-access memories (RAM's) should be made. The terminology "random-access memory" is unfortunate; far more explicit would be "read/write memory." This is because the memory consists of electrical storage circuits which can be addressed by appropriate "write" input signals so as to set and reset each of them selectively[43,44,46]. This set/reset pattern may subsequently be read out as an output word by applying appropriate "read" input signals. Thus the

RAM is a memory device which is fully programmable and erasible by *normal logic input signals,* which the PROM is not, but it does have one further fundamental disimilarity. This is that the storage in a RAM is volatile; that is if the d.c. power is removed from the package then the storage pattern is lost, exactly as in any normal shift register for example. Hence a RAM can never be used as an equivalent for hardwired logic as can the ROM and PROM families.

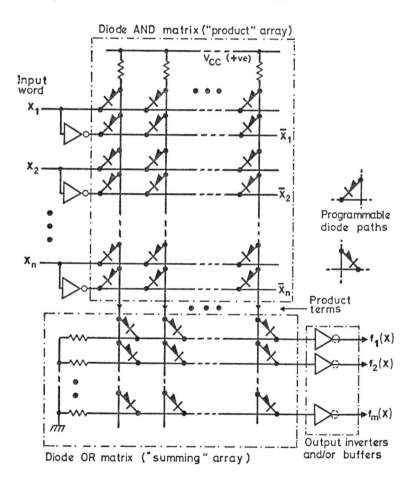

Fig. 5.36. Schematic arrangement of a PLA. (The diodes shown in the AND and OR matrices may be transistor links in specific cases and input buffers may also be present on the non-complemented input lines.)

5.5.2. Programmable Logic Arrays (PLA's)

As we have now seen, PROM's may be used to realize random-logic requirements by realizing a sum-of-minterms expansion for any given function $f(x)$. The programmable logic array which we shall now consider may be applied to the same purposes, but implements a sum-of-prime-implicants expression for $f(x)$ rather than the minterm expansion. Hence minimization of $f(x)$ is necessary, or at least desirable, for PLA realizations.

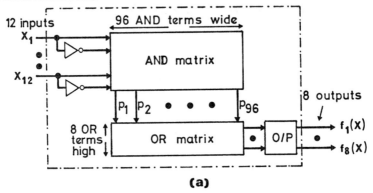

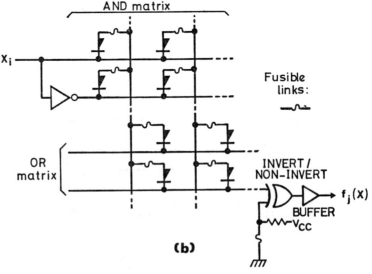

(continued)

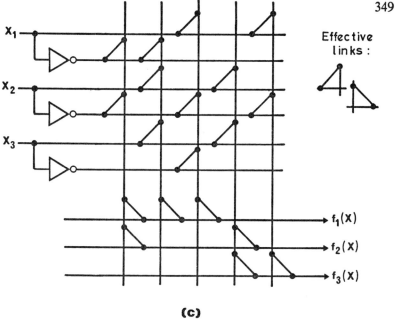

Fig. 5.37. PLA details: (*a*) the size of a typical PLA, which may be packaged in a 24-pin pack; (*b*) the concept of fusible links in each section, which may be programmed to blow open-circuit; (*c*) the paths required in a simple PLA for the three functions

$f_1(x) = [\bar{x}_1\bar{x}_2 + \bar{x}_1 x_2 x_3 + x_1 \bar{x}_2 \bar{x}_3]$
$f_2(x) = [\bar{x}_1\bar{x}_2 + x_2 x_3]$
$f_3(x) = [x_2 x_3 + x_1 \bar{x}_2]$.

The PLA differs fundamentally from the PROM in that the x_i input data is not internally decoded by fixed decoding down to minterm signal level, but instead the input signals are fed to an AND matrix, which can be programmed to produce any prime-implicant terms of the input variables, including of course the minterm-level AND prime implicants if necessary. The AND prime implicants are then OR-ed by a separate programmable OR matrix, so as to produce sum-of-product terms, these sum-of-product signals then being processed by a final output inversion/buffer stage. Figure 5.36 illustrates a typical complete structure.

A production PLA may have, say, 12 x_i inputs, provision for 96 AND terms, and 8 outputs, as shown in Fig. 5.37*a*. The programming of the AND and OR matrices may be done by mask programming, to connect the appropriate paths of the circuit during the final production stage of the package, or by fusible links within the structure, as illustrated in Fig. 5.37b. In the latter case we also have field-programmable capability, and hence the terminology FPLA may be encountered[48]. Figure 5.37c shows the diode paths required for a simple output requirement.

It is clear from the preceding illustrations that it is necessary to minimize the expressions for any required set of functions $f_1(x)$, $f_2(x), \ldots$, so as not to have more product terms in total than the capacity of the AND matrix. However, there is no particular advantage in minimizing to the theoretical minimum number of irredundant prime implicants, once we are within the capacity of the PLA. Notice also that the number of variables in each AND product term is irrelevant—only the total number of product terms and not their individual complexity is a limitation.

If a comparison between the logical capacity of a PLA and a PROM is made, it will be found that the PLA provides greater flexibility and can handle larger functions in almost every case[43,46,49,50]. Where the PLA may fail in comparison with the PROM is where little classical minimization of the required functions is possible, for example in strongly dependent exclusive-OR/NOR functions; in such cases more products may be present than the capacity of the AND matrix, and therefore the full minterm capacity of a PROM may be required.

Another facet of PROM *versus* PLA operation may also be mentioned, this being the question of logic hazards (see §4.3.3, Chapter 4). As the PROM fundamentally provides a minterm realization, then logic hazards may occur on any input change, no masking of such hazards being possible in the PROM. Further the propagation delay through the PROM for different input minterms is dissimilar, owing to different length logic paths through the internal decoders, and therefore this again leads to transitory wrong outputs ("glitches") during input data changes. The PLA, on the other hand, has (ideally) equal input–output delays since dissimilar path lengths through the AND/OR matrix are not involved, and in addition can be arranged to have overlapping prime-implicant coverage to eliminate possible logic hazards, provided of course that sufficient AND matrix capacity is available. Hence it may well be preferable to use PLA's rather than PROM's in situations where transitory wrong outputs may be troublesome; certainly in asynchronous circuits the use of PROM's should be avoided if possible.

Design Using Spectral and Other Techniques

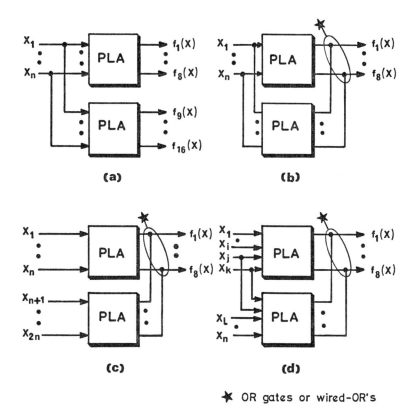

Fig. 5.38. Multiple PLA arrangements: (*a*) expansion of output function capacity; (*b*) expansion of AND matrix capacity; (*c*) doubling of input capacity; (*d*) expansion of input capacity with part sharing of input variables between the two PLA's.

It is also possible to use multiple PLA's to extend system capability, although certain restrictions arise. Expansion of the number of outputs from, say, 8 per single PLA to 16 may be made by paralleling two 8-output PLA's, as shown in Fig. 5.38*a*. However, the total number of different product terms which may be handled by the AND matrices has not necessarily been doubled, as identical product terms which are required in any of the first eight outputs and also in any of the second eight outputs will have to be generated in each PLA. Only if all the product terms in the two PLA's can be arranged to be dissimilar is the total AND matrix capacity doubled.

To increase the AND matrix capacity without requiring an increase in number of output functions, two PLA's in parallel with OR-ed outputs can be used, as shown in Fig. 5.38b. The prime-implicant terms programmed into each AND matrix are now dissimilar, but collectively provide the required sum-of-products for each $f(x)$ output. A "half-way" stage between Figs. 5.38a and b is also possible, where prime implicants from the lower PLA are OR-ed with certain outputs of the upper PLA, whilst other outputs from upper and lower PLA require no sharing of the PLA outputs.

Input expansion of PLA input capacity is also possible, but more troublesome to design. In the simple parallel arrangement shown in Fig. 5.38c the input-variable capacity has been doubled, but the prime-implicant AND terms which can be generated are only those which contain either the top set of input variables or lower set of input variables but not any mixed set of variables. If the required output functions cannot be expressed in this way, then some sharing of inputs between the two PLA's is necessary, with a resultant reduction in the total input capacity of the pair. Figure 5.38d illustrates this principle.

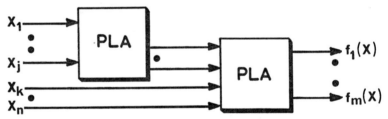

Fig. 5.39. A PLA multi-rail cascade for which no complete design algorithms are yet available.

Cascading of PLA's is also possible, for example as illustrated in Fig. 5.39. However, like the cascaded PROM arrangement illustrated in Fig. 5.34c, no really satisfactory design techniques are yet available to exploit the considerable possibilities of such arrangements. The present theories of multi-rail cascades, which will be found well reviewed and referenced by Elspas[51], are entirely inadequate to cater for the full possibilities of Fig. 5.39; possibly the spectral and symmetry considerations considered earlier in this chapter may be of future help.

The use of PLA's in synchronous applications may be done in a similar way to that shown in outline in Fig. 5.35 for PROM's. However, for reasons which have already been mentioned, the use of

Design Using Spectral and Other Techniques

PLA's may be more satisfactory in sequential work than PROM's and in addition the considerable logic capacity of the PLA's allows, for example, complex sequence generation and decoding to be jointly performed. For synchronous operation the schematic arrangement of Fig. 5.35a is still valid, although frequently the clocked bistable elements (the "delays") are included in the feedback paths to the PLA and the outputs from the PLA used directly as the network outputs. This is illustrated in Fig. 5.40a. The application of PLA's to the generation of a 4-bit cyclic code plus a seven-segment read-out is shown in Fig. 5.40b. Further details of these possibilities will be found published[43].

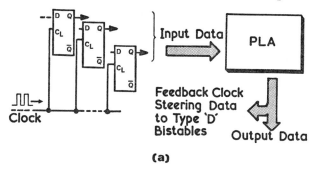

(a)

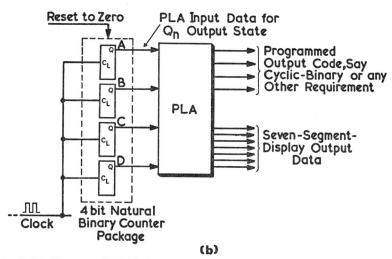

(b)

Fig. 5.40. The use of PLA's in sequential operations: (a) a typical schematic, using say type D bistables; (b) the use of a PLA plus type JK bistables as a code convertor; if the PLA also provides the clock-steering gating for the JK circuits then the whole assembly is a PLA-based code generator.

As a final note of caution on both PROM's and PLA's, it must be mentioned that careful testing of a programmed device is essential to ensure that it truly realizes the required output(s), and does not generate more minterms or prime-implicant outputs than it should. It is relatively easy to check that the outputs are correct when the apropriate input words are present, but this is no check that the outputs are correct on all other input conditions. Hence it is necessary to verify the device programming by cycling the inputs *through all 2^n input words*, checking for correct response on all of them. The long-term reliability of the internally programmed paths of these devices is yet another consideration which the practising engineer must by no means overlook.

5.5.3. Further Possibilities

Whilst the PROM or PLA circuit currently represents the best choice of l.s.i. circuit for limited-quantity random-logic applications, other l.s.i. possibilities are available, which we shall briefly review.

The first alternative is the *uncommitted gate array* ("UGA") or *uncommited logic array* ("ULA"). This consists of an integrated-circuit chip which has been made with a matrix of individual gates or individual active devices and resistors. No permanent interconnections between the devices, however, are initially made, and the production silicon slices are therefore "uncommitted" for any particular duty. To cater for a particular customer requirement the final production-stage interconnection (metalization) mask has to be made, so as to interconnect the chip to perform the required combinatorial and/or sequential duty. The device therefore is a mask-programmable device, the final packaged circuit being the exact customer requirement[52,53].

There are many pros and cons of this approach. Firstly each customer's requirements have to be manufacturer-designed to fit the uncommited chip, with optimum design being desirable. However, the design is not constrained to sum-of-minterms or sum-of-products realizations as is the case with PROM and PLA realizations, and exclusive-OR relationships, for example, may be considered at the design stage. Hence good computer-aided design techniques by the manufacturer are desirable as well as his CAD of the final metalization mask for the production line. Notice that once a customer design has been commited, no circuit changes are generally possible; hence this approach is not suitable for applications where customer changes of mind are likely. Secondly, the number of gates which can be accommodated per chip in a ULA or UGA layout is generally considerably less than the number employed in, say, a PROM or PLA or other l.s.i. circuit. This arises because isolation between all items laid down on the ULA chip has to be maintained, and room left available to connect them in whatever configurations are finally required. Thus utilization of silicon area for logic purposes is not as high as in other l.s.i. areas where optimum layout can be achieved. Finally, the manufacturer's cost of design and individual metalization mask for each customer requirement is considerable, possibly in the region of $2000 to $10,000 depending upon complexity. Hence the ULA/UGA approach is not commercially viable for very small quantity requirements, but may find its application where, say, thousands of a special circuit are required. The full reliability of a normal integrated circuit should be available from a ULA/UGA package, without the query of the long-term reliability of the PROM and PLA approach.

Another possible l.s.i. approach to custom requirements is the use of a calculator chip. This is particularly and obviously relevant where the custom requirement is a number-processing application, such as in data handling, on-line control systems with A–D/D–A conversions, and the like. Unfortunately publicity on the availability, structure, and application of calculator chips is not as widespread as data on other l.s.i. products, such as PROM's, multiplexers, microprocessors (see later), etc. Nevertheless owing to their high-quantity production, standard calculator chip costs should be minimal.

The calculator chip by itself will only serve to perform arithmetic operations on binary numbers presented to it, these numbers being in binary, BCD, or octal format. To produce a fully operational system the calculator chip needs to be surrounded by appropriate input and output registers, together with program memory and program sequence control. Notice that all this will be a normal hardware logic-design exercise, not involving the considerable software emphasis of a microprocessor or microcomputer system. Also no device programming of the calculator chip itself is involved, the l.s.i. device programming remaining in the program-memory (PROM or PLA) part of the system.

A complete system to perform some arithmetic process therefore may be as illustrated in schematic form in Fig. 5.41. The precise schematic will largely depend upon the requirements of the particular calculator chip, as well as on the particular arithmetic processes which are required in the system. The instructions which the calculator chip obeys may for example be as shown in Table 5.7, but this again will vary with the type of chip chosen for the application. It will be ap-

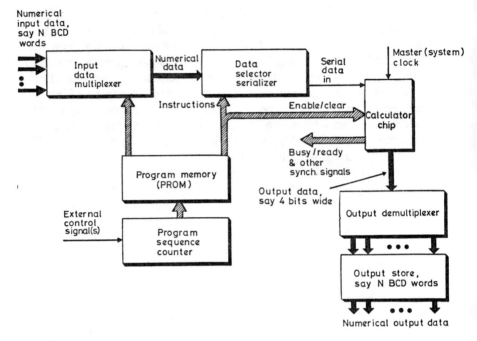

Fig. 5.41. A schematic arrangement for an arithmetic-type system using an l.s.i. calculator chip as the main logic processor. The necessary synchronizing signals between blocks are not shown in detail.

preciated from Fig. 5.41 that the program memory which controls the sequence of operation of the calculator chip may itself be a large but basically simple logic assembly and hence ideally realized with a PROM package. The multiplexers and demultiplexers which also are necessary around the periphery of the calculator chip are standard m.s.i./l.s.i. packages, and hence a complete and relatively complex system is partitioned into appropriate packages, each of which utilizes l.s.i. technology resources. However, this is not to say that the complete system could not be redesigned to use considerably less logic if a sufficiently powerful design philosophy was available to consider globally the total input–output requirements of the system. Currently this is not a possibility, and hence partitioning of the system down into recognizable and handleable subsections is our only presently viable technique. Global symmetry considerations may be of future use, however, in this area.

Table 5.7. Typical calculator chip instructions

Instruction	Possible input instruction code	Description of instruction
Clear	10000	Clear all internal data registers and stored instructions
Multiply	10010	Multiply output register by last data entry; answer to output register
Divide	10011	Divide output register by last data entry; answer to output register
Add	10100	Add last data entry to contents of output register
Subtract	10110	Subtract last data entry from contents of output register
Increment	10101	Add 1 to output register
Decrement	10111	Subtract 1 from output register
Add till overflow	11000	Continuously increment until output register overflows
Subtract till zero	11001	Continuously decrement until output register zero
Shift right	11010	Move output register contents one place towards least-significant place
Shift left	11011	Move output register contents one place towards most-significant place
Exchange	11100	Interchange the number stored in the two data registers
Equals	10001	Execute the last instruction

Finally, a brief mention should be made of microprocessors, although like the calculator chip the microprocessor chip is standardized and the data/instructions fed to it are the variables. The microprocessor package currently represents the ultimate in packaging of logic on a single l.s.i. chip. Unlike the calculator chip, however, the microprocessor chip is designed to perform a comprehensive set of instructions on the digital data presented to it, including making its own inter-

358 The Logical Processing of Digital Signals

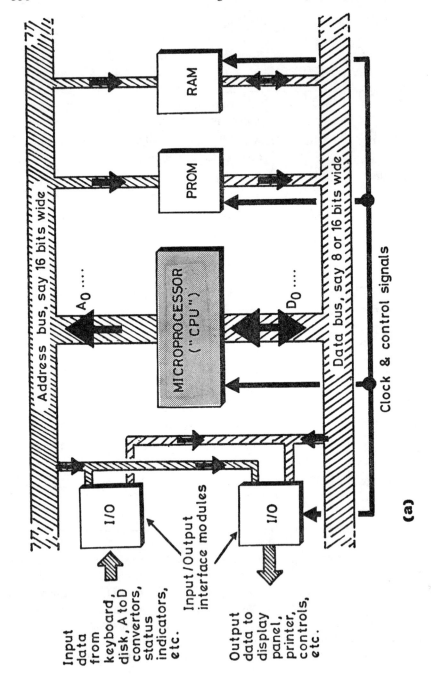

(a)

Design Using Spectral and Other Techniques

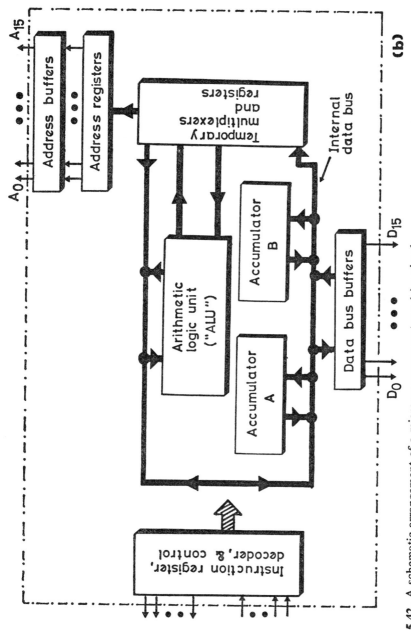

Fig. 5.42. A schematic arrangement of a microprocessor system: (*a*) a simple system; (*b*) the basic microprocessor elements.

nal decisions when comparing its internally stored information[54,55]. Figure 5.42 illustrates in broad outline the system architecture of present microprocessor packages.

However, it is now not feasible for a system designer to consider the individual digital signals which have to be fed into the microprocessor in order to execute a particular requirement, as the complexity of the total input requirements and the step-by-step execution becomes too great to handle mentally. Instead we have to resort to software programming as with any digital computer, the software then being verified by program verifiers and other aids provided by the microprocessor manufacturer. When the software has been verified as correct, then it is transferred into the permanent memory of the system, which normally will be ROM's or PROM's external to the microprocessor. Notice in Fig. 5.42a that the PROM provides the permanent memory of the program sequence for the particular application, whilst the RAM provides the temporary memory for changing data during the operation of the system.

The microprocessor has been hailed as the perfect solution for original-equipment-manufacturer ("o.e.m.") use when designing digital systems. Certainly it may be applied to both simple and complex logic requirements, but there is a great danger that it may provide a more sophisticated solution than a system requires. Where it may be of use is where the system logic requirements change, as now a reprogrammed ROM rather than any wiring changes may cater for the changed requirements. Equally a PLA rather than a microprocessor-based system might cater for such contingencies. Where system requirements are fixed, however, then the original equipment manufacturer must carefully consider the real economics of microprocessor system realization, including total system design costs, total system documentation, and on-site maintenance and diagnosability of his end product.

5.6. Universal Logic Modules

In the preceding sections we considered logic circuits which may be programmed in appropriate ways so as to realize specific requirements, which in all cases finally came down to altering electrically the internal paths of an l.s.i. circuit or circuits. However, let us now briefly look at possible circuits which have a fixed internal interconnection pattern, but which may be used for more than one specific logic requirement by selectively using the inputs provided. Such circuits may be said to be "programmable" in the sense that we may program their use by hard-

Design Using Spectral and Other Techniques

wired connections to them, but we shall not use this term in this context.

5.6.1. Multiplexers as Universal Logic Modules

Multiplexer circuits, which conceptually may be regarded as single-pole, multi-position switches as suggested in Fig. 5.43a, may be used as combinatorial universal logic modules ("ULM's").

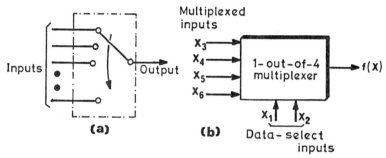

Fig. 5.43. Multiplexer principles: (a) mechanical switch analogy; (b) a 1-out-of-4 multiplexer, $f(x)$ = either x_3, or x_4, or x_5, or x_6.

Consider the simple case of a 1-out-of-4 multiplexer, as shown in Fig. 5.43b. When employed as a conventional multiplexer, the independent $\{0,1\}$ logic signals present on inputs x_3, x_4, x_5, and x_6 are effectively transferred to the multiplexer output as the data-select inputs x_1, x_2 cycle through 00, 01, 10 and 11 respectively. The output equation therefore is

$$f(x) = [\bar{x}_1\bar{x}_2(x_3) + \bar{x}_1 x_2(x_4) + x_1\bar{x}_2(x_5) + x_1 x_2(x_6)].$$

However, if we consider the case where the multiplexed inputs x_3, x_4, x_5, and x_6 are not separate independent binary input signals, but instead are either 0, 1, x_3, or $\bar{x}_3$ in any order, then with these four input possibilities on each of the four multiplexed inputs we have a total of $4^4 = 256$ different input patterns which x_1, x_2 can multiplex to the output. However, 256 is the total number of possible combinatorial functions of $n = 3$, and hence the 1-out-of-4 multiplexer may be used as an $n = 3$ ULM.

More formally we may state that the multiplexer output used in this way is

$$f(x) = [\bar{x}_1\bar{x}_2 f_0(x_3) + \bar{x}_1 x_2 f_1(x_3) + x_1\bar{x}_2 f_2(x_3) + x_1 x_2 f_3(x_3)]$$

where $f_0(x_3),\ldots,f_3(x_3) \in \{0, 1, x_3, \bar{x}_3\}$.

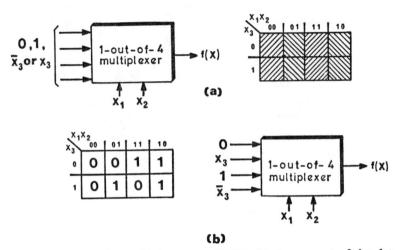

Fig. 5.44. The use of a multiplexer as a ULM: (a) the concept of the data-select inputs x_1, x_2 selecting the columns of an $f(x_1, x_2, x_3)$ Karnaugh map; (b) multiplexer realization of $f(x) = [\bar{x}_1 x_2 x_3 + x_1 \bar{x}_3 + x_1 \bar{x}_2]$.

Effectively the x_1, x_2 data-select inputs are now selecting the columns of the $n = 3$ Karnaugh map shown in Fig. 5.44a, and hence the realization of any required $n = 3$ function $f(x)$ readily follows as illustrated in Fig. 5.44b.

In a similar manner a 1-out-of-8 multiplexer may be employed as an $n = 4$ ULM. To express this the other way round, n-variable ULM capability is provided by a 1-out-of-2^{n-1} multiplexer circuit. Notice that in all cases the $n - 1$ data-select inputs multiplex $\{0, 1, x_n, \bar{x}_n\}$ signals from the remaining inputs, and hence in all cases are progressively selecting pairs of adjacent minterms of the required function $f(x)$. Further, it may be seen that although both x_n and $\bar{x}_n$ may be required on the multiplexed inputs, with possibly a considerable fan-in requirement on each signal, the data-select inputs require only the true and never the complemented signal of their inputs to be applied.

Where the number of input variables n is within the capacity of a single multiplexer unit, then the best design technique is to plot the required function $f(x)$ on a Karnaugh map layout, from which the required $0, 1, x_n, \bar{x}_n$ signals multiplexed by the remaining data-select inputs may be read off. Figure 5.45a illustrates this procedure for an $n = 5$ function. However, the best choice of which variables $x_1, \ldots, x_{n-1}$ to use as the data-select inputs and which remaining one x_n becomes the multiplexed input variable is not always clear. If the data-select inputs

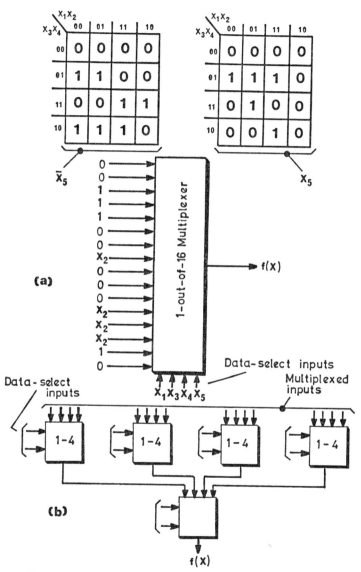

Fig. 5.45. Function realization with multiplexers: (*a*) a five-variable function $f(x_1,x_2,x_3,x_4,x_5)$ with x_1,x_3,x_4,x_5 chosen as the data-select inputs and $\{0,1,\bar{x}_2,x_2\}$ as the multiplexed inputs, with $\bar{x}_2$ not required in this particular example; (*b*) the use of 1-out-of-4 multiplexers in cascade; note the data-select input signals to the first-level multiplexers need not be all the same but may be varied to produce some optimum realization.

are chosen such that (*a*) a maximum number of 0 or 1 signals are multiplexed and/or (*b*) the complement of the multiplexed x_n input is not required, then this will probably be the best choice of input variable selection. This has been attempted in the function shown in Fig. 5.45*a*, where x_1, x_3, x_4, x_5 have been chosen as the $n - 1$ data-select inputs, with x_2 as the remaining multiplexed input variable.

The realization of a large-*n* function may require the use of multilevel multiplexer networks, as illustrated in Fig. 5.45*b*. However, we now have an even more difficult problem in how best to partition the data-select variables and the multiplexed variables at each level of processing, a problem which can be aggravated if both $f(x)$ and $\overline{f(x)}$ are available from each multiplexer. Currently there are no known methods to optimize this problem[1,4,44], and hence intuitive or trial-and-error methods become the usual approach.

5.6.2. Universal Cellular Arrays

The previous section introduced the concept of variable input connections to a module so as to provide universality, that is the hard wiring to the module was the "programming" for a specific application. We also saw that it was assumed that at least one input x_i and its complement $\bar{x}_i$ was freely available, together with the steady binary signals 0 and 1. These concepts will be encountered again in some of the further "universal" structures we shall consider in due course.

The definition *cellular array* refers to networks composed of some regular interconnection of logic cells. These arrays may be either one-dimensional, as shown in Fig. 5.46*a*, two-dimensional, as shown in Fig. 5.46*b*, or theoretically of any higher dimension of three or more. Practical considerations, however, usually constrain circuit layouts to the one- or two-dimensional case. Arrays may be further classified as *unilateral arrays,* in which the signal flow is consistently from one cell to adjacent cell(s), or *bilateral arrays*, in which signals are fed back between cells as well as forward.

The number of cascade feedforward signal paths is yet a further defining parameter which may be used, more specifically for unilateral arrays. A single signal path as shown in Fig. 5.46*a* is termed a *single-rail cascade,* with Fig. 5.46*c* illustrating a two-rail cascade. In general we may have *multi-rail cascades* between cells.

When the cells in any form of cellular array are identical, then we have an *iterative* array structure. The advantages of all cells being the

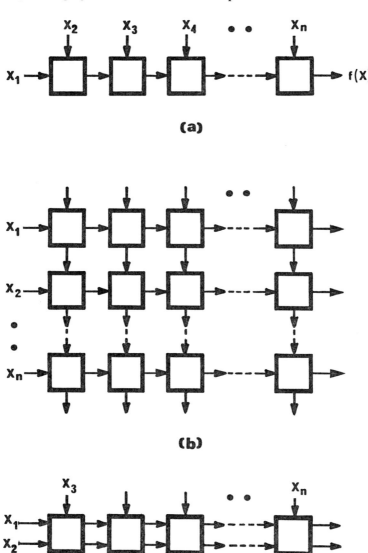

Fig. 5.46. The structures of cellular arrays: (*a*) a one-dimensional single-rail cascade; (*b*) a two-dimensional single-rail array, with inputs conventionally shown on the left-hand side; (*c*) a one-dimensional cascade, but two-rail interconnection between cells.

same for l.s.i. realization are obvious, although this advantage may be more than offset by the necessary complexity then required in each cell. In the following considerations we look at iterative and non-iterative structures, but in both cases we shall tend to confine our discussions to arrays with unilateral signal flows—the additional complexity involved by considering bilateral signal flows is in general an unsolved situation.

The simple one-dimensional single-rail array shown in Fig. 5.46a may readily be shown to be inadequate to realize all possible combinatorial functions $f(x)$ of n input variables, irrespective of the length of the array[56–59]. Fundamentally there is a completely inadequate information capacity available in the single binary path linking the cells to build up other than a very restricted range of functions. This is true whether the cells are all identical (an iterative cascade) or dissimilar.

A two-rail cascade, however, may be shown to be functionally complete, but only with functionally dissimilar cells and not with an

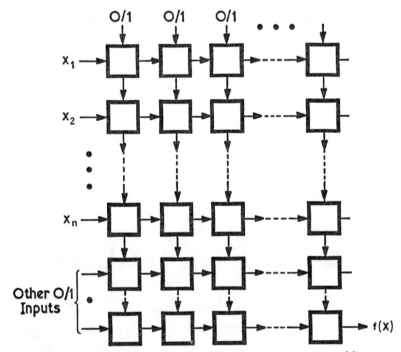

Fig. 5.47. A cellular fixed-interconnection rectangular array with uncomplemented x_1 to x_n inputs.

iterative cascade⁽⁵⁸⁾. If we consider multi-rail cascades, with more than two signal paths between cells, then we are entering the same area of consideration as we briefly noted in connection with Figs. 5.34 and 5.39, this being an area of very great difficulty for which no ready synthesis methods are yet available[51,58]. If the additional constraint of making all cells in the one-dimensional multi-rail cascade logically identical is imposed, then this adds yet another order to the difficulties of synthesis.³ In theory a multi-rail iterative cascade can realize any required function $f(x)$, but the determination of optimum cell requirements and number of cells in cascade is currently unsolvable.

Two-dimensional arrays, however, have received much greater attention, as functionally complete structures are more readily achieved. With such arrays we most commonly have the x_i input variables applied to the left-hand edge of the array, with binary 0,1 signals as the top inputs, the final output function $f(x)$ being produced at the bottom right-hand corner of the array. The general arrangement of such "edge-fed" two-dimensional arrays is illustrated in Fig. 5.47.

Let us consider first arrays where the input and cell interconnection wiring is fixed, but cell functions are not all identical. One such family of rectangular arrays which has received considerable theoretical attention is where each cell is as illustrated in Fig. 5.48a, the x input being a straight-through bus connection and the output $f(x,y)$ being any one of the 16 possible functions of the two cell inputs x,y. However, developments by Matra, Minnick, and others[56,59–61] have shown that only six two-input–one-output cell functions are necessary, assuming only uncomplemented x_i inputs available as cell inputs, giving complete array structures capable of realizing any function $f(x)$ as shown in Figs. 5.48b and c. In Fig. 5.48b we are generating a sum-of-products output for $f(x)$, whilst in Fig. 5.48c we are generating a products-of-sum output. Such structures are frequently referred to as *cutpoint cellular arrays*.

The difficulty, both theoretical and practical, of this approach is that for the minimum synthesis of any required function $f(x)$ the individual cells of the array (except for the lowest row) have to be individually determined and realized, and hence the simplicity of the fixed cell-interconnection wiring is more than offset by synthesis, minimiza-

---

³ It may be recalled from §5.1 of this chapter that Boolean matrices enable us to analyze a cascade of any multi-input–multi-output cells by matrix multiplication, but what was not possible was to determine the individual cell requirements by finding the mth root of the overall function matrix, where m is the number of cells in cascade.

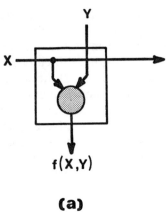

(a)

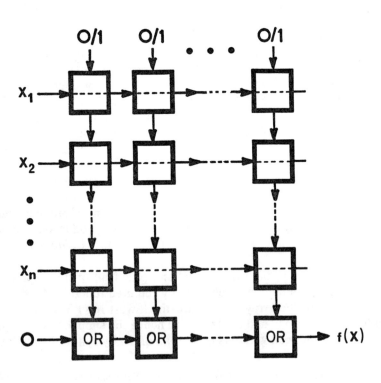

(b)

(continued)

Design Using Spectral and Other Techniques

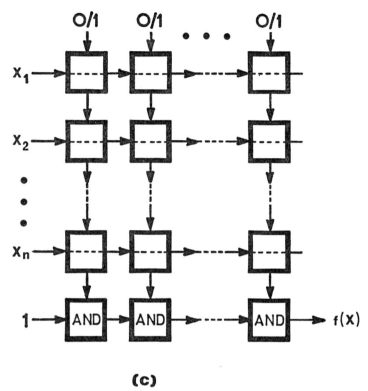

(c)

Fig. 5.48. "Cutpoint" cellular arrays: (a) the "cutpoint" cell; (b) sum-of-products cutpoint array; (c) product-of-sums cutpoint array. Note the cell functions in (b) and (c) are not all the same, but are from the set

$$f(x,y) = \left\{[y], [x + y], [xy], [\bar{x} + y], [\bar{x}y], \text{ and } [\bar{x} \oplus y]\right\}.$$

tion, and realization complexities. Papakonstantinou's consideration of the synthesis problem represents a state-of-the-art publication[61].

An alternative approach to this problem is as follows. Here standardization on cells and a fixed interconnection pattern between cells has been achieved in the main part of the array, but closed or open vertical paths to the lowest (summing) rows of cells are required. Hence the full array is not completely iterative or fixed wired, but the majority portion of it is. The general arrangement is illustrated in Fig. 5.49a.

The realization of the final output is given by a sum-of-minterms expansion of $f(x)$, no minimization to sum-of-product form being available. The minterms, however, are formed by AND/exclusive-OR cells, as illustrated in Fig. 5.49b, the horizontal output to the next-right cell being $[x \oplus y]$, sometimes therefore referred to as the half-adder

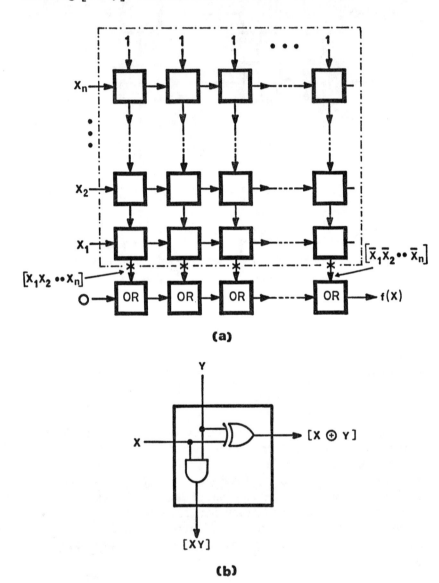

(a)

(b)

(continued)

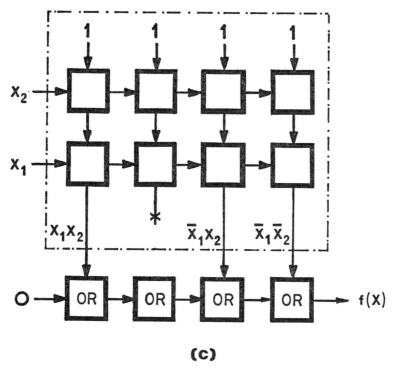

Fig. 5.49. The AND/exclusive-OR minterm generator cellular array: (a) the general arrangement, with the final vertical links marked x made or broken as required; (b) the iterative cell; (c) cellular realization of $f(x) = [\bar{x}_1 + x_1 x_2]$.

cell output, the vertical output to the cell below being the AND output $[xy]$. It may be quickly verified that the final vertical outputs from the iterative array to the bottom row of OR cells are the minterms $x_1 x_2 \ldots x_n, x_1 x_2 \ldots \bar{x}_n, \ldots, \bar{x}_1 \bar{x}_2 \ldots \bar{x}_n$, reading from left to right in Fig. 5.49a. Notice that in the topmost row of cells the horizontal feedforward signal is alternately $\bar{x}_n, x_n$; in the second row the horizontal feedforward signal is either $\bar{x}_{n-1}$ or x_{n-1} depending upon the vertical input signal, and so on. Hence to realize any function $f(x)$ the lowest vertical connections corresponding to 1-valued minterms must be included, whilst those corresponding to 0-valued minterms must be removed or omitted. Figure 5.49c illustrates the realization for a trivial three-variable function. Notice that a total of $n(2^n)$ AND/exclusive-OR cells plus 2^n OR cells are necessary in this type of array for the realization of any n-variable function $f(x)$.

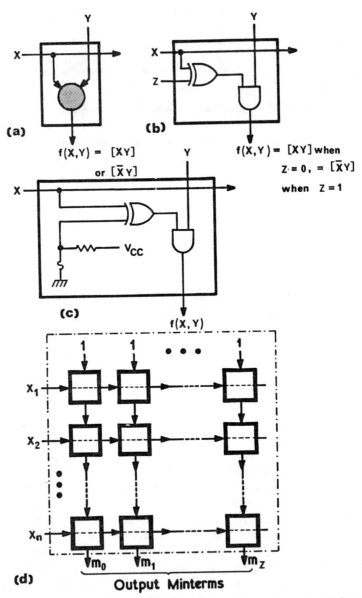

Fig. 5.50. Simple programmable cell for an iterative universal cellular array: (a) the cell logic requirements; (b) cell output programmed by an additional input z; (c) cell output programmed by some internal mechanism such as fusible link; (d) the rectangular iterative array for minterm generation.

Design Using Spectral and Other Techniques

The search for a "best" compromise between (i) variable hard-wiring on the x_i inputs and/or between cells and (ii) non-standardization of cell functions inevitably leads on to the consideration of some form of programming of the cells, so that they may be fabricated as identical layouts with a fixed interconnection pattern but then programmed in some manner for specific duties.

The simplest concept in this area is still to consider the realization of any function $f(x)$ by a sum-of-minterms expansion. If we now consider a cell as shown in Fig. 5.50a, where the horizontal input x is bussed through unchanged but the vertical output can be programmed to $[xy]$ or $[\bar{x}y]$, then we have the makings of a universal minterm generator for any required function $f(x)$. The selection between cell output $[xy]$ or $[\bar{x}y]$ could be made by some method such as shown in Figs. 5.50b and c. If an array of such cells is now made as in Fig. 5.50d then clearly each column output can be arranged to be any chosen minterm of the required function, ranging from $\bar{x}_1\bar{x}_2\ldots\bar{x}_n$ if all the column cells are programmed to the $[\bar{x}y]$ state, through to $x_1x_2\ldots x_n$ if all cells are in the $[xy]$ state. The final summation of these minterm functions to produce $f(x)$ now requires to be made by a lowest-level OR summation, which if necessary could be done by using the same construction cells, firstly inverting each minterm output $m_0, m_1, \ldots$ by a next row of cells, and then forming the final output in the form

$$f(x) = [\overline{\overline{m_0}\,\overline{m_1}\ldots\overline{m_z}}]$$

in a final row of cells. Alternatively merged minterm and summation cells may be suggested[62].

At this stage the reader may have feelings of being on ground which has already been covered, but before we pause for reflection let us continue a little further with theoretical developments. These further considerations may also serve as a useful basis for certain ternary developments covered in §6.6 of the following chapter.

Instead of a sum-of-minterms expansion for any function $f(x)$, we may also consider a generalized Reed–Muller exclusive-OR expansion, which we first encountered in Chapter 1. It will be recalled that this expansion is the modulo-2 addition

$$f(x) = [a_0 \oplus a_1x_1 \oplus a_2x_2 \oplus \ldots \oplus a_{2^n-1}x_1x_2\ldots x_n]$$

where $a_j, j = 0$ to $2^n - 1$, are 0 and 1 constants operating on all possible different combinations of $x_i, i = 1\ldots n$, taken one at a time, two at a time, up to n at a time. Hence a rectangular array may be suggested, which can generate all the $2^n - 1$ product terms of the x_i's, which

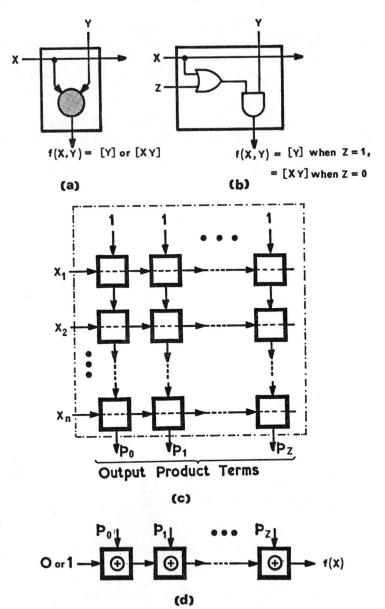

Fig. 5.51. The Reed–Muller universal cellular array: (*a*) the cell logic requirements; (*b*) cell output programmed by an additional input z; (*c*) the rectangular iterative array for product generation; (*d*) the final modulo-2 row addition.

terms are finally exclusive-OR'ed by an appropriate bottom-level row of cells. Notice that no complements of any of the x_i inputs are required in this Reed–Muller expansion.

The requirements from each of the cells in the product generator array is shown in Fig. 5.51a. Clearly where a particular x_i input variable is not required in a product term, then the appropriate cell is programmed so that its output is merely y, but where the variable has to be included then the cell is set to $[xy]$. This is illustrated by Figs. 5.51b and c. The final row below the output of the matrix of Fig. 5.51c is a cascade of exclusive-OR gates performing the modulo-2 addition, cascaded to give the final output $f(x)$ as indicated in Fig. 5.51d. Notice that like the previous minterm array where only the non-zero-valued minterms required to be generated and OR'ed, here only the products whose value in the Reed–Muller expansion is non-zero have to be generated and added modulo-2.

Like the previous minterm array case, it is possible to suggest a common cell configuration which will act both as the vertical-column product generator and also as the final-row exclusive-OR cascade. However, the circuitry required per cell is now rather excessive, and hence to standardize on this one type of cell for the whole structure becomes very wasteful. For further details of both these arrays and other developments reference may be made to the comprehensive survey of Kautz[62].

Finally in these universal rectangular array considerations let us look at an entirely different approach, that of Akers[63]. Consider the cell shown in Fig. 5.52a, having three separate cell inputs and two logically identical cell outputs. To tie up with the ternary developments of Chapter 6, let us label the inputs p, q, and e rather than x, y, and z as often used elsewhere, where e represents the "external" variable input signal to each cell. A complete iterative array of such cells is shown in Fig. 5.52b.

The truth table for the cell action is as follows, where — indicates a don't-care condition.

| Cell inputs | | | Cell output | Comments (see later in text) |
p	q	e	f	
0	0	—	0	Topmost row of cells redundant
1	1	—	1	Leftmost column of cells redundant
1	0	0	0	Variable conditions
1	0	1	1	

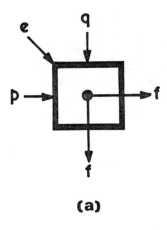

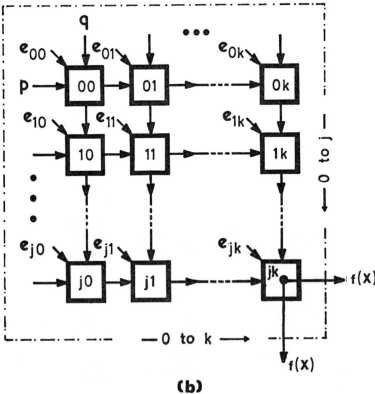

Fig. 5.52. Akers' iterative cellular array: (*a*) cell input–output designations; (*b*) the rectangular array.

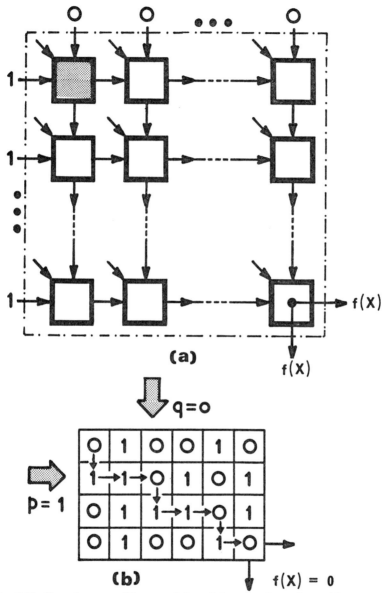

Fig. 5.53. Boundary conditions and 0 and 1 paths through an Akers array: (a) the 1 and 0 boundary signals on p and q, respectively; (b) the 0 and 1 paths through an example 6 × 4 array, where the 0 and 1 entries shown are the e cell input signals.

Consider the situation where all the p inputs to the leftmost column of cells are made logic 1, and all the q inputs to the topmost row of cells are made logic 0, as illustrated in Fig. 5.53a. Notice that the only cell whose three inputs p, q, and e are all initially known will be the top left-hand corner cell, shown shaded in Fig. 5.53a. It will now be seen from the above truth table that if the e input to this initial cell is 0 then the cell output f to its adjoining cells will be 0, whilst if e is 1 then output f will be 1. However, when this cell output is 0 it will further be seen from the truth table that this 0-valued output signal cascades through all the topmost layer of cells, giving $p = 0$, $q = 0$, and hence output $f = 0$ conditions in every cell of this row. Hence the topmost row of cells may effectively be removed, or considered redundant, as logic 0 is now the vertical q input to all the second row of cells in the array. Conversely, when the initial cell input and therefore output is 1, it will be seen that this 1-valued signal cascades down all the leftmost column of cells, and hence all leftmost cell outputs are 1. Thus the leftmost column of cells may in this case be considered redundant. Hence the action of each top left-hand corner input e is to reduce the remaining effective cell array size, reducing its *height* by one row when the input e is 0, and reducing its *width* by one column when this input e is 1.

If we consider the action of the particular array inputs shown in Fig. 5.53b, then it will readily be seen that 0 and 1 signal paths are formed by the e inputs as shown, giving a final cell output value 0. Thus it is a trivial exercise to determine the final cell output, knowing all the external cell input signals e. It may be visualized that this form of array can be made into an array to generate any specific function $f(x)$, the input variables x_i and $\bar{x}_i$, $i = 1 \ldots n$, and logic 0 and 1, forming the e cell inputs, or into an universal array to generate any function $f(x)$, although in both cases the necessary cell inputs are not immediately obvious. Details of how to allocate the 0, 1, x_i and $\bar{x}_i$ input signals to the e inputs may be found[2,63]. The universal array is based upon the expansion of any function $f(x)$ into

$$f(x) = [\bar{x}_2\bar{x}_3\ldots\bar{x}_n f_0(x_1) + \bar{x}_2\bar{x}_3\ldots x_n f_1(x_1) + \ldots$$
$$+ x_2 x_3 \ldots x_n f_{2^n-1}(x_1)]$$

where $f_0(x_1), f_1(x_1), \ldots \in \{0, 1, x_1, \bar{x}_1\}$.

Figure 5.54 illustrates the e-cell inputs for an $n = 4$ universal array; it is left as an instructive exercise for the reader to confirm that the final cell

Design Using Spectral and Other Techniques

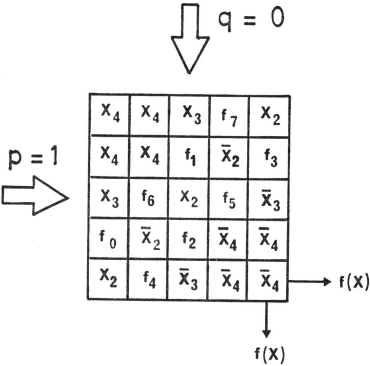

Fig. 5.54. The required e inputs for an $n = 4$ universal Akers array, where $f_0 = f_0(x_1), f_1 = f_1(x_1)$, etc.

output is correctly $f_0(x_1)$, $f_1(x_1)$, etc. depending upon the remaining inputs x_2, x_3, x_4.

All the arrays so far considered have been rectangular arrays. However, let us finally look at one non-rectangular array before we collectively summarize this present discussion. Consider the cell illustrated in Fig. 5.55a. This is in effect a change-over switch which multiplexes its two left-hand outputs to the right-hand output. Now let us form these cells into an iterative triangular array, as shown in Fig. 5.55b. The action of this triangular array is simple; the extreme left-hand inputs a_0, a_1, etc. are multiplexed through to the output as the x_i inputs cycle through 000, 001, 010, etc. Therefore to realize any $n = 3$ function $f(x)$, the a_i inputs require to be set to 0 or 1 as detailed in the truth table for $f(x)$. Note that, with this type of array for any n-input function $f(x)$ there are n levels in the realization and 2^n left-hand input signals of 0 or 1. However, for any specific application if both left-hand inputs to a cell are identical then the cell is redundant, and if two (or more) cells at any level of realization have identical inputs then these cells may be combined into a single cell with the same inputs.

Looking back at the array structures that we have now considered in this section, we may compile and compare certain statistics as detailed in Table 5.8. In compiling these figures maximum possible array size has been assumed, for example the full 2^n cell width of minterm generating arrays, and therefore no possible array minimization techniques have been taken into consideration.

Points which can be made collectively surveying this area are as follows.

(i) In all the universal rectangular arrays a considerable area of the array is performing no useful purpose when the array is generating any single given function $f(x)$.

(ii) A very considerable number of cells are required in all the rectangular arrays, and the possible input–output cell propagation time may be long in comparison with a custom-designed realization for the same output function.[4]

(iii) The triangular array is superior on cell count and propagation path lengths to the rectangular arrays where single-output functions are concerned, as the logic information content is converging on the one-output cell without involving the unused areas of array of the rectangular formats.

However we have previously considered in this chapter PROM's, PLA's, and multiplexers as universal logic devices. How should they be compared with these results?

Firstly let us consider PROM's. Looking back at Fig. 5.32a, for example, it will be recalled that the PROM is effectively a programmed minterm generator, with the programmed minterms finally OR'ed to give the required output $f(x)$. Only n inputs x_1 to x_n with no complements of any of these inputs are, however, necessary owing to the internal decoding. Hence the PROM provides an ideal sum-of-minterms solution, implementing the same principles as a sum-of-minterms cellular array, although it is not usually referred to as a "cellular array" device.

---

[4]. This of course depends upon what consitutes a "cell." If a cell can be made as a straight-through bus connection, with wired-AND's or wired-OR's to other bus lines[64], then these points may not be significant.

Design Using Spectral and Other Techniques

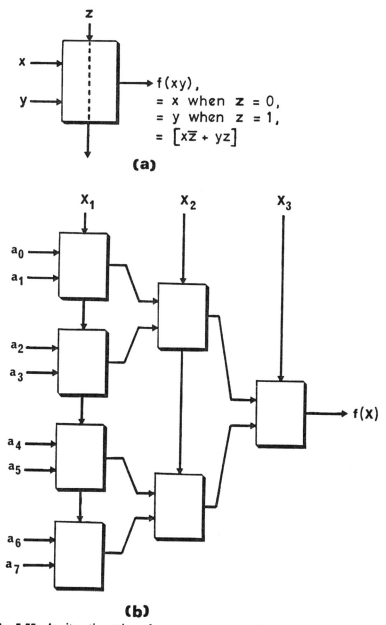

Fig. 5.55. An iterative triangular array: (*a*) the cell logic requirements; (*b*) a universal $n = 3$ array;

Table 5.8. A comparison of the various arrays for single-output universal n-variable function realization

Type of array	Absolute max. number of cells	No. of input connections including variable 0, 1 connections	Complements of x_i required	Max. no. of cells in series	Iterative or dissimilar cells	Fixed interconnection pattern
Cutpoint arrays, Fig. 5.48	$(n + 1)2^{n-1}$ *	$n + 2^{n-1}$ *	No	$n + 2^{n-1}$	Dissimilar	Yes
Sum-of-minterms array, Fig. 5.49	$(n + 1)2^n$	n	No	$n + 2^n$	Main array iterative, OR lowest row	Not to lowest row
Programmable sum-of-minterms array, Fig. 5.50	$(n + 1)2^n$	n, excluding the possible 2^{n+1} cell programming inputs	No	$n + 2^n$	Main array iterative, OR lowest row	Yes
Programmable Reed-Muller array, Fig. 5.51	$(n + 1)(2^n - 1)$	n, excluding the possible $n(2^n - 1)$ cell programming inputs	No	$n + (2^n - 1)$	Main array iterative, X-OR lowest row	Yes
Akers programmable array, Fig. 5.52	$(2^{n-2} + 1)^2$	$(2^{n-2} + 1)^2$	Yes	$(2^{n-1} + 1)$	Iterative	Yes
Triangular array, Fig. 5.55**	$(2^n - 1)$	$n + 2^n$	No	n	Iterative	Yes

* See (58).
** Universal $(n + 1)$ variable capability (see later in text).

Design Using Spectral and Other Techniques

Secondly let us consider PLA's. From Fig. 5.37 it will be recalled that the PLA is a sum-of-products generator, with multi-output capability within the capacity of its product generator. Hence it realizes in an eminently practical way the objectives of the output cellular array concepts of Matra, Minnick, and others, without the theoretical difficulties of the general cellular array approach.

Thirdly let us consider multiplexers. The similarity of the triangular array concept of Fig. 5.55 to the multiplexer development of Fig. 5.43 etc. should be obvious. However, in Fig. 5.55 we multiplexed 0, 1 signals only to the output $f(x)$ under the control of the input variables, but in the earlier work we multiplexed $\{0, 1, x_i, \bar{x}_i\}$ inputs to the output. This additional capability is of course readily adopted with the array of Fig. 5.55, thus giving this array $(n+1)$-variable universal-function capability. Hence there is no functional difference whatsoever between a multiplexer package used as a universal function generator and this iterative triangular array.

This leaves us with two rectangular categories in Table 5.8 which do not at present appear to have production l.s.i. counterparts, namely the Reed–Muller exclusive-OR type of array, and the Akers programmable array. If we consider both of these, there is no apparent attraction in the complex Akers array in comparison with PROM, PLA, or multiplexer solutions; the rectangular Reed–Muller array likewise has no immediately obvious attraction, particularly as each final (output) row of this array requires exclusive-OR rather than simple inclusive OR summation. The general conclusion of all this general array theory, therefore, is that currently simple unilateral single-rail sum-of-minterms or sum-of-products rectangular arrays or multiplexers represent the best practical solutions. However, these in their turn to not necessarily represent the most efficient use of silicon area to realize specific output requirements or provide the highest speed capability. Bilateral and multi-rail cell interconnections raise still unsolved problems, but are potentially more powerful means of synthesis.

5.6.3. Other Cellular Arrays

Whilst we are still considering cellular arrays, let us briefly detour to refer to a particular area of application where arrays have received very great attention and have useful potentialities. This is not for universal application, as previously, but instead is specifically for binary arithmetic purposes.

It may be intuitively suggested that some repetitive combinatorial logic structure may be suitable for the asynchronous addition or other mathematical manipulation of binary numbers, and indeed this proves to be the case. A very large number of papers have been published in this area, and many ingenious iterative array structures proposed. Cell interconnections are often multi-rail and bilateral, although it is significant that no formal set-theoretic means are used for the logic design and cell requirements of the proposed structures. Instead cell duties and interconnections are intuitively built up from consideration of the arithmetic requirements, modeling the required shift, add, transfer, etc. bit requirements of the arithmetic process.

Amongst the published arrangements will be found iterative structures, usually rectangular, for binary multiplication, binary division, binary logarithm and antilogarithm generation, binary comparators, square and square-root generation, BCD multiplication, and others. Considerable literature in this area is available[65-71].

As a single example of the possible array complexity, the circuit shown in Fig. 5.56 is that of a floating-point multiplier with variable dynamic range[69]. The basic cell of this array contains a full-adder circuit, but with additional logic to shift the bit patterns to the left so as to loose least-significant digits, this shift left being controlled both by

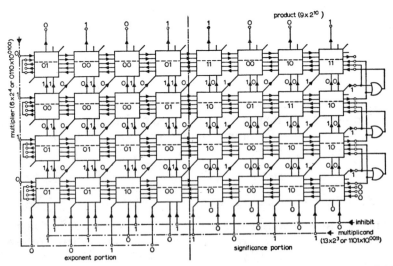

Fig. 5.56. Edwards's iterative floating-point multiplier array, with variable dynamic range.

exponent-length control inputs and digit overflow. Hence maintenance of the length of the output product to within a chosen number of digits is achieved, less significant digits being lost, with a record of exponent count being given by the remaining array outputs. In the example shown in Fig. 5.56 we have the inhibit lines set to give a dynamic range in the output product of four bits (0–15), with an exponent range also of four bits (2^0–2^{15}). The numbers being multiplied are as follows.

Multiplicand: $13 \times 2^3 = 104$ (denary)
                applied as 1101×0011

Multiplier: $6 \times 2^4 = 96$ (denary)
                applied as 0110×0100

giving the product output 9×2^{10}, read out as 1001×1010

Note that the possible error on this output product may lie in the range $+0, -(2^{10} - 1)$, the maximum undervalue in the output being when all 1's have been discarded in the less-significant digits.

This example typifies what may be achieved with iterative arrays for special-purpose applications. Further examples in yet other areas may be suggested[72,73]. Unfortunately, however, once we move into special-purpose areas global demand for such arrays is not great, and therefore the very high cost of translating such ideas into l.s.i. packages tends to discourage their manufacture and general marketing.

5.6.4. Other Universal Modules

In addition to the consideration of complete arrays which may be used for the realization of individual or multiple functions of n input variables, $n \geq 2$, there have been several disclosures for basic universal modules which can realize any combinatorial function of $n = 2$. These modules may in turn be built up into assemblies for the realization of any function of $n > 2$. The terminology "universal function generator," "universal logic circuit," or "universal logic module" may be found used for these basic configurations[74-78]. Here we shall use the latter term, which is usually abbreviated to "ULM."

The majority of proposed ULM's for $n = 2$ consist of a circuit configuration with more than two input terminals. Any function $f(x_1, x_2)$ is realized by appropriately connecting $0, 1, \dot{x}_1, \dot{x}_2$ to the inputs, where $\dot{x}_1, \dot{x}_2$ is x_1, x_2 true or complemented. The configurations are therefore hard-wired programmable circuits. What we shall here consider, however, is an algebraic approach to ULM's, and show that this approach will result in the same circuit configurations which have been previously proposed, often by intuitive means.

Consider any $n = 2$ function $f(x)$. This may be expanded into its standard sum-of-minterms form as

$$f(x_1,x_2) = [a_0\bar{x}_1\bar{x}_2 + a_1\bar{x}_1x_2 + a_2x_1\bar{x}_2 + a_3x_1x_2]$$

where a_i, $0 \leq i \leq 2^n - 1$, $\in \{0,1\}$. Factorizing we obtain

$$f(x_1,x_2) = [\bar{x}_2(a_0\bar{x}_1 + a_2x_1) + x_2(a_1\bar{x}_1 + a_3x_1)]. \qquad (5.5)$$

Considering all the possible combinations of a_0,a_2 and a_1,a_3 in each inner bracket, each may take the value $0, x_1, \bar{x}_1,$ or 1. Hence we may re-express equation (5.5) as

$$f(x_1,x_2) = [\bar{x}_2 p + x_2 q] \qquad (5.6)$$

where $p,q \in \{0,1,x_1,\bar{x}_1\}$. This final expression will once again be recognized as that of a simple multiplexer circuit or electromechanical change-over contact, which multiplexes its two data inputs p and q to the output under the control of its data-select input x_2.

However, as we are dealing with universal arrays we may complement and rearrange equation (5.6) in several alternative ways, for example

$$f(x_1,x_2) = [\overline{\bar{x}_2 p + x_2 q}] \qquad (5.7)$$

$$f(x_1,x_2) = [(\bar{x}_2 + p)(x_2 + q)] \qquad (5.8)$$

and

$$f(x_1,x_2) = [\overline{(\bar{x}_2 + p)(x_2 + q)}] \qquad (5.9)$$

Equations (5.6) to (5.9) therefore are all equations for $n = 2$ ULM's, although it must be appreciated that for any particular function $f(x)$ the p,q values may be dissimilar in the different realizations. Figure 5.57 illustrates these four basic possibilities. Notice also that p,q can be made $\{0,1,x_2,\bar{x}_2\}$ with x_1 as the data-select input, if desired, and that x_1,x_2 may be freely negated in all these expressions.

Extending the universal sum-of-minterms expansion to the $n = 3$ case, we obtain after factorizing

$$f(x_1,x_2,x_3) = [\bar{x}_3(a_0\bar{x}_1\bar{x}_2 + a_2\bar{x}_1x_2 + a_4x_1\bar{x}_2 + a_6x_1x_2)$$
$$+ x_3(a_1\bar{x}_1\bar{x}_2 + a_3\bar{x}_1x_2 + a_5x_1\bar{x}_2 + a_7x_1x_2)] \qquad (5.10)$$

whence it may readily be seen that each inner bracket is the previous general term $[\bar{x}_2 p + x_2 q]$. Hence we have now algebraically realized

Design Using Spectral and Other Techniques 387

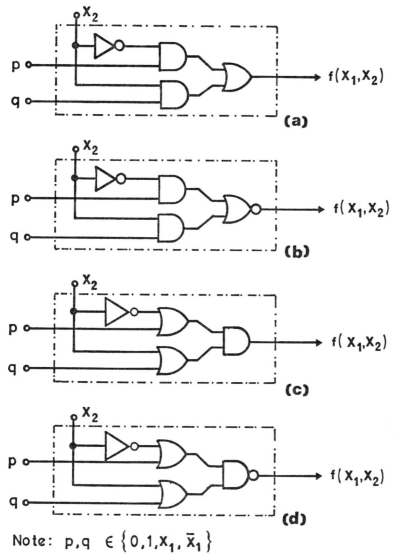

Note: $p, q \in \{0, 1, x_1, \bar{x}_1\}$

Fig. 5.57. $n = 2$ ULM's derived from the sum-of-minterms expansion of $f(x)$: (a) equation (5.6); (b) equation (5.7); (c) equation (5.8); (d) equation (5.9).

the triangular-array multiplexer circuit originally illustrated in Fig. 5.55, with the four circuits of Fig. 5.57 as possible ULM's for this array.

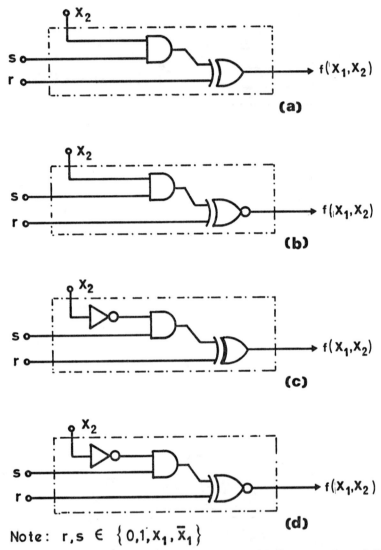

Note: $r, s \in \{0, 1, x_1, \bar{x}_1\}$

Fig. 5.58. $n = 2$ ULM's derived from the Reed–Muller expansion of $f(x)$: (a) equation (5.12) (b) equation (5.13); (c) equation (5.14); (d) equation (5.15).

In place of the previous sum-of-minterms expansion for any function $f(x)$, we may choose a Reed–Muller exclusive-OR expansion. For $n = 2$ we have

$$f(x_1, x_2) = [a_0 \oplus a_1 x_1 \oplus a_2 x_2 \oplus a_3 x_1 x_2] \tag{5.11}$$

where $a_i, 0 \leq i \leq 2^n - 1, \in \{0,1\}$. Factorizing we obtain

$$f(x_1,x_2) = [(a_0 \oplus a_1x_1) \oplus x_2(a_2 \oplus a_3x_1)].$$

Considering as before all possible combinations of the a_i's, each bracket in this expansion may take the value.

$0 \oplus 0 = 0$

$0 \oplus x_1 = x_1$

$1 \oplus 0 = 1$

or

$1 \oplus x_1 = \bar{x}_1$

and therefore we may re-express the function as

$$f(x_1,x_2) = [r \oplus x_2 s] \tag{5.12}$$

where $r, s \in \{0, 1, x_1, \bar{x}_1\}$. The circuit for this alternative $n = 2$ universal logic module, which differs from those of Fig. 5.57, is shown in Fig. 5.58a. This turns out to be the ULM configuration recently proposed by Murugesan[78], a possible circuit arrangement for which will be mentioned in Chapter 7.

Again, for universal purposes we may propose variations upon equation (5.12) without impairing its universality. Simply negating equation (5.12) we obtain the exclusive-NOR variant

$$f(x_1,x_2) = [r \overline{\oplus} x_2 s]. \tag{5.13}$$

whilst if the original Reed–Muller expansion of equation (5.11) had been expressed with either or both of its x_i's complemented we may obtain further variants, such as

$$f(x_1,x_2) = [r \oplus \bar{x}_2 s] \tag{5.14}$$

and

$$f(x_1,x_2) = [r \overline{\oplus} \bar{x}_2 s]. \tag{5.15}$$

Equations (5.13), (5.14), and (5.15) are illustrated in Figs. 5.58b–d.

Again x_1, x_2 may be freely interchanged without impairing the module universality.

Extending the universal Reed–Muller expansion to the $n = 3$ case we obtain after factorizing:

$$f(x_1,x_2,x_3) = [\{(a_0 \oplus a_1x_1) \oplus x_2(a_2 \oplus a_3x_1)\} \\ \oplus x_3\{(a_4 \oplus a_5x_1) \oplus x_2(a_6 \oplus a_7x_1)\}] \tag{5.16}$$

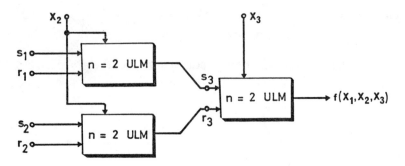

Fig. 5.59. The $n = 3$ triangular array of equation (5.16)

whence the universal realization using $n = 2$ ULM's as shown in Fig. 5.59 follows. Clearly there is a triangular array structure for any n similar to that shown in Fig. 5.55b.

If we map any of the functions expressed by equations (5.6) to (5.9) or (5.12) to (5.15) we shall find that they are all characterized by equal count true/false minterm patterns. Figure 5.60 illustrates this feature. Thus their Rademacher–Walsh spectrum has the feature that the zero-ordered coefficient r_0 is always zero. Further details of their

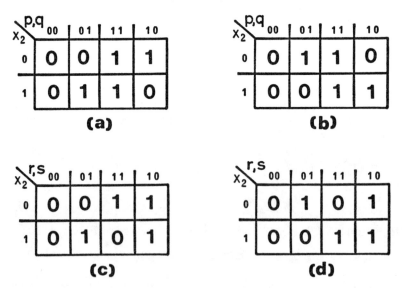

Fig. 5.60. Mapping of some of the $n = 2$ ULM proposals: (a) Fig. 5.57a; (b) Fig. 5.57c; (c) Fig. 5.58a; (d) Fig. 5.58c.

interrelationships may be found published[79]. Thus whilst a number of apparently different ULM's for $n = 2$ can be proposed, which may then be built up into universal arrays for $n \geq 3$, upon closer examination it will be found that all stem from the fundamental algebraic developments we have just covered. The ULM variants which require the complement of one input variable only to be made available, with no internal generation of any other complements, may possibly be the better from a circuit realization point of view.

5.7. Sequential Topics

It may be considered unusual that in this book relatively little mention has been made of sequential topics, and yet the majority of standard logic textbooks devote a high proportion of their pages to sequential design. As an explanation of this apparent discrepancy the following factors may be advanced for the readers' consideration.

(a) There have been few recent new developments in the theory of sequential logic design—conventionally developments tend to take place in the combinatorial area and at a later stage be extended if appropriate to the sequential field. Hence existing material on sequential logic design is admirably covered in many readily available sources[1-5,80,81] and elsewhere, this material being largely based upon the work of Huffman, Mealy, and Moore[82-84].

(b) Industrial sequential design requirements frequently have a rigorous structural requirement, such that a network realization follows directly without the need for sophisticated design techniques or algorithms. Indeed it is possible to overdesign using theoretical concepts such that the final network realization is more complex than some simple iterative realization.

(c) The advent of microelectronics has reduced the need for maximum theoretical minimization of the design; instead maximum simplification of network interconnection pattern or network testability may be of more practical significance.

To illustrate some of these latter views, and to indicate possible changes of emphasis in the sequential logic area, let us consider some fundamental points and simple problems in the paragraphs immediately following.

The classic representation of a sequential system has been illustrated in Fig. 4.33 of the previous chapter when dealing largely with threshold-logic realizations, and also in specific form in Fig. 5.35 of this

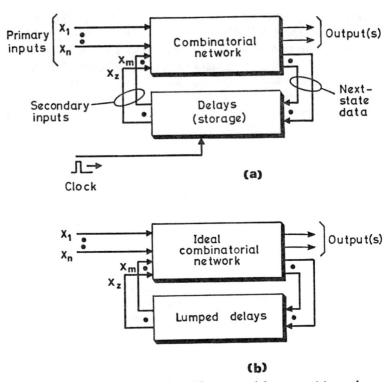

Fig. 5.61. The classic representations of a sequential system: (*a*) synchronous (clocked); (*b*) asynchronous (unclocked).

chapter. These basic concepts are repeated here in Fig. 5.61. Notice in a Mealy model that each output is considered as a function of the inputs $\{x_1,\ldots,x_n,x_m,\ldots,x_z\}$, but in a Moore model each output is considered to be a function of the inputs $\{x_m,\ldots,x_z\}$. In both cases, however, the next-state data is in general a function of the whole input set $\{x_1,\ldots,x_n,x_m,\ldots,x_z\}$.

In synchronous logic the change-over of the output(s) in response to changing input data is under the rigorous control of the clock signal, and hence provided the clock repetition rate is such that all circuit changes have settled down between clock pulses no critical logic races are encountered. Network design therefore is straightforward from this point of view. The penalty which arises with a synchronous realization, however, is that the input information is sampled at definite intervals of time by the clock pulses, and hence the overall response time of the network is governed by the clock repetition frequency. The absolute

Design Using Spectral and Other Techniques

minimum possible response time which the gates could provide therefore is not available. However, if the incoming data is itself say a serial stream of data under the control of a master clock, i.e. "synchronous data," then this is no hardship.

Consider a very trivial problem, that of recognizing an input sequence of 101 in a clocked serial stream of data. One of the simplest intuitive realizations for this problem is shown in Fig. 5.62a, this being particularly valid for monolithic circuit layout owing to its regular construction. A classic sequential design procedure for the same problem, however, would go through the standard steps of compiling state diagrams, state tables, reduced (minimized) state tables, and eventually to a realization based upon the final tables, and for this particular problem may yield a solution such as further illustrated in Fig. 5.62.[5] Notice what happens in designs based upon such classical approaches, namely the number of storage elements (bistable circuits) is normally reduced, but at the expense of additional combinatorial logic surrounding the storage elements. Hence minimization of state tables *does not guarantee in any way a minimum overall realization,* a feature which has been previously publicized[85] but not always academically appreciated. Finally from the maintainability and testing point of view a realization such as Fig. 5.62a may be preferable to that of Fig. 5.62d.

In asynchronous working the advantage gained is that of higher speed; the network does not have to await the arrival of a clock pulse to respond to any change of primary input data. Against this must be weighed the problem of ensuring that there are no critical race hazards in the network design which may give rise to incorrect network response. This is the greatest problem with asynchronous working, one which makes original equipment manufacturers prefer synchronous (clocked) systems whenever possible. A further disadvantage with asynchronous working is that it cannot be slowed down for fault-diagnosis purposes; a very powerful technique with most clocked systems is to replace the normal clock with a one-shot clock pulse, so that step-by-step operation of the system may be monitored.

However, where asynchronous operation is necessary, there again we may find that classic design techniques do not necessarily yield a final network realization which is preferable to alternative solutions. For example, consider again our previous problem of recognizing a 101

---

[5] Although there is no critical race condition present in this standard textbook solution, nevertheless a transitory spike ("glitch") may appear at the output on receipt of the first 1 input should the second bistable circuit set faster than the first bistable.

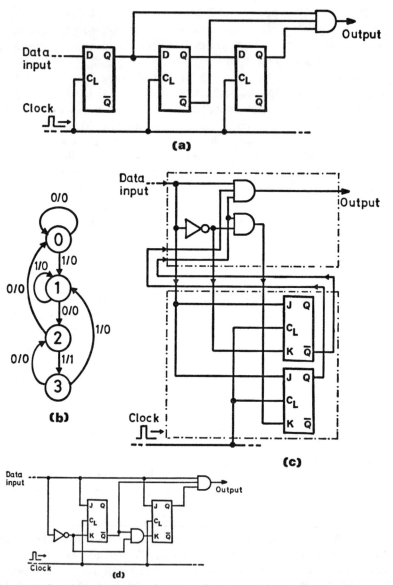

Fig. 5.62. Synchronous realizations for 101 sequence recognition, omitting any initial resetting requirements: (*a*) cascaded D type bistable realization; (*b*) the Mealy state diagram of the requirements; (*c*) a textbook realization based upon classical design theory drawn in the classic "two-box" representation; (*d*) a re-draw of (*c*). (Note: RS type bistables instead of JK are applicable in solutions (*c*) and (*d*).)

input sequence. Notice that in completely asynchronous working there is no such input situation as, say, 1101 or 1001, etc. as no time dependency is present. Hence what this problem now involves is the detection of the second 1 input signal. Thus a trivial asynchronous solution is as shown in Fig. 5.63a, which may be compared with a classically minimized hazard-free solution as in Fig. 5.63b.

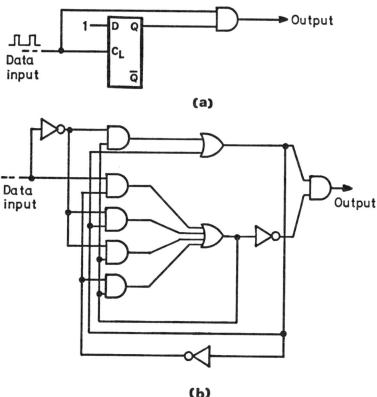

Fig. 5.63. Asynchronous realizations for 101 sequence recognition omitting any initial resetting requirements: (a) an asynchronous counter solution, assuming the D type bistable circuit triggers on the 1 to 0 trailing edge; (b) a textbook realization based upon classical design theory.

What we are capitalizing upon in the apparently simpler realizations of Figs. 5.62a and 5.63a is the availability of more complex modules, i.e. the standard D type bistable circuit in these cases, in our realization. However, when we consider more complex requirements than these trivial examples more formal design techniques are clearly necessary.

Currently there is no fully established theory to enable complex modules to be accommodated in sequential synthesis, or to specify what complex module(s) may be most profitably used. A strong trend, however, is towards the one-to-one translation of a directed control graph into a network realization, using one complex module per node of the graph *without any attempt at overall system minimization*.

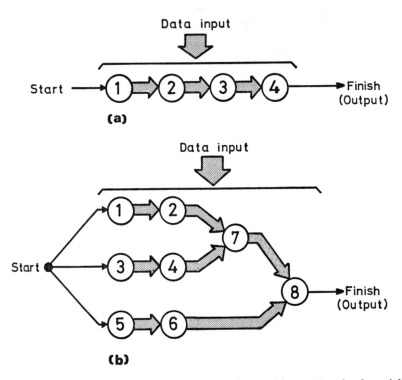

Fig. 5.64. Directed control graphs for sequential machines: (*a*) a simple serial machine; (*b*) a machine with parallel operating branches.

Figure 5.64*a* illustrates a directed control graph for a simple serial machine, the nodes in this diagram representing the sequence of separate events which are to be implemented. A more complex machine with parallel branches is illustrated in Fig. 5.64*b*, in which certain sequential operations can take place simultaneously in the parallel branches. In both diagrams each node represents one unique state in the machine sequence—the parallel-branch machine may be considered or even represent separate serial machines operating in parallel, their final outputs being combined as and when required.[6]

The serial control graph can be directly translated into a synchronous realization by employing clocked bistable modules at each node, provided the worst-case completion times for the operation of each nodal requirement are known. If parallel branches and/or feedback loops are present then the problem may become more difficult, and dummy stages have to be included to achieve synchronization between various branch signals. This technique is particularly successful for two-phase and four-phase custom-designed MOS circuits, where the logic action is in any case under the control of the system clocks ϕ_1, ϕ_2 or $\phi_1, \phi_2, \phi_3, \phi_4$. However, high-speed capability is not available.

Where highest possible speed of operation is essential, or where the completion time of the various nodal operations are unknown or vary widely, then asynchronous operation becomes necessary. However, now the problem of critical race hazards arises. Assuming that exact timing of signals and completion of operations cannot be calculated, as is often the case in a complex system, then we require logic modules which we can use as the control graph nodes to have the following basic control properties.

(i) On completion of its operation, each nodal module should initiate a "request transfer" signal to its successor(s) to say that it has completed its operation and has onward data awaiting.

(ii) If the receiving module is in an idle state, that is it is not busy on a previous operation or awaiting some other control input, it should accept this request input control signal and the revised input data.

(iii) On completion of its operation on the revised input data, the module should return an "accepted" signal to its predecessor(s) to release them for further activity.

This arrangement is shown schematically in Fig. 5.65. Variations on this basic concept may be proposed within this general framework of status control signals between the nodes. This concept is the familiar *request–acknowledge* or *handshake* technique which arises widely in many fields of science and engineering. In the digital logic area it has

---

[6] If the information coming into the machine is in completely serial form, as for example in a control system which has to await the completion of each operation before going on to the next, clearly this is represented by a serial control graph. On the other hand if parallel information is simultaneously available, then parallel operation with its faster overall response capability is available to the designer.

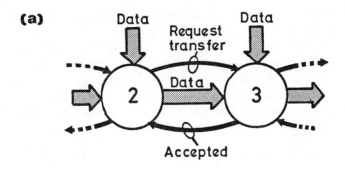
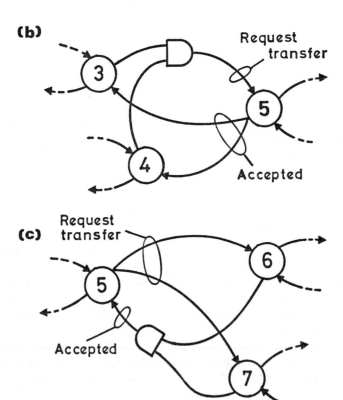

Fig. 5.65. The handshake control signals between nodes: (*a*) simple serial machine; (*b*) convergence of two parallel branches (data signals omitted for clarity); (*c*) divergence into two parallel branches (data signals omitted for clarity).

Design Using Spectral and Other Techniques

been pursued in several guises with (unfortunately) dissimilar terminologies being used by different authorities. Among the main published contributions in this area may be mentioned the Petri network concepts of Petri and others[44,86–89], the LOCOS digital system structure of Heath et al.[90–92], and other contributions in this area[85,93–96]. The fundamental feature of all approaches, however, is the separation of the data signal paths and the control signal paths in the overall system structure.

As far as possible circuits for asynchronously operating nodal modules are concerned, there have been several published suggestions[85,98–100]. Walker[85] suggests a single "node element" with somewhat restricted capability, but Howard[98] proposes a range of "control modules" suitable for both simple serial and more complex parallel control graph realizations. Development work in this area has perforce to be undertaken with s.s.i. and discrete-component assemblies, but for economic exploitation m.s.i. realizations are clearly necessary.

Much more work in the area of sequential machine design is necessary, not only from the point of view of establishing the best machine structure for more complex sequential requirements, but also from the latter point of view of establishing the most useful range of modules for the subsequent hardware realization. The concept of programmable modules for sequential applications, for example[62], is an almost completely unexplored field for the future.

5.8. Chapter Summary

The length of this chapter has reflected the considerable developments ongoing in the binary logic area, but still largely outside everyday use. Basically there are two trains of development, which influence each other, these being as follows: (i) newer methods of expressing and manipulating binary design data, for example, the spectral techniques and symmetry considerations; (ii) more complex logic modules which now become economically viable owing to the development of monolithic m.s.i. and l.s.i. technologies.

The emphasis of logic design is no longer primarily one of reduction of the total number of, say, AND or OR gates or bistable circuits in the system, this being a necessary requirement in discrete-component or discrete-gate days, but now the emphasis is more one of simplification of the resultant network topology and interconnection complexity. This is a much more difficult goal to quantify unless a known range of complex logic modules is available.

This change of emphasis will be strongly felt in the sequential design area, where the classic techniques involving state minimization which occupy a large part of many present-day logic textbooks will no longer remain of prime significance. Newer modular design methods and more powerful means of globally considering large system requirements are wanted, but once a system structure is established subsequent realization with appropriate complex modules should be straightforward.

Programmable devices have been introduced in this chapter. Without doubt the future of programmable devices is assured, so that future original equipment manufacturers may be able to specify the bare minimum of packages for their system realization, subsequently programming them for their particular requirements. Here again, however, we have an interesting problem of precisely how complex a programmable module and what range of such modules should be available—combinatorial, synchronous sequential, asynchronous sequential, or the whole gamut in one package for the users' delight?

REFERENCES

1. Lewin, D. *Logical Design of Switching Circuits*. Elsevier/North-Holland, New York; Nelson, London, 1974.

2. Friedman, A.D., and Menon, P.R. *Theory and Design of Switching Circuits*. Computer Science Press, Calif., 1975.

3. Bannister, B.R., and Whitehead, D.G. *Fundamentals of Digital Systems*. McGraw-Hill, Maidenhead, U.K., 1973.

4. Barna, A., and Porat, D.I. *Integrated Circuits in Digital Electronics*. John Wiley Interscience, New York, 1973.

5. Kostopoulos, G.K. *Digital Engineering*. John Wiley Interscience, New York, 1975.

6. Zissos, D. *Logic Design Algorithms*. Oxford University Press, Oxford, 1972.

7. Morris, R.K. and Miller, J.R. *Designing with TTL Integrated Circuits*. McGraw-Hill, New York, 1971.

8. Campeau, J.O. *The Synthesis and Analysis of Counters in Digital Systems by Boolean Matrices.* M.S. Thesis, University of California, June 1955.

9. Campeau, J.O. The synthesis and analysis of digital systems by Boolean matrices. *Trans. IRE* **EC.6,** 231–41, Feb. 1957.

10. Edwards, C.R. The logic of Boolean matrices. *Computer J.* **15** (3), 247–53, 1972.

11. Edwards, C.R. *Matrix Methods in Combinational Logic Design.* PhD. Thesis, University of Bath, U.K., 1973.

12. Page, E.W. Matrix methods for systematising sequential-curcuit design. *Electron. Lett.* **12,** 49–50, Jan. 1976.

13. Hurst, S.L. The application of Chow parameters and Rademacher–Walsh matrices to the synthesis of binary functions. *Computer J.* **16,** 165–73, May 1973.

14. Edwards, C.R. The application of the Rademacher–Walsh transform to Boolean classification and threshold-logic synthesis. *Trans. IEEE* **C24,** 48–62, Jan. 1975.

15. Dertouzos, M.L. *Threshold Logic Synthesis: A Synthesis Approach.* MIT Press, Cambridge, Mass., 1965.

16. Lechner, R.J. Harmonic analysis of switching functions. In *Recent Developments in Switching Theory* (Ed. A. Mukhopadhyay). Academic Press, New York, 1971.

17. Stone, H.S. Universal logic modules. In *Recent Developments in Switching Theory* (Ed. A. Mukhopadhyay). Academic Press, New York, 1971.

18. Hurst, S.L. Improvements for general-purpose s.s.i. logic packages and m.s.i./l.s.i. logic subsystems. *Electron. Lett.* **11,** 78–81, Feb. 1975.

19. Dietmeyer, D.L. *Logic Design of Digital Systems.* Allyn and Bacon, Boston, 1971.

20. McClusky, E.J. Detection of group invariance or total symmetry of a Boolean function. *Bell Syst. Tech. J.* **35,** 1445–53, Nov. 1956.

21. Marcus, M.P. The detection and identification of symmetric switching functions with the use of tables of combinations. *Trans. IRE* **EC5,** 237–9, Dec. 1956.

22. Mukhopadhyay, A. Detection of total or partial symmetry of a switching function with the use of decomposition charts. *Trans. IEEE* **EC12,** 553–7, Oct. 1963.

23. Das, S.R., and Sheng, C.L. On detecting total or partial symmetry of switching functions. *Trans. IEEE* **C20**, 352–5, March 1971.

● 24. Edwards, C.R. The design of easily-tested circuits using mapping and spectral techniques. *Radio Electron. Eng.* **47**, 321–42, July 1977.

25. Hurst, S.L., and Edwards, C.R. Preliminary considerations of the design of combinatorial and sequential digital systems under symmetry methods. *Int. J. Electron.* **40** (5), 499–507, 1976.

26. Hurst S.L. The detection of symmetries in combinatorial functions by spectral means. *IEE, Electron. Circuits and Systems.* **1**, Sept. 1977, 173–80.

27. Gibbs, J.E. and Ireland, B. Walsh functions and differentiations. *Proc. Symp. on the Applications of Walsh Functions, Washington, D.C., March 1974*, 147–76.

28. Gibbs, J.E. *Walsh Spectrometry: A Form of Spectral Analysis Well Suited to Binary Digital Computation.* Internal Rep. National Physical Lab., U.K., 1967.

● 29. Chang, H.Y., Manning, E.G., and Metze, G. *Fault Diagnosis in Digital Systems.* John Wiley Interscience, New York 1970.

● 30. Friedman, A.D. and Menon, P.R. *Fault Detection in Digital Circuits.* Prentice-Hall, Princeton, N.J., 1971.

● 31. Breuer, M.A., and Friedman, A.D. *Diagnosis and Reliable Design of Digital Systems.* Computer Science Press, Calif., 1975.

32. Susskind, A.K. Diagnostics for logic networks. *IEEE Spectrum* **10**, 40–7, Oct. 1973.

33. Seshu, S. On an improved diagnosis programme. *Trans. IRE* **EC14**, 69–76, Jan. 1965.

34. Roth, J.P. Diagnosis of automata failure: a calculus and a method. *IBM J. Res. Dev.*, **10**, 278–91, 1966.

35. Kautz, W.H. Fault testing and diagnosis in combinational digital circuits. *Trans. IEEE* **C17**, 352–66, April 1968.

36. Sellers, F.F., Hsiao, M.Y., and Bearnson, L.W. Analysing errors with the Boolean difference. *Trans. IEEE* **C17**, 676–83, July 1968.

37. Bennetts, R.G., Brittle, D.C., Prior, A.C., and Washington, J.L. A modular approach to test sequence generation for large digital networks. *Digital Process.* **1** (1), 3–24, 1975.

38. Prior, A.C., and Bennetts, R.G. The application of the Boolean difference technique to sequential logic. *Electron. Lett.* **10,** 486-8, Nov. 1974.

39. Bennetts, R.G. and Lewin, D.W. Fault diagnosis of digital systems—a review. *Computer J.* **14** (2), 199-206, 1971.

40. Bennetts, R.G., and Scott, R.V. Recent developments in the theory and practice of testable logic design. *IEEE Comput.* **9,** 47-62, June 1976.

41. Elphick, R. Deluge of l.s.i. circuits may cause logjams in system design. *Electron. Des.* **24** (1) 26-8, Jan. 1976

42. Green, R., and House, D. *Designing with Intel PROM'S and ROM'S.* Application Note AP-6, Intel Corp., Calif., April 1975.

43. Carr, W.N., and Mitz, J.P. *MOS/LSI Design and Application.* McGraw-Hill, New York, 1972.

44. Lewin, D. Outstanding problems in logic design. *Radio Electron. Eng.* **44,** 9-17, Jan. 1974.

45. Proiste, J.E. Replacing sequential logic with read-only memories. *Electron. Eng.* **48,** 35-8, Jan. 1976.

46. Blakeslee, T.R. *Digital Design with Standard MSI and LSI.* John Wiley Interscience, New York, 1975.

47. Uimari, D. PROM'S—a practical alternative to random logic. *Electron. Products* 75-91, Jan. 1974.

48. Cavlan, N. Field-PLA's simplify logic designs. *Electron. Des.* **23** (18) 84-90, Sept. 1975.

49. National Semiconductor Corporation *How to Design with Programmable Logic Arrays.* Application Note AN.89, National Semiconductor Corp., Calif., 1973.

50. Hemel, A. The PLA: a "different kind" of ROM. *Electron. Des.* **24** (1) 78-87, Jan. 1976.

51. Elspas, B. The theory of multirail cascades. In *Recent Developments in Switching Theory* (Ed. A. Mukhopadhay). Academic Press, New York, 1971.

52. Grundy, D.L., Bruchez, J., and Down, B. Collector diffusion isolation packs many functions on a chip. *Electronics* **45** (14), 96-104, July 1972.

53. Ferranti Ltd. *The Uncommitted Logic Array—A CDI Standard Product.* Publication ESB 620274, Ferranti Ltd., U.K., 1974.

54. McGlyn, D.A. Microprocessors: technology, architecture and application. John Wiley Interscience, New York, 1976.

55. Osborne, A. An introduction to microcomputers, Vols. 1 and 2, Adam Osborne and Associates, Berkeley, Calif., 1977.

56. Maitra, K.K. Cascaded networks of two-input flexible cells. *Trans. IRE* **EC11**, 136–43, April 1962.

57. Stone, H.S., and Korenjak, A.J. Canonical form and synthesis of cellular cascades. *Trans. IEEE* **EC14**, 852–62, Dec. 1965.

58. Mukhopadhay, A., and Stone, H.S. Cellular logic. In *Recent Developments in Switching Theory* (Ed. A. Mukhopadhay). Academic Press, New York, 1971.

59. Minnick, R.C., Short, R.A., Goldberg, J., Stone, H.S., Green M.W., Yeoli, M., et al. *Cellular Arrays for Logic and Storage*. Stanford Research Institute Final Rep. AFCRL Contract AF19(628)-4233, 1966.

60. Weiss, C.D. The characterization and properties of cascade-realizable switching functions. *Trans. IEEE* **C18**, 624–33, July 1969.

61. Papakonstantinou, G.K. A synthesis method for cutpoint cellular arrays. *Trans. IEEE* **C21**, 1286–92, Dec. 1972.

62. Kautz, W.H. Programmable cellular logic. In *Recent Developments in Switching Theory* (Ed. A. Mukhopadhay) Academic Press, New York, 1971.

63. Akers, S.B. A rectangular logic array. *Trans. IEEE* **C21**, 848–57, Aug. 1972.

64. Heutink, F. Implications of busing for cellular arrays. *Computer Des.* **13**, 95–100, Nov. 1974.

65. White, G. A cellular 8421 BCD multiplier. *Radio Electron. Eng.* **40**, 321–2, Dec. 1970.

66. Majithia, J.C., and Katai, R. An iterative array for multiplication of signed binary numbers. *Trans. IEEE* **C20**, 214–6, Feb. 1971.

67. Deverell, J. Multiplication of complex numbers using iterative arrays. *IEE Electron. Lett.* **7**, 205–7, May 1971.

68. Guild, H.H. Fast versatile binary comparator array. *IEE Electron. Lett.* **7**, 225–6, May 1971.

69. Edwards, C.R. Floating-point cellular-logic multiplier with variable dynamic range. *IEE Electron. Lett.* **7**, 747–9, Dec. 1971.

70. Majithia, J.C. Cellular array for extraction of squares and square roots of binary numbers. *Trans. IEEE* **C21**, 1023–4, Sept. 1972.
71. Deverell, J. The design of cellular arrays for arithmetic. *Radio Electron. Eng.* **44**, 21–6, Jan. 1974.
72. Bandyopadhyay, S., Basu, S., and Choudhury, A.K. A cellular permuter array. *Trans. IEEE* **C21**, 1116–9, Oct. 1972.
73. Dean, K.J. Non-arithmetic cellular arrays. *Proc. IEE,* **119**, 785–9, July 1972.
74. Yau, S.S. and Tang, C.K. Universal logic modules and their application. *Trans. IEEE* **C19**, 141–9, Feb. 1970.
75. Sobocinski, B. On a universal decision element. *J. Computer Syst.* **1**, 71–5, 1953.
76. Butter, J.T., and Breeding, K.J. Some characteristics of universal cell networks, *Trans. IEEE* **C22**, 897–903, Oct. 1973.
77. Muzio, J.C. Particular universal function generator. *IEE Electron. Lett.* **11**, 429, Sept. 1975.
78. Murugesan, S. Universal logic gate and its applications. *Int. J. Electron.* **42** (1), 55–63, 1977.
79. Edwards, C.R., and Hurst, S.L. An analysis of universal logic modules. *Int. J. Electron.* **41** (6), 625–28, 1976.
80. Unger, S.H. *Asynchronous Sequential Switching Circuits.* John Wiley Interscience, New York, 1969.
81. Caldwell, S.H. *Switching Circuits and Logical Design.* John Wiley, New York, 1958.
82. Huffman, D.A. The synthesis of sequential switching circuits. *J. Franklin Inst.* **257**, 161–90, March 1954; 275–303, April 1954.
83. Mealy, G.H. A method of synthesizing sequential circuits. *B.S.T.J.* **34**, 1045–79, 1955.
84. Moore, E.F. Gedanken-experiments on sequential machines. *Ann. Math. Stud. Princeton Univ.* **34**, 129–53, 1955.
85. Walker, B.S. The design of sequential circuits. *Radio Electron. Eng.* **44**, 45–9, Jan. 1974.
86. Petri, C.A. *Kommunication mit Automaten.* Ph.D. Thesis, University of Bonn, 1963.

87. Patel, S.S., and Dennis, J.B. The description and realization of digital systems, *Proc. IEEE 6th Annual Computer Society Int. Conf.*, 1972, pp. 223–6.

88. Misunas, D. Petri nets and speed-independent design. *A.C.M. Commun.* **16,** 474–81, 1973.

89. Keller, R.M. Towards a theory of universal speed-independent modules. *Trans. IEEE* **C23,** 21–33, Jan. 1974.

90. Heath, F.G. The LOCOS system. *IEE Conf. on Computer Aided Design,* pp. 225–30. IEE Conf. Publ. No. 86, 1972.

91. Rose, C.W., and Bradshaw, F.T. The LOCOS representation system. *Proc. IEEE 6th Annual Computer Society Int. Conf., Sept. 1972,* 187–90.

• 92. Heath, F.G., and Howard, B.V. Asynchronous control modules for directed graph realization of parallel digital machines. *IEE Conf. on Computer Systems Technology,* 130–7. IEE Publ. No. 121, 1974.

• 93. Karp, R.M., and Miller, R.E. Parallel program schemata. *J. Computer Syst. Sci.* **3,** 147–95, 1969.

• 94. Miller, R.E. A comparison of some theoretical models of parallel computation. *Trans. IEEE,* **C22,** 710–7, 1973.

95. Slutz, D.R. Flow graph schemata. *Project MAC Conf. on Concurrent Systems and Parallel Computation, ACM Records,* 129–41, 1970.

96. *Project MAC Conf. on Concurrent Systems and Parallel Computation, ACM Records,* 183–99, 1970.

97. Peatman, J.B. *The Design of Digital Systems.* McGraw-Hill, New York, 1972.

98. Howard, B.V. Parallel computation schemata and their hardware implementation. *Digital Process.* **1,** 183–206, 1975.

99. Perry, P.J. A high-voltage industrial logic family. *Tech. digest.* Plessey (U.K.) Ltd., 1976.

100. Bell, C.G., Eggert, J.L., Granson, J., and Williams, P. The description and use of register transfer modules (RTMS). *Trans. IEEE* **C21,** 495–500, May 1972.

Chapter 6

The Design of Ternary Logic Networks

Introduction

In the practical design of any digital system, whether binary or higher ordered, an early decision has to be made concerning the types of logic gate or circuit configurations which will be used in the final equipment. In the normal binary case we may make the decision to restrict ourselves to NAND gates only for the combinatorial requirements, or alternatively to allow more complex gates such as we have been considering in previous chapters of this book. Clearly the "best" system realization may differ markedly depending upon such initial constraints.

In the design of ternary logic networks the same considerations arise, but with even greater impact. We have already seen in Chapter 1 that there are several dissimilar algebras which may be used to define a ternary combinatorial network, each of which imply a particular type of gate or range of gates for a practical implementation of the chosen algebra. Thus in this ternary case even more so than in the binary case we must decide what gates we shall be finally using, as this will dictate the appropriate algebra or relevant design procedure.

However, the whole question of ternary system design is still largely in the theoretical arena, owing to the non-availability of any commercial solid-state ternary building blocks. In this respect it is interesting to note that logic designers had greater scope in pre-solid-state days, as three-position relays were commercially available and were used in industrial control applications. Very attractive circuit realizations could be produced with such relays in many practical situations. However, now we possess more theoretical know-how, but no current state-of-the-art hardware for its commercial realization!

Nevertheless it is instructive in a book of this nature, which is ranging fairly widely over ideas, concepts, and principles, to review what information we do have which is relevant to ternary network or system design; such information will stand us in good stead irrespective of the direction of any subsequent commercial development of ternary circuits.

6.1. Ternary Truth Tables and Algebraic Minimization

Chapter 1 has already introduced truth tables for ternary functions. It will be recalled that the truth values for a ternary system may be denoted by, say, 0, 1, and 2, or by -1, 0, and $+1$, or indeed by any other choice of three symbols. In the following pages we shall generally adopt the usual choice of 0, 1, and 2 to represent our three signal values. Precisely what 0, 1, and 2 may be in any circuit realization will not be our present concern; this will form a consideration in the following chapter, where we shall be looking at possible circuit configurations of non-standard logic gates.

In the pages immediately following we shall not confine ourselves to any specific algebra which may be associated with one proposed family of ternary logic gates. Instead we shall concern ourselves with general expressions and truth tables. Finally, the term "minterm" will be used as in the normal binary case to denote a fundamental logical AND input combination containing all the input variables of the given system.

Let us start by considering a very simple two-variable ternary combinatorial function $f(X)$, whose truth table is as follows.

Inputs		Output
X_1	X_2	$f(X)$
0	0	2
0	1	0
0	2	0
1	0	2
1	1	1
1	2	1
2	0	2
2	1	0
2	2	0

The information content of this table may also be expressed by the three following statements:

The Design of Ternary Logic Networks

(1) output $f(X) = 0$ when $[(X_1 = 0$ AND $X_2 = 1)$
  OR $(X_1 = 0$ AND $X_2 = 2)$
  OR $(X_1 = 2$ AND $X_2 = 1)$
  OR $(X_1 = 2$ AND $X_2 = 2)]$

(2) output $f(X) = 1$ when $[(X_1 = 1$ AND $X_2 = 1)$
  OR $(X_1 = 1$ AND $X_2 = 2)]$

(3) output $f(X) = 2$ when $[(X_1 = 0$ AND $X_2 = 0)$
  OR $(X_1 = 1$ AND $X_2 = 0)$
  OR $(X_1 = 2$ AND $X_2 = 0)]$.

If we represent the juxtaposition of terms to represent the logical AND operation and + to represent the logical OR operation, then we may express the above three clauses by the more compact expressions following where the symbol $=_0$ means "equals 0 when," $=_1$ means "equals 1 when," and $=_2$ means "equals 2 when":

(1) $f(X) =_0 [X_1^0 X_2^1 + X_1^0 X_2^2 + X_1^2 X_2^1 + X_1^2 X_2^2]$

(2) $f(X) =_1 [X_1^1 X_2^1 + X_1^1 X_2^2]$

(3) $f(X) =_2 [X_1^0 X_2^0 + X_1^1 X_2^0 + X_1^2 X_2^0]$.

The reader must be careful to appreciate that these are not mathematical equations with numerical equality always holding on each side of the equals sign. Instead each individual expression only summarizes when $f(X) = 0$, 1, or 2, respectively, and gives no precise information about the value of $f(X)$ when its right-hand conditions are absent[1].

Now in a binary network, where the output $f(x)$ is either 0 or 1, it is sufficient to define the input conditions which are required to give $f(x) = 1$ (or 0); when these input conditions are not present then the output $f(x)$ must take the alternative value. In the ternary case, however, two of the three groups of input conditions are necessary in order to define fully the function output value $f(X) = 0$, 1, or 2; in the absence of the two chosen then the third output value of $f(X)$ must be present. However, when minimization of the minterms into a more compact prime-implicant form is sought, which in the binary case involves the classic procedures of mapping, or Quine–McClusky, or similar procedures[1-5], there arises the question which is the "best" set

---

[1]The reader must also be careful to distinguish between logical AND and "the minimum value of," and between logical OR and "the maximum value of," these being synonymous in the binary case but not in higher-valued situations. In this ternary field we shall use the previously defined symbols of v for "the maximum value of" and & for "the minimum value of" when required later on in this chapter.

of minterms to minimize. It will be appreciated that in the binary case the $f(x) = 0$ set of minterms may for certain functions minimize better than the $f(x) = 1$ minterms; a similar situation may arise in the ternary case, as we shall illustrate.

Consider the same ternary function $f(X)$ which we have already tabulated. If we plot this function on a two-variable ternary Karnaugh map, as originally introduced in Chapter 3, the resultant plot is as shown in Fig. 6.1a. Minimization using this ternary Karnaugh map involves searches for three adjacent same output-value minterms. In this case it is obvious that $f(X) = 2$ is produced by the input signal X_2^0, that is the minterm expression (3) above minimizes to

$$f(X) =_2 [X_2^0].$$

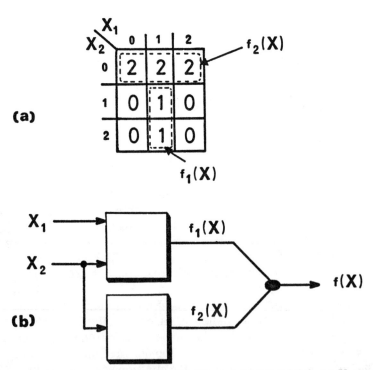

Fig. 6.1. A simple ternary function $f(X_1, X_2) = 2,0,0,2,1,1,2,0,0$: (a) Karnaugh map plot of $f(X)$; (b) possible schematic realization, where $f_1(X)$ is a network realizing the $f(X) = 1$ output requirements and $f_2(X)$ is a network realizing the $f(X) = 2$ output requirements, these two being appropriately combined to give $f(X)$.

The Design of Ternary Logic Networks 411

There is no other immediate minimization of the minterms of either of the other two output levels; the expression for $f(X) = 1$, however, contains only two minterms compared with the four minterms in $f(X) = 0$, and therefore would be preferable should a 2-out-of-3 choice be relevant for the final system realization. This is illustrated in Fig. 6.1b, without so far considering precisely how we combine the two signals $f_1(X)$ and $f_2(X)$ to give the overall function $f(X)$.

The simple Karnaugh map of Fig. 6.1a is ideal for the trivial case of only two ternary input variables ($n = 2$). For three ternary input variables ($n = 3$) the map construction begins to be inconvenient, as we have examined in Chapter 3, and a preferable geometric construction for simple algebraic minimization purposes will be found to be the $n = 3$ hypercube construction. Consider, for example, the three-variable function following. For ease of presentation we shall abbreviate our previous minterm designations so as to drop the X_i, leav-

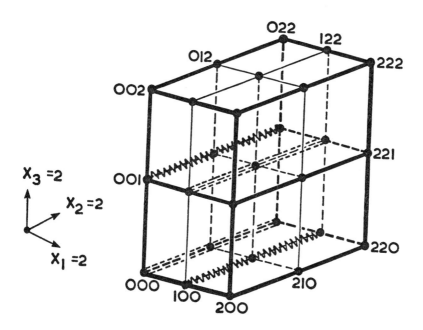

Fig. 6.2. The hypercube plot of the given $n = 3$ ternary function: ∿∿∿ $f(X) = 1$ minterms; ==== $f(X) = 2$ minterms; $f(X) = 0$ minterms are left unmarked. (Note: considerably greater clarity in these constructions can be achieved by the use of key color-coding on the hypercube.)

ing only the truth-value designation to be written, i.e. for $X_1^0 X_2^1 X_3^1$ we shall merely write 011, for $X_1^2 X_2^0 X_3^2$ we shall write 202, and so on, noting that the order of writing the three truth values is always X_1 followed by X_2 followed by X_3.

Example $n = 3$ ternary function:

$$f(X) =_1 [100 + 011 + 110 + 021 + 120 + 001]$$

$$f(X) =_2 [000 + 020 + 101 + 121 + 010 + 111]$$

$$f(X) =_0 [\text{all other minterms}].$$

This function is shown plotted on the $n = 3$ hypercube construction in Fig. 6.2. Notice that in this diagram we have the following features: (*a*) any straight line of three points represents a prime-implicant term of two input variables only, and (*b*) any complete surface of the hypercube represents a prime-implicant term of one input variable only. A Hamming distance of 1 also exists between the corresponding nodes on opposite faces of the hypercube, but this is of no particular significance in the minimization procedures.

From Fig. 6.2 it may be seen that the minterms which collectively give output $f(X) = 1$ minimize to two straight-line prime implicants, giving

$$f(X) =_1 [1\text{--}0 + 0\text{--}1]$$
$$=_1 [X_1^1 X_3^0 + X_1^0 X_3^1]$$

whilst the minterms which give output $f(X) = 2$ likewise minimize to

$$f(X) =_2 [0\text{--}0 + 1\text{--}1]$$
$$=_2 [X_1^0 X_3^0 + X_1^1 X_3^1].$$

However, when we examine the $f(X) = 0$ minterms, we see that they precisely occupy two faces of the hypercube, and hence we have that $f(X) = 0$ minimizes to

$$f(X) =_0 [2\text{--}\text{--} + \text{--}\text{--}2]$$
$$=_0 [X_1^2 + X_3^2].$$

Clearly it is worth considering the $f(X) = 0$ term as a part of the final realization if the type of gates available allows this to be done.

In the same way that the Karnaugh map becomes decreasingly useful above $n = 4$ in the binary case, so the hypercube and other geometric constructions become decreasingly useful above $n = 3$ in

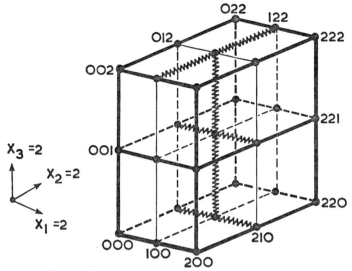

Fig. 6.3. The hypercube plot of the given $f(X) = 2$ minterms which minimize to four "lines", the 11– line, however, being redundant.

this ternary area. Above $n = 3$ we must resort to algebraic means to reduce ternary truth tables or other given forms of data to a minimum algebraic form. However, before we leave the hypercube which is ideal for the $n = 3$ case, let us use it to illustrate a well-known feature which occurs in binary minimization and which arises also in higher-radix working. This is the feature of redundant and irredundant prime-implicant terms[2-5].

Consider the case where a function output was, say,

$$f(X) =_2 [010 + 011 + 110 + 102 + 122 + 111 + 112 + 210 + 211].$$

These minterms will be found to minimize to the four prime-implicant terms

$$f(X) =_2 [-11 + -10 + 1-2 + 11-],$$
$$=_2 [X_2^1 X_3^1 + X_2^1 X_3^0 + X_1^1 X_3^2 + X_1^1 X_2^1].$$

However, when we examine the hypercube plot of this information, as shown in Fig. 6.3, it will be seen that three of these four prime-implicant terms are "essential" or "irredundant," that is none of them can be discarded without losing one or more of the required minterms, but one prime implicant is redundant, as its component minterms are jointly covered by the other three prime implicants. Hence we may discard this term, giving the minimum irredundant result of

$$f(X) =_2 [-11 + -10 + 1-2]$$
$$=_2 [X_2^1 X_3^1 + X_2^1 X_3^0 + X_1^1 X_3^2].$$

This of course is exactly the same as occurs in binary minimization, and which requires the Quine–McClusky or equivalent procedure to sort out the redundant terms from the irredundant ones.

A formal algebraic technique, directly analogous to the binary Quine–McClusky technique can be formulated for the ternary case. In the binary situation it is often only applied to, say, the $f(x) = 1$ minterms, although strictly speaking both the $f(x) = 0$ and the $f(x) = 1$ minterms should be minimized and compared, but in the ternary case at least two of the three sets of minterms must be minimized, say the $f(X) = 1$ set and the $f(X) = 2$ set. The technique is best illustrated by a numerical example before any formal procedure is stated.

Consider a four-variable function and consider, say, the $f(X) = 2$ output state:

$$f(X) =_2 [2\text{--}11 + -0\text{--}2 + 102\text{--} + 10\text{--}0 + 1001 + 1\text{--}11$$
$$+ -2\text{--}1 + 11\text{--}1].$$

Let us proceed as follows.

(a) Expanding these terms into minterm form, we have

2–11	+ –0–2	+ 102–	+ 10–0	+ 1001	+ 1–11	+ –2–1	+ 11–1
2011	0002	1020	1000	1001	1011	0201	1101
2111	1002	1021	1010		1111	1201	1111
2211	2002	1022	1020		1211	2201	1121
	0012					0211	
	1012					1211	
	2012					2211	
	0022					0221	
	1022					1221	
	2022					2221	

(b) Add together the value of each minterm and then tabulate in "total value" order.

The Design of Ternary Logic Networks

Total value	0	1	2	3	4	5	6	7	8
Minterm population	—	1000	1010 0002 1001	1020 1002 0012 1020* 1011 0201 1101	2002 1012 1021 1111 1201 1111* 0022 0211 2011	2012 1022 1211 2201 1121 1022* 1211* 0221 2111	2022 2211 1221 2211	2221	—

(c) Delete all duplications in this tabulation (those indicated by an asterisk in this example).

(d) Compare each minterm in the tabulation with all other minterms in the *next two* columns, looking for any variable which takes all three values 0, 1, and 2, the remaining $n - 1$ variables being identical. This gives us the following.

```
1000   1000   0002   0002   1010
1001   1010   1002   0012   1011
1002   1020   2002   0022   1012
100–   10–0   –002   00–2   101–

1001   1001   1020   1002   0012
1101   1011   1021   1012   1012
1201   1021   1022   1022   2012
1–01   10–1   102–   10–2   –012

1011   0201   0201   1101   2002
1111   0211   1201   1111   2012
1211   0221   2201   1121   2022
1–11   02–1   –201   11–1   20–2

1021   1201   0022   0211   2201
1121   1211   1022   1211   2211
1221   1221   2022   2211   2221
1–21   12–1   –022   –211   22–1

0221   2011
1221   2111
2221   2211
–221   2–11
```

(e) Print out any minterms not combining to form a triplet of terms, i.e. essential minterms in the final minimization, equal to NONE in this particular example.

(f) Tabulate the resulting terms from (d) in – order.

xxx–	xx–x	x–xx	–xxx
100–	10–0	1–01	–002
101–	00–2	1–11	–012
102–	10–1	1–21	–201
	10–2	2–11	–022
	02–1		–211
	11–1		–221
	20–2		
	12–1		
	22–1		

(g) Delete any redundancies (none in this example), and then compare each term in each column with all other terms in the *same* column, looking for complete variable values. This gives us the following.

```
100–    10–0    00–2    10–1
101–    10–1    10–2    11–1
102–    10–2    20–2    12–1
10--    10--    –0–2    1--1

02–1    1–01    –002    –201
12–1    1–11    –012    –211
22–1    1–21    –022    –221
–2–1    1--1    –0–2    –2–1
```

(h) Print out any prime implicants from table (f) which do not combine in (g) i.e. term 2–11 only in this particular example.

(i) Tabulate the resulting terms from (g) in appropriate columns.

xx--	x–x–	–xx–	–x–x	x--x	--xx
10--			–0–2	1--1	
			–2–1		

(j) Repeat procedure (g). In this particular example no further triplet terms can be formed from this last table. This therefore is the last stage of minimization, giving the result

$$f(X) =_2 [2\text{–}11 + 10\text{--} + \text{–}0\text{–}2 + \text{–}2\text{–}1 + 1\text{--}1]$$
$$=_2 [X_1^2 X_3^1 X_4^1 + X_1^1 X_2^0 + X_2^0 X_4^2 + X_2^2 X_4^1 + X_1^1 X_4^1].$$

The Design of Ternary Logic Networks

(k) Should *four or more* prime implicants be printed out at any stage of minimization after the first minterm stage, a search must be made to check whether one (or more) prime implicant is redundant (cf. Fig. 6.3). Any prime implicant found to be redundant should be discarded, leaving only the irredundant prime implicants in the final solution. (This situation does not arise in this example we have just worked through.)

The formal flow diagram for this procedure is as detailed in Table 6.1. Computer programming of this procedure is entirely straightforward, although, like the binary Quine–McClusky procedure, a large volume of intermediate information is generated during the procedure before minimization down to the final prime-implicant terms occurs.

Table 6.1. The flow chart to produce a minimized algebraic expression for any ternary output $f(X)$

1. Establish and list all minterms of the function output $f(X) = 0$ (or 1, or 2) in numerical form, e.g. 20101, etc. ↓
2. Look for and identify all triplets of minterms which contain a variable value 0, 1, and 2, all remaining $n - 1$ variables being identical, e.g. 20101, 21101, and 22101, = 2–101 ↓ → Print out all minterms not combining to form triplets, i.e. irredundant minterms
3. Look for and identify all triplets of prime implicants which contain a variable value 0, 1, and 2, all remaining $n - 1$ variables being identical, e.g. 2–10– ↓ → Compare all prime implicants not combining to form triplets for redundancy; delete any redundant and print out irredundant remainder
4. Repeat this search for triplets of prime implicants up to a maximum of n times → Compare and delete any redundant terms at each stage, printing out all irredundant terms ↓ Final minimized solution is the sum of all the printed-out irredundant terms

This procedure should be applied for $f(X) = 0, f(X) = 1$, and $f(X) = 2$.

Further details relating to the elimination of redundant prime implicants produced during such a procedure may be found published[6]. It may also be noted that this procedure may be generalized to the minimization of any R-valued system, $R \geq 2$, of which the ternary case just considered is merely one particular case[7].

So far these ternary procedures, like the usual binary ones, are based upon a "sum-of-products" starting point, that is the logical OR of the input AND minterms. If the given data was in a "product-of-sums" form, that is the logical AND of input OR maxterms, then multiplication of this starting point into the more amenable "sum-of-products" form can be undertaken.

For example, supposing we were given that a function output was

$$f(X) = 2 \quad \text{when} \quad [(X_1 = 1 \text{ OR } X_2 = 1 \text{ OR } X_3 = 0)$$
$$\text{AND } (X_1 = 2 \text{ OR } X_2 = 1 \text{ OR } X_3 = 0)$$
$$\text{AND } (X_1 = 0 \text{ OR } X_2 = 2 \text{ OR } X_3 = 2)]$$

Expressing this in more compact form and then multiplying out the right-hand side we obtain the following:

$$f(X) =_2 [(X_1^1 + X_2^1 + X_3^0)(X_1^2 + X_2^1 + X_3^0)(X_1^0 + X_2^2 + X_3^2)]$$
$$=_2 [(X_1^1 X_1^2 + X_1^1 X_2^1 + X_1^1 X_3^0 + X_2^1 X_1^2 + X_2^1 X_2^1 + X_2^1 X_3^0$$
$$+ X_3^0 X_1^2 + X_3^0 X_2^1 + X_3^0 X_3^0)(X_1^0 + X_2^2 + X_3^2)]$$

which may be simplified to[2]:

$$[(X_1^1 X_2^1 + X_1^1 X_3^0 + X_1^2 X_2^1 + X_2^1 + X_2^1 X_3^0 + X_1^2 X_3^0 + X_3^0)(X_1^0 + X_2^2 + X_3^2)].$$

Multiplying again, and immediately discarding any inadmissible terms such as $X_1^0 X_1^1 X_2^1$ and writing terms such as $X_3^0 X_2^2 X_2^2$ as $X_2^2 X_3^0$, we have

$$f(X) =_2 [X_1^0 X_2^1 + X_1^0 X_2^1 X_3^0 + X_1^0 X_2^1 X_3^0 + X_1^0 X_3^0 + X_1^1 X_2^2 X_3^0$$
$$+ X_1^2 X_2^2 X_3^0 + X_2^2 X_3^0 + X_1^1 X_2^1 X_3^2 + X_1^2 X_2^1 X_3^2 + X_2^1 X_3^2]$$
$$=_2 [X_1^0 X_2^1(1 + X_3^0 + X_3^0) + X_1^0 X_3^0 + X_2^2 X_3^0(X_1^1 + X_1^2 + 1)$$
$$+ X_2^1 X_3^2(X_1^1 + X_1^2 + 1)]$$
$$=_2 [X_1^0 X_2^1 + X_1^0 X_3^0 + X_2^2 X_3^0 + X_2^1 X_3^2].$$

---

[2] The laws which apply for such ternary expressions may be found published[6]; they follow closely the more familiar Boolean identities.

The Design of Ternary Logic Networks

Hence the more convenient sum-of-products expression for the ternary output value can be obtained.

However, it must be emphasized again that so far we have not considered the implications of any form of circuit realization upon our algebra; all we have at present done is to show how Boolean-like logic expressions which represent the truth table of a ternary system may be independently minimized to an irredundant sum-of-product expression. Also the expressions we have been using to define $f(X) = 0$, 1, or 2 are merely logical expressions and not precise mathematical equations. Nevertheless such expressions giving the output requirements in a compact algebraic form can form the basis for a realization, as we shall subsequently show.

6.2. The effect of "maximum-of" and "minimum-of" operators on algebraic minimization procedures

In any circuit realization of a function $f(X)$, the output must of course be 0, 1, or 2 in accordance with its truth table. The previous algebraic minimization of the minterms which separately define $f(X) = 0$, $f(X) = 1$, and $f(X) = 2$ is *not* a final realization for $f(X)$, but merely a possibly more compact way of expressing the three differing output states. In Fig. 6.1 we illustrated a possible structure for generating the final function $f(X)$. Let us now consider the implications of this structure more closely.

Consider the network whose output we labeled $f_2(X)$, whose duty was to provide an output signal, say logic 2, whenever $f(X) = 2$. When $f(X) \neq 2$ then the output from this network may be 0 or 1. Let us assume it is 0 under such circumstances, and hence let us relabel its output $f(X)_{02}$. Similarly, considering the network whose output we labeled $f_1(X)$, assume it provides an output 1 when $f(X) = 1$, and that its output is also 0 when $f(X) \neq 1$. Therefore in order finally to realize the complete function $f(X)$, the final-level combining of these two output signals must be such as to select the *maximum* of each signal. When $f_2(X) \neq 2$ and $f_1(X) \neq 1$, then the final output $f(X)$ will be 0 as required. Thus $f(X)$ may be expressed as

$$f(X) = [f(X)_{02} \vee f(X)_{01}]$$

where $\vee \triangleq$ maximum-of. However, if we now consider the output $f(X)_{01}$, because the final-level decision element selects the maximum signal from $f(X)_{02}$ and $f(X)_{01}$, output $f(X)$ will remain functionally cor-

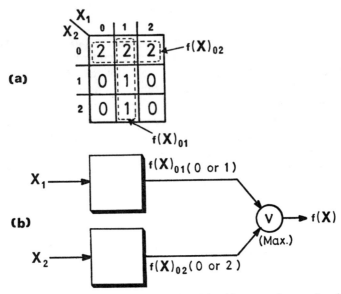

Fig. 6.4. Repeat of the example of Fig. 6.1 with a maximum-signal output decision element: (*a*) the Karnaugh map with revised minimized cover for $f(X) = 1$; (*b*) the realization.

rect if $f(X)_{01}$ is 1 for any or all of the $f(X) = 2$ minterms in addition to its essential $f(X) = 1$ minterms, provided of course that the $f(X)_{02}$ output is precisely correct. If we look again at the example originally mapped in Fig. 6.1*a*, it is now no longer necessary to take the $f(X) = 1$ minterms as $[X_1^1 X_2^1 + X_1^1 X_2^2]$, but instead the minimization to merely X_1^1 is entirely satisfactory. This is indicated in Fig. 6.4*a*, giving the final realization shown in Fig. 6.4*b*.

A similar procedure is also possible if the final-level decision element selects the minimum instead of the maximum of its input signals.[3] The function $f(X)$ is now given by

$$f(X) = [f(X)_{20} \, \& \, f(X)_{21}]$$

where $\& \triangleq$ minimum-of. This is as illustrated in Fig. 6.5. Notice that now the $f(X)_{21}$ network output can overlay the $f(X) = 0$ minterms if desired, but these $f(X) = 0$ minterms must be precisely covered by the $f(X)_{20}$ network output.

However, we are still only considering basic principles which may be applied to the simplification of "Boolean-like" algebraic ex-

---

[3] The corresponding Boolean concepts are of course the sum-of-products and the product-of-sums methods of realizing any given binary function $f(x)$.

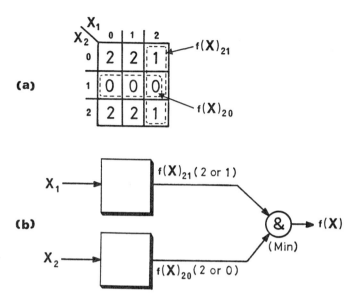

Fig. 6.5. A further simple example with a minimum-signal output decision element: (a) the Karnaugh map of the required function; (b) the realization.

pressions used to define $f(X)$. So far we have not considered other possible algebras, such as were breifly mentioned in Chapter 1, or the circuit implications of the realizations shown schematically in Figs. 6.4 and 6.5. Let us continue with further considerations along these paths.

6.3. Realizations based upon "maximum-of" and "minimum-of" operators

In the synthesis of binary networks it is possible to standardize upon one type of logic gate which provides functional completeness, for example the NAND gate, or to employ a larger family of gates which together allow a more compact network realization to be achieved. The same philosophy is true for ternary networks. As was mentioned in Chapter 1, the adoption of one simple functionally complete ternary operator, for example the Webb operator[8,9], frequently produces inordinately long algebraic expressions, and correspondingly large ternary network realizations.

For these reasons the majority of ternary research which has practical network realization in mind has not concentrated upon single functionally complete gates, but rather has used sets of gates which collectively provide functional completeness. This has resulted in many proposals which basically employ (i) single-input–single-output "conversion" functions together with (ii) multi-input–single-output "maximum-of" ("OR") and "minimum-of" ("AND") functions. Examples of this approach will be found published by many authors[10-21]. Minimization procedures similar to those introduced in the preceding sections of this chapter will also be found in certain of these publications and also in the work of Su and Cheung and others[22-23].

The single-input functions ("unary" functions) frequently consist of a family of gates which each accept a ternary input signal of value 0, 1, or 2, but which provide a two-valued output signal, say 0 and 2 only. The multi-input maximum-of and minimum-of functions of course provide a three-valued output signal of 0, 1, or 2, depending upon their input signals.

The choice of which single-input functions to employ with the multi-input maximum-of or minimum-of functions is still a matter of debate. A choice may be based upon producing a minimum set of gates necessary to achieve functional completeness, but non-minimal sets may be advantageous in that a smaller total number of gates may be required per network[17,19]. The ease or otherwise of the circuit realization of the gates themselves is also a very pertinent point, one which we shall consider in more detail in the following chapter.

Consider the minimum set of single-input gates defined by Mühldorf (and others), which with the logic 1 signal and the maximum-of operator provides functional completeness. We have already introduced these gates in Chapter 1, §1.8.4, where it was shown that the minterms of any given function $f(X)$ can form the basis of a direct synthesis of $f(X)$ using the Mühldorf operators. It will be recalled that the single-input operators are the three j_k operators

$$j_k(X_i) = 0, X_i \neq k$$
$$= 2, X_i = k$$

giving the following full tabulation.

X_i	$j_0(X_i)$	$j_1(X_i)$	$j_2(X_i)$
0	2	0	0
1	0	2	0
2	0	0	2

The functional completeness given by the j_k operators enables all the 27 possible single-variable functions of X_i to be generated, as detailed in Table 6.2. Functional completeness for any multi-variable function $f(X_1, \ldots, X_n)$ may be shown to follow[16].

The Design of Ternary Logic Networks

Table 6.2. The generation of all possible single-variable ternary functions, see Chapter 1, Table 1.6, using Mühldorf's functions plus logic 1

	Realization for $f(X)$	X		
		0	1	2
1	j_0X & j_1X	0	0	0
2	j_2X & 1	0	0	1
3	j_2X	0	0	2
4	j_1X & 1	0	1	0
5	$(j_1X \lor j_2X)$ & 1	0	1	1
6	$j_2X \lor (j_1X$ & 1$)$	0	1	2
7	j_1X	0	2	0
8	$j_1X \lor (j_2X$ & 1$)$	0	2	1
9	$j_1X \lor j_2X$	0	2	2
10	j_0X & 1	1	0	0
11	$(j_0X \lor j_2X)$ & 1	1	0	1
12	$j_2X \lor (j_0X$ & 1$)$	1	0	2
13	$(j_0X \lor j_1X)$ & 1	1	1	0
14	1	1	1	1
15	$j_2X \lor 1$	1	1	2
16	$j_1X \lor (j_0X$ & 1$)$	1	2	0
17	$j_1X \lor 1$	1	2	1
18	$j_1X \lor j_2X \lor 1$	1	2	2
19	j_0X	2	0	0
20	$j_0X \lor (j_2X$ & 1$)$	2	0	1
21	$j_0X \lor j_2X$	2	0	2
22	$j_0X \lor (j_1X$ & 1$)$	2	1	0
23	$j_0X \lor 1$	2	1	1
24	$j_0X \lor j_2X \lor 1$	2	1	2
25	$j_0X \lor j_1X$	2	2	0
26	$j_0X \lor j_1X \lor 1$	2	2	1
27	$j_0X \lor j_1X \lor j_2X$	2	2	2

However, instead of a network realization using the Mühldorf set of operators and based upon the minterms of the required function $f(X)$, as we initially illustrated in Chapter 1, it is equally possible and indeed preferable to apply the Mühldorf set to a minimized sum-of-products expression for $f(X)$. Consider again the very simple problem shown in Fig. 6.4.

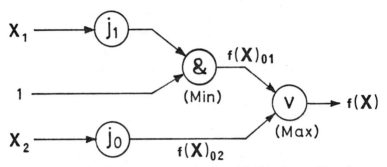

Fig. 6.6. The circuit of Fig. 6.4 realized with a Mühldorf set of functions.

The function output $f(X)_{02}$ of this problem was required to be 2 when its input X_2 was 0, and to be 0 in all other cases. This is directly given by using a single j_0 gate, which will provide the logic 2 output signal when required. The output $f(X)_{01}$ has to be 1 when its input was 1, and 0 otherwise. This is not directly available from a single j_k gate, but from Table 6.2 or by intuition such a requirement is realized by the two-level "minimum-of" network $[j_1(X_i) \ \& \ 1]$. The full network realization for this little problem is therefore as illustrated in Fig. 6.6. The algebraic equation for this realization is

$$f(X) = [j_0(X_2) \ \vee \ \{1 \ \& \ j_1(X_1)\}].$$

Notice that now we have a single mathematically correct equation for the final output $f(X)$, and not an algebraic expression for the individual conditions which have to be present to make the output 0, or 1, or 2, respectively.

The maximum-of and minimum-of operators, however, also allow another form of minterm grouping to be used in network realizations. Consider the situations shown in Fig. 6.7. In map (a) we have a normal minimization by the elimination of one variable. However in map (b) as well as minimizing the center row into $\{1 \ \& \ j_1(X_2)\}$ we also see that the $X_1 = 2$ column entries of the map are identical to the X_2 input values. Therefore we can realize all these column entries by the one expression $\{j_2(X_1) \ \& \ X_2\}$, the $j_2(X_1)$ term being zero valued except when $X_1 = 2$, the & operator then ensuring that the value of the whole expression follows the value of X_2 itself. Hence the realization for the function shown in Fig. 6.7b is

$$f(X) = [\{j_2(X_1) \ \& \ X_2\} \ \vee \ \{1 \ \& \ j_1(X_2)\}].$$

The Design of Ternary Logic Networks

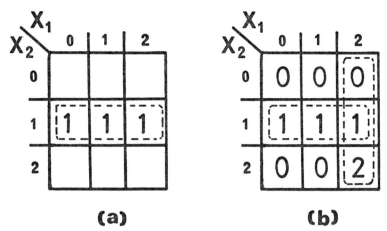

Fig. 6.7. Further map minterm groupings: (*a*) the previous elimination of one variable in a group of same-output-value minterms; (*b*) a further grouping of minterms with the output values of 0, 1, and 2.

As a further illustration of these basic techniques, let us consider a final simple problem, that of a ternary full-adder circuit, with three inputs A, B, and carry C_{in}, and two outputs, the sum output S_{out} and the carry output C_{out}. One feature of this problem should be noted at the start; this is that although the two inputs A and B can each take the three values 0, 1, or 2, the carry-in signal C_{in} can only be 0 or 1 and never 2, assuming of course that this carry-in signal is the same as the carry-out signal C_{out} from a preceeding less significant stage of addition. Hence all the $C_{in} = 2$ states in the full-adder design may be taken as "don't cares," and as a result a triple of $n = 2$ ternary Karnaugh maps rather than an $n = 3$ hypercube construction becomes adequate to illustrate the design requirements.

The requirements for S_{out} and C_{out} are shown in Fig. 6.8. Taking the S_{out} output first, no useful minimization can be found owing to the very nature of the arithmetic requirements. However, some 0, 1, 2 groupings are useful as indicated in Fig. 6.8*a*, giving a possible realization for S_{out} of

$$S_{out} = \Big[\big(j_0 C_{in} \& \{ (j_0 A \& B) \vee (j_0 B \& A) \vee (j_1 A \& j_1 B) \\ \vee \{1 \& (j_2 A \& j_2 B)\}\} \big) \\ \vee \big(j_1 C_{in} \& \{ (j_2 A \& B) \vee (j_2 B \& A) \vee (j_1 A \& j_0 B) \\ \vee (j_0 A \& j_1 B) \vee \{1 \& (j_0 A \& j_0 B)\}\} \big) \Big].$$

The C_{out} requirements have a certain minimization possible, as the $C_{in} = 0$ 1-valued minterms can also be included in the $C_{in} = 1$ and 2 maps. One realization for C_{out} is therefore

$$C_{out} = \left[1 \ \& \ \{(j_2A \ \& \ j_1B) \vee (j_2A \ \& \ j_2B) \vee (j_1A \ \& \ j_2B) \\ \vee (j_1C_{in} \ \& \ \{j_2A \vee j_2B \vee (j_1A \ \& \ B)\})\}\right].$$

Notice that in this particular realization we have treated all the C_{out} minterms as though they were value 2, except for minterm $A^1B^1C_{in}^1$, which we have taken as 1. It is immaterial whether we consider all these minterms as 1 or 2, as the overall "1 &" operation will reduce them all to the value 1 in the final realization. The full circuit realization for the full-adder is therefore as illustrated in Fig. 6.8c.

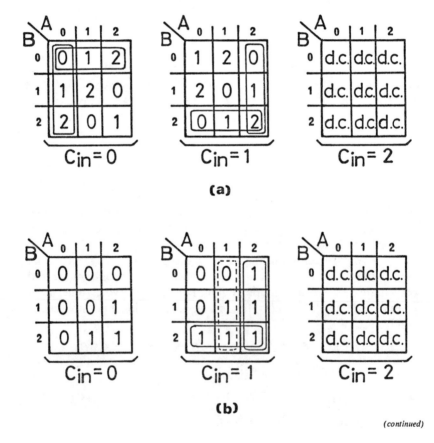

(a)

(b)

(continued)

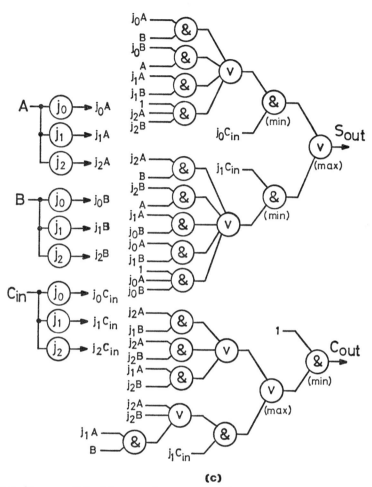

Fig. 6.8. Ternary full-adder requirements: (*a*) the S_{out} output sum requirements; (*b*) the C_{out} output carry requirements; (*c*) the full realization. (Note: The C_{in} signal never takes the value $C_{in} = 2$.)

The relative complexity of these realizations for S_{out} and C_{out} will be obvious, even though the Mühldorf set of operators and variations on them are frequently considered to be one of the more promising set. Even more complex are the realizations for this problem which do not consider the $C_{out} = 2$ situation to be a "don't-care" situation.[24]

Algebraic rearrangements of realizations such as above are possible, for example to convert the sum-of-products form we have been using into an equivalent product-of-sums form (minimum-of-maximum terms). Full details of such algebraic manipulations, which rely upon the commutative and associative properties of the algebra, will be found well detailed in Nutter and elsewhere[19,20,22,24–26]. The ternary equivalent of DeMorgan's law may also be introduced into these procedures. However, in general a sum-of-product form (maximum-of-minimum terms) such as we have illustrated is usually the more readily understandable, as indeed is frequently the case in binary working.

Before we leave this example of the ternary full-adder circuit, it may be recalled from Chapter 1, §1.7 that for the radix-3 number representation it is also possible to use a signed numbering system, using the three numbers $-1, 0, +1$, or written more conveniently $-, 0, +$. With this numbering system the truth tables for the full adder are not the same as for the 0, 1, 2 ternary numbering system. Also all these carry signals of $-, 0$, and $+$ may arise, unlike the "don't-care" $C_{out} = 2$ case in our previous example. It is left as a useful exercise for the reader to formulate the precise truth table for the $-, 0, +$ full-adder, and thence to attempt a realization using the Mühldorf set of functions; the $-, 0, +$ signals may be treated as equivalent to 0, 1, and 2 respectively for the purpose of applying the j_k and maximum-of and minimum-of operators. Some initial guidance may be found in this area[27], but suffice it to say here that no compact Mühldorf realization results for this alternative full-adder specification.

To summarize this section, therefore, we may state that whilst the maximum-of and minimum-of operators together with Mühldorf's three j_k operators and logic 1 enable us to generate any ternary function $f(X)$, nevertheless a considerable network of such operators rapidly builds up, even for quite simple problems. Various authorities have augmented and/or suggested alternatives to the j_k operators[11,12,19], but still relatively large assemblies continue to be necessary.

6.4. Network Realizations Using Other Algebraic Operators

The maximum-of and minimum-of operators used in the preceding sections are frequently quoted as a natural choice of operators, as they can "be easily realized with electronic circuits." For example, simple passive diode gates can be proposed which pass the maximum (or minimum) of the gate input voltages and which would be functionally perfect if ideal diodes existed. However, the realities of diode forward volt-drop, for example, may be significant, and complex voltage-level-restoring circuits may become necessary in practice. Such circuit problems will come into our considerations in Chapter 7.

Therefore other possible operators should not at this stage be dismissed out of hand. Referring back to Chapter 1 it will be recalled that we introduced several other possibilities, amongst them being (i) the Post operators and algebra, (ii) the modulo-3 operators and algebra of Bernstein, (iii) the single "Sheffer-stroke" type operator and algebra of Webb, and (iv) the "T-gate" operator and algebra of Lee and Chen.

We have already indicated in Chapter 1 that the Post operators do not prove to be a very convenient set to adopt for ternary realizations. Associative and commutative algebraic rules do not hold, and inordinately long algebraic expressions and hence practical network realizations result. No authority concerned with eventual network realization has found sufficient justification for their sole adoption and use.

Similar remarks apply to the single operator of Webb. This therefore leaves us with modulo-3 operators and the T-gate operator, both of which certainly possess attractions in allowing meaningful networks for any given function $f(X)$ to be readily derived. An example of the design procedure using modulo-3 arithmetic operators has been shown in Chapter 1. The key procedure is the solution of the set of 3^n simultaneous modulo-3 additions of the coefficient values, which have to total to the output value $f(X)$ at each minterm. Any final rearrangement of the coefficients is a normal arithmetic rearrangement.

It is an intructive exercise for the reader to apply this procedure in detail to the design of the ternary full-adder which we have previously considered. Assuming the $C_{in} = 2$ situation is still taken to be a "don't care" situation, then we have a total of 18 different input conditions and a total of 18 mod_3 coefficients, namely

$$a_{000}, a_{001}, a_{002}, a_{010}, \ldots, a_{112}, a_{120}, a_{121}, a_{122}$$

where the coefficient subscripts are in the order representing C_{in}, A, and B respectively. The 9 remaining coefficients $a_{200}, a_{201}, \ldots, a_{222}$ lie in the "don't-care" region.

However, when we solve these 18 unknown coefficient values, we shall find that for the sum output all are zero-valued except the a_{001}, a_{010}, and a_{100} coefficients; the latter are each unity valued, giving the realization

$$S_{out} = \{1 \cdot A + 1 \cdot B + 1 \cdot C\}_{mod_3}$$
$$= \{A + B + C\}_{mod_3}.$$

This of course must be the realization from the very definition of the problem!

The carry output C_{out} is not quite so straightforward a solution. The solution of the 18 coefficient equations shows that 10 of the 18 are zero valued, the 8 remaining being as follows:

$a_{011} = 2$
$a_{012} = 2$
$a_{021} = 2$
$a_{101} = 1$
$a_{102} = 2$
$a_{110} = 1$
$a_{111} = 1$
$a_{120} = 2.$

Hence the realization is

$$C_{out} = \{2AB + 2AB^2 + 2A^2B + C_{in}A + 2C_{in}A^2 + C_{in}B$$
$$+ C_{in}AB + C_{in}B^2\}_{\text{mod}_3}$$
$$= \{2A(B + B^2 + AB) + C_{in}(A + B + AB + 2A^2 + 2B^2)\}_{\text{mod}_3}.$$

This is illustrated in Fig. 6.9.

This realization for C_{out} using the modulo-3 operators is a rather fascinating result. Clearly when C_{in} is zero, it is necessary for the first bracket $2A(B + B^2 + AB)$ to produce the $C_{out} = 1$ output when required, which it duly does whenever the sum of A plus B equals or exceeds the value 3. The second bracket adds nothing in these cases, as it is multiplied by $C_{in} = 0$. However, when $C_{in} = 1$, it may be thought at first sight that the second bracket now provides the overall result for $C_{out} = 1$ when necessary, but this is not so, as the first bracket is still operative. In fact the second bracket only provides its contribution of 1 for the three input conditions of C_{in}, A, and B equal to 111, 120, and 102, respectively, and correctly contributes zero to the total when C_{in}, A, and B are 112, 121, and 122. A glance back at the C_{out} Karnaugh map of Fig. 6.8b will illustrate this nice distinction in this modulo-3 realization.

Modulo-3 synthesis is therefore a perfectly straightforward exercise, but of course remains entirely an academic one unless we can construct the modulo-3 gates in a viable manner.

Turning finally to the T-gate operator of Lee and Chen, there is very little which can be added to the information given in Chapter 1 and the typical realization illustrated in Fig. 1.22. Circuit design using the

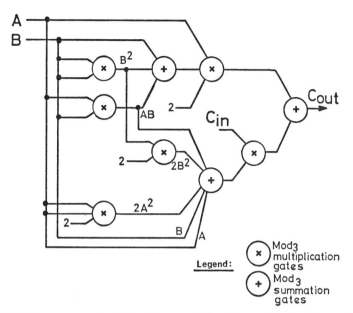

Fig. 6.9. The carry output of a ternary full-adder, realized with modulo-3 gates.

T-gate operator is a direct realization of the truth table of the required function, giving an initial n-level realization for any n-variable function $f(X)$. A certain amount of gate sharing and gate elimination may be possible to reduce the final number of gates used per realization.

For illustration, Fig. 6.10 illustrates how the basic synthesis of the sum output of our full-adder requirement would be built up using T-gates; the next stage to that shown would be to look for duplicated T-gates and combine them if present, and also to look for any T-gate with all identical left-hand input signals X_1, X_2, X_3, which of course is entirely redundant. The latter situation does not arise in this example, but the total number of first-level gates can be reduced to half the number shown in Fig. 6.10.

6.5. Further Realizations Based upon Basically Binary Operators

The biggest question in ternary network realization remains that of how many functions to propose. We have already seen that a minimum set necessary to achieve functional completeness may result in lengthy

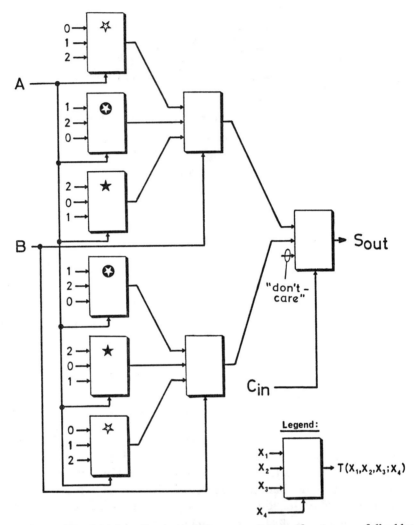

Fig. 6.10. The initial realization of the sum output of a ternary full-adder realized with T-gates. The gates shown with different asterisk identifications are duplicates, and therefore can be combined to produce a minimal final realization.

expressions for $f(X)$ and complex realizing networks, and therefore a larger range of ternary functions in general enables fewer functions (gates) to be used for any given problem. All this of course assumes that we can in fact make all such functions that may be suggested without excessive difficulty, which may or may not be the case.

The Design of Ternary Logic Networks

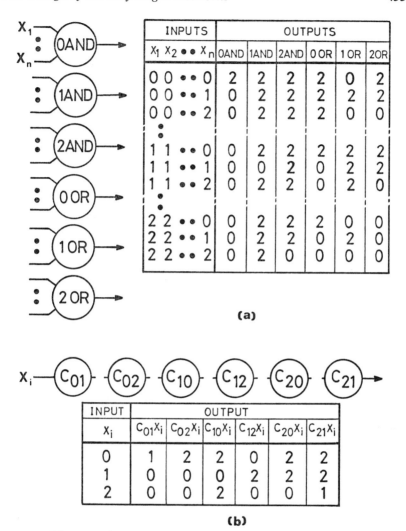

Fig. 6.11. The non-minimal set of ternary AND, OR, and conversion functions: (*a*) the multi-input AND, OR set; (*b*) the unary conversion set.

One set which has already been introduced in Table 1.13 of Chapter 1 suggests a total of six multi-input operators together with six unary operators. The philosophy behind this choice was as follows.

(i) To enable the simultaneous presence (conjunction) of any one chosen signal value to be detected, i.e. logical AND of 0 signals, logical AND of 1 signals, and logical AND of 2 signals.

(ii) To enable the independent presence (disjunction) of any chosen signal value to be detected, i.e. logical OR of 0 signals, logical OR of 1 signals, and logical OR of 2 signals.

(iii) To have the facility of being able to convert each of the signal values 0, 1, and 2 into another signal value.

The complete set of these functions is illustrated in Fig. 6.11. It will be noted that all these functions have binary outputs. However, unlike the Mühldorf j_k functions, the unary conversion functions do not all have 0, 2 output signals. A discussion on the choice of these functions and their particular output signals may be found published[17,28].

The synthesis techniques using this range of functions are basically to realize two of the three output levels of the required function $f(X)$, and then finally to combine them by means of an appropriate maximum-of operator. In this respect the final output gate, which is in addition to the 12 combinatorial gates listed in Fig. 6.11, is the same as in most of the realizations covered in earlier sections of this chapter. The synthesis therefore requires a minimized algebraic expression for two of the three output cases

$$f(X) = 2$$
$$f(X) = 1$$
or
$$f(X) = 0$$

which are then realized and finally combined. Notice that should we, for instance, choose to realize the minimized expression for $f(X) = 1$ and $f(X) = 2$ and finally combine them by the maximum-of operator, then the $f(X) = 1$ minimization may intrude into the $f(X) = 2$ minterm space, if advantageous.

However, in considering the circuit implications of all these proposed functions, it has been suggested that functions which have to operate on the center 1 signal level may be more complex circuitwise than those which operate on the 0 or 2 signal level. Therefore it may be preferable to choose the two sets of algebraic expressions which contain the fewest logic 1 signals in their expressions. Let us look back by way of illustration at the problem originally introduced and minimized in §6.1 and illustrated in Fig. 6.2. The three output conditions of this particular function $f(X)$ were

$$f(X) =_0 \left[X_1^2 + X_3^2 \right]$$
$$f(X) =_1 \left[X_1^1 X_3^0 + X_1^0 X_3^1 \right]$$
$$f(X) =_2 \left[X_1^0 X_3^0 + X_1^1 X_3^1 \right].$$

The Design of Ternary Logic Networks

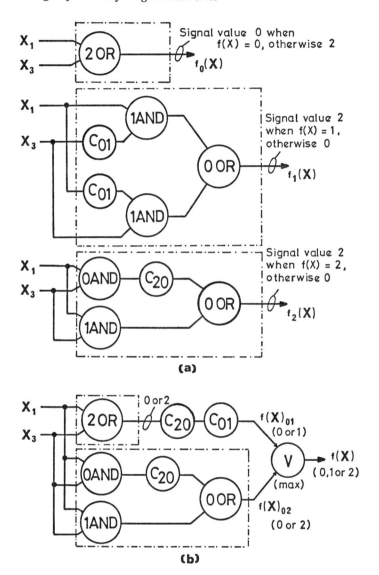

Fig. 6.12. The realization of the ternary function originally plotted in Fig. 6.2. (a) the individual realizations selecting $f(X) = 0$, 1, and 2 respectively; (b) the final realization, choosing

$$f(X) = [f(X)_{01} \vee f(X)_{02}].$$

(Notice that all the gates in (a) are basically binary gates, in comparison with the full ternary gates of, for example, Fig. 6.10.)

Now these expressions may each be formed by the networks shown in Fig. 6.12a. However, it must be noticed that each network within its dotted boundary will detect the required input minterm conditions and given a unique output signal when they are present, but this signal *is not necessarily the correct signal value* for the final $f(X)$ output.

In considering which two of the three networks to finally employ, the $f_1(X)$ network as well as being the largest also contains the greatest number of functions involving 1 signal levels. It is therefore advantageous to discard this one and to form the final realization for $f(X)$ from the $f_0(X)$ and $f_2(X)$ networks, see Fig. 6.12b. Notice particularly that the two conversion functions in cascade placed at the output of the $f_0(X)$ network will finally provide 0 and 1 logic signals only[4]. Thus the final output $f(X)$ will correctly take the values 0, 1, or 2 as required.

Network synthesis using this range of gates is therefore seen to be a straight realization of the chosen algebraic expressions defining $f(X)$, using the appropriate conversion and logical operators for each $f(X)$ output state. In this respect the synthesis procedure mirrors conventional binary practice with AND/OR/INVERTER vertex gates. The viability of this ternary approach, however, hinges entirely upon (*a*) the generation of minimal algebraic expressions for the $f(X)$ output values and (*b*) the possibility of the successful circuit realization of the very many operators used.

Several other approaches to ternary function realization using basically binary operators have been suggested. Here we shall mention just two further published possibilities, those of Birk and Farmer[20] and of Pugh[14]. A comprehensive work of Duncan (as yet unpublished) is also in this area[29].

Birk and Farmer's synthesis technique is based upon earlier publications, and suggests the use of logic-level ("threshold") detectors as first-level gates, followed by AND/OR configurations of normal binary gates. A final single multi-valued output gate, realizing the maximum-of its input signals, provides the final $f(X) = 0$, 1, or 2 output signal. Notice that the term "threshold" used by authors in this context refers not to any form of threshold-logic function, but merely to a *signal-level detection,* as provided for example by a Schmitt trigger circuit.

The basis of Birk and Farmer's development is the "literal" operators $^aX_i^b$, where for a ternary system $^aX_i^b$ is defined as

---

[4] The original paper which discussed these functions had additional unary functions, one of which would realize this required conversion directly. We have not, however, considered all these variations in this text.

The Design of Ternary Logic Networks

$$^aX_i^b, a,b \in \{0,1,2\} = 2 \text{ if } a \leq X_i \leq b$$
$$= 0 \text{ otherwise.}$$

The complement of this operator is also available, namely

$$\overline{^aX_i^b} = 0 \text{ if } a \leq X_i \leq b$$
$$= 2 \text{ otherwise.}$$

The truth table for all the possible ternary $^aX_i^b$ literal operators is therefore as detailed in Table 6.3. Note that the literal operators are not exclusively ternary operators, but may be extended to any R-valued system, where $a,b \in \{0, 1, \ldots, R - 1\}$[22,30].

Table 6.3. The ternary literal $^aX_i^b$ operators.

X_i	$^0X_i^0$	$^0X_i^1$	$^0X_i^2$	$^1X_i^1$	$^1X_i^2$	$^2X_i^2$
0	2	2	2	0	0	0
1	0	2	2	2	2	0
2	0	0	2	0	2	2

The six $\overline{^aX_i^b}$ operators are similar but with the 0 and 2 output values interchanged.

Of these six possible ternary operators and their complements, the last two only are suggested as being the most useful, as these effectively detect where the ternary input signal equals or exceeds the signal level 1 or 2, respectively. We shall here, therefore, modify the previous designations and henceforth use those shown in Table 6.4. Also, as the outputs from these gates will be used in following-level AND/OR gates, we shall use the binary designations 0 and 1 rather than 0 and 2 to signify their output states.

Table 6.4. The final two literal operators of Table 6.3 and their complements, redesignated as binary-output signal-level detectors

X_i	^1X_i	$\overline{^1X_i}$	^2X_i	$\overline{^2X_i}$
0	0	1	0	1
1	1	0	0	1
2	1	0	1	0

If we now consider a simple two-variable ternary function $f(X)$ as shown in Fig. 6.13a, we may group the $f(X) = 2$ and the $f(X) = 1$

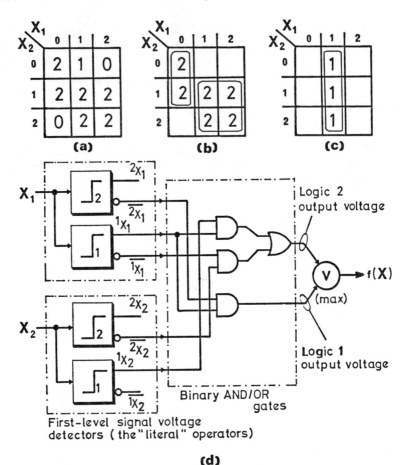

Fig. 6.13. Ternary synthesis using first-level signal-voltage-detector functions: (a) a given function $f(X)$; (b) $f(X) = 2$ minterm minimization; (c) $f(X) = 1$ minterm minimization; (d) final realization of $f(X)$.

minterms as shown in Fig. 6.13b and c, assuming we are using an output maximum-of operator. Now the $f(X) = 2$ blocks of minterms may be selected by the binary AND/OR detection

$$[^1X_1{}^1X_2 + \overline{^1X_1}\,\overline{^2X_2}]$$

whilst the $f(X) = 1$ block of minterms may be selected by the single binary AND detection

$$[^1X_1\overline{^2X_1}].$$

The Design of Ternary Logic Networks 439

Therefore the final overall realization using the first-level signal-level detectors becomes as shown in Fig. 6.13*d*.

Birk and Farmer do not publish any detailed circuit considerations, but instead concentrate largely upon minimization techniques for the $f(X) = 2$ and $f(X) = 1$ requirements. However, they suggest that the final AND/OR gates before the output maximum-of operator may have their supply voltages adjusted such that their output signals are effectively logic 0/1 or logic 0/2 in a ternary case, as required. This is as indicated at the outputs of the final binary gates in Fig. 6.13*d*.

The binary-type synthesis philosophy of Pugh, whilst still very largely based upon the employment of binary gates, departs somewhat from our previous methods of treating the ternary logic signals in two distinct parts by proposing a "negative-logic" half corresponding to the 0 and 1 logic levels and a "positive-logic" half corresponding to the 1 and 2 logic levels. Pugh's work, however, used the symmetrical set of ternary truth values $-1, 0, +1$ instead of 0, 1, 2, and hence considered the 0, -1 levels as the negative-logic half of the synthesis and the 0, $+1$ levels as the positive-logic half. This is illustrated in schematic form in Fig. 6.14. Notice that "translating circuits" may be required to interconnect the two halves of the synthesis. The two halves of the syn-

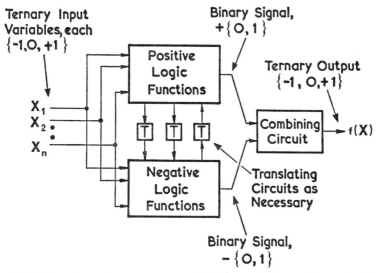

Fig. 6.14. The schematic arrangement of Pugh's positive- and negative-logic quasi-binary ternary synthesis.

thesis are each assumed to be sensitive to their own polarity signals only, but can tolerate opposite polarity signals without response. For example the positive-logic gates will be responsive to +1 logic signal inputs but not to 0 or −1 signals, and conversely for the negative-logic gates. There is also redundancy in the overall system in that if the two halves were entirely separate and able independently to provide their 0, 1 binary output signals, then there would be quaternary system capability, with the outputs from the two systems providing the following four possibilities.

Upper output	Lower output
0	0
0	1
1	0
1	1

However, only the first three of these states are present in ternary synthesis, representing the ternary conditions shown in Table 6.5.

Table 6.5. The positive and negative quasi-binary system of representing the three ternary states

Upper (positive) output	Lower (negative) output	Ternary output	
		$\{-1,0,+1\}$ system	$\{0,1,2\}$ system
0	0	0	1
0	1	−1	0
1	0	+1	2

The published synthesis techniques of this approach involve the conversion of the required ternary function $f(X)$ into a quasi-binary input–output truth table. This involves dividing each input X_i and the output $f(X)$ into two parts, the positive part X_i^+, $f(X)^+$, and the negative part X_i^-, $f(X)^-$, in accordance with the relationships detailed in Table 6.5. Each half is then basically a straightforward binary system synthesis, the output from each half being finally combined in the output-level "combining" circuit. Notice that unlike most previous cases this output-level gate is not a maximum-of (or minimum-of) gate, but instead is a unique circuit configuration for this particular system.

As an example of this synthesis technique let us take again the simple ternary function used in Fig. 6.13. Converting the previous $\{0,1,2\}$ ternary data into $\{-1,0,+1\}$ and then splitting this $\{-1,0,+1\}$

The Design of Ternary Logic Networks

data into quasi-binary data, we obtain the new truth table for $f(X)$ as follows.

Inputs X_1, X_2								Output $f(X)$		
{0,1,2}		{−1,0,+1}		Quasi-binary				{0,1,2}	{−1,0,+1}	Quasi-binary
X_1	X_2	X_1	X_2	X_1^+	X_1^-	X_2^+	X_2^-	$f(X)$	$f(X)$	$f(X)^+ f(X)^-$
0	0	−1	−1	0	1	0	1	2	+1	1 0
0	1	−1	0	0	1	0	0	2	+1	1 0
0	2	−1	+1	0	1	1	0	0	−1	0 1
1	0	0	−1	0	0	0	1	1	0	0 0
1	1	0	0	0	0	0	0	2	+1	1 0
1	2	0	+1	0	0	1	0	2	+1	1 0
2	0	+1	−1	1	0	0	1	0	−1	0 1
2	1	+1	0	1	0	0	0	2	+1	1 0
2	2	+1	+1	1	0	1	0	2	+1	1 0

Should we be working continuously in the $\{-1, 0, +1\}$ domain, there would of course be no need to cross-refer at all to the 0, 1, 2 system, and hence we would tabulate the above function as follows, with asterisks to indicate all the redundant situations which cannot occur owing to the ternary system limits.

Quasi-binary inputs				Quasi-binary outputs	
X_1^+	X_1^-	X_2^+	X_2^-	$f(X)^+$	$f(X)^-$
0	0	0	0	1	0
0	0	0	1	0	0
0	0	1	0	1	0
0	0	1	1	*	*
0	1	0	0	1	0
0	1	0	1	1	0
0	1	1	0	0	1
0	1	1	1	*	*
1	0	0	0	1	0
1	0	0	1	0	1
1	0	1	0	1	0
1	0	1	1	*	*
1	1	0	0	*	*
1	1	0	1	*	*
1	1	1	0	*	*
1	1	1	1	*	*

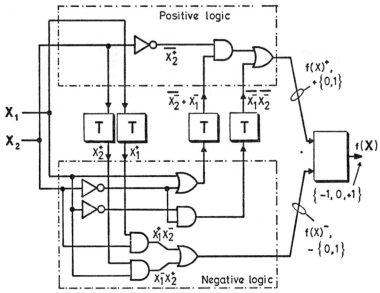

Fig. 6.15. Synthesis of the same ternary function of Fig. 6.13 using positive and negative quasi-binary synthesis.

If these outputs $f(X)^+$ and $f(X)^-$ are separately minimized, say by normal Karnaugh mapping techniques, it will be found that they minimize to

$$f(X)^+ = [\overline{X_2^+ X_2^-} + X_1^- \overline{X_2^+} + \overline{X_1^- X_2^-}]$$
$$= [\overline{X_2^+}(\overline{X_2^-} + X_1^-) + \overline{X_1^- X_2^-}]$$

and

$$f(X)^- = [X_1^+ X_2^- + X_1^- X_2^+].$$

Full use of the seven asterisk conditions can be made in these minimizations.

However, it may now be seen that in the $f(X)^+$ output equation we have X_i^- terms present, and similarly in the $f(X)^-$ output equation we have X_i^+ terms, but as the logic gates in the positive half will not operate on the X_i^- inputs, and *vice versa*, we therefore have a number of translations necessary for this particular ternary requirement. Figure 6.15 illustrates the direct synthesis of the above equations for $f(X)^+$ and $f(X)^-$ into a network realization for $f(X)$.

The original work on this method of ternary synthesis covers appropriate rearrangements of the algebraic expressions which occur in the syntheses. The above problem may be revised into alternative synthesis by such means, although not a great deal of overall simplification results for this particular problem. In general it may be said that this approach to ternary network synthesis does not yield such good results as some of the other approaches we have previously considered, particularly if there is a scatter of $f(X)^+$ output terms determined by X_i^- input signals, and/or $f(X)^-$ output terms determined by X_i^+ input signals. Its main claim for consideration lies in the possibility of being able to make the required logic gates readily using complementary transistor circuit configurations—this, however, is somewhat irrelevant in the present days of monolithic integrated circuit technology.

To summarize this section, therefore, we may conclude by saying that although the concept of making ternary networks from assemblies of binary or quasi-binary gates is attractive from the point of view of possible circuit realization, as conventional circuit techniques then become relevant, it is seen that large networks of such gates frequently become necessary for quite trivial ternary functions. Fundamentally it cannot be a maximally efficient approach, as gate interconnections carrying only two and not three signal levels predominate, thus requiring correspondingly more signal paths in a realizing network to realize the overall ternary requirements.

6.6. Ternary Cellular Arrays

The theoretical attractions of arrays of identical cells which are capable of realizing any required combinatorial function has been introduced for the binary case in Chapter 5. Similar theoretical attractions may be claimed for the ternary case, although it must be stated at the outset that the circuit complexity per ternary cell is likely to prove disadvantageous. Nevertheless a pursuit of such techniques is a viable exercise in order to examine the possibilities and extend our knowledge in this area.

The first published material specifically on universal ternary arrays is possibly that of Yoeli in 1968[31]. More recent disclosures are those of Miller and Muzio[32] and of Kopanapani and Setlur[33], both in 1974. As these latter two disclosures employ noticeably different approaches, we shall briefly look at each and consider their main features in the following pages.

Miller and Muzio's approach is to extend the two-valued universal array techniques of Akers[34] to the ternary case, although in so doing many of the binary results do not generalize to the three-valued case. In particular Akers' 0 and 1 paths through the array structure have been replaced by a path of what is termed "pivot" cells, around which certain planes of cells become redundant. It is shown that for any n-variable ternary function $f(X)$ at most 2^{3n} cells are required, whilst for completely symmetric functions it is conjectured that at most $\{2^n - (n-1)^3\}$ cells are required.

The disclosed array basically consists of a three-dimensional block of identical cubic cells as shown in Fig. 6.16a and b. The cell input and output designations are those of the present author; notice that the three outputs f are all identical signals. The truth table for each cell is as follows, where — indicates a "don't-care" condition.

p	q	r	e	Cell outputs f	Comments (see later in text)
0	0	—	—	0 ⎫	Backmost layer of cells
0	—	0	—	0 ⎭	redundant
1	1	—	—	1 ⎫	Leftmost layer of
—	1	1	—	1 ⎭	cells redundant
2	—	2	—	2 ⎫	Topmost layer of
—	2	2	—	2 ⎭	cells redundant
0	1	2	0	0 ⎫	
0	1	2	1	1 ⎬	First pivot cell conditions
0	1	2	2	2 ⎭	

The p, q, and r inputs of each cell within the array are respectively connected to the output f of the cells immediately behind, to the left, and above the cell in question. At the edges of the complete array, however these inputs are energized as follows: (i) the p inputs of all cells in the backmost layer are connected to 0; (ii) the q inputs of all cells in the leftmost layer are connected to 1; (iii) the r inputs of all cells in the topmost layer are connected to 2. This is illustrated in Fig. 6.17 for a 2×2 array, without showing the fourth input e to each cell in this illustration. The fourt input e to each cell is the variable or "external" cell input, which may be either a constant 0, 1, or 2, or an input variable X_i, or a rotation of an input variable $\overline{X_i}$ or $\overline{\overline{X_i}}$. The chosen rotation of an

The Design of Ternary Logic Networks

X_i is defined as follows, being similar to the Post operator defined in Chapter 1.

X_i	$\overline{X_i}$	$\overline{\overline{X_i}}$
0	2	1
1	0	2
2	1	0

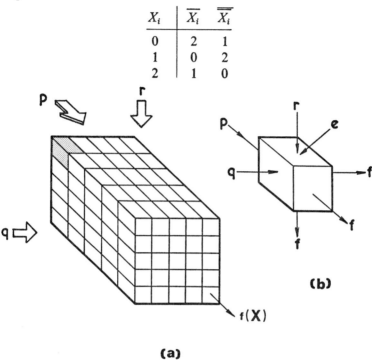

(a)

Fig. 6.16. Miller and Muzio's ternary cellular array: (*a*) a 5 × 5 × 5 cubic array; (*b*) individual cell input–output details.

The cubical format shown in Fig. 6.16 is not a practical layout for showing all the individual cell input and output signals. The 2 × 2 cell array may be redrawn as shown in Fig. 6.18, which represents more realistically the possible circuit layout of such an array, assuming the circuit realization was confined to a usual single-layer circuit construction. The constant $p = 0$, $q = 1$, and $r = 2$ inputs are included in this diagram. In considering the action of a complete array, all the cell inputs p, q, r, and e are only known for the cell shown shaded in Fig. 6.16. This is termed the "first pivot cell." If now the input e to this first pivot cell is 0, then the cell output f is 0, which is the input p of the cell in front, input q of the cell to the right, and input r of the cell below. Reference back to the truth table for the cells will now show that the cell below will have an output 0, and in fact it may be quickly con-

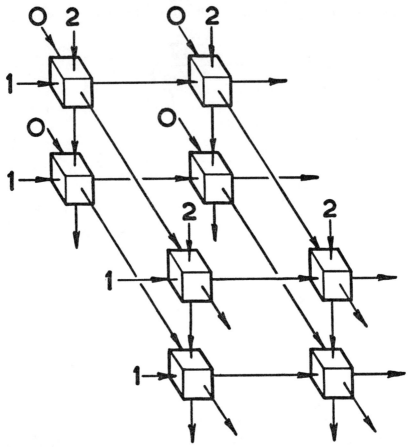

Fig. 6.17. A 2 × 2 array showing the constant inputs but omitting the variable cell inputs e.

firmed that output $f = 0$ will be present at all cells in the back vertical layer of the array. This back layer of cells is therefore completely redundant in such circumstances as all their outputs are the same value as the original input e, and thus may be discounted. Similarly, if input e to the first pivot cell is 1, then the cell output f will be 1. By the same reasoning it may be shown that this output $f = 1$ propagates to all the leftmost layers of cells, and hence this layer may now be considered redundant. Finally if input e to the first pivot cell is 2, then the cell

The Design of Ternary Logic Networks

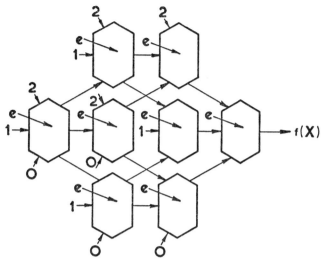

Fig. 6.18. The 2 × 2 array redrawn as a two-dimensional circuit diagram.

output will be 2, and all the top layer of cells may now be considered as redundant.

Hence the action surrounding each pivot cell is to reduce the remaining effective cell array size, the input to this pivot cell explicitly defining a smaller array with a new pivot cell and associated input e. The final output $f(X)$ of the complete array may thus be found by progressively working through the array until finally only the one output cell or layer remains.

To facilitate the analyses of these proposed arrays, a matrix terminology has been suggested, the signal e of each cell constituting the matrix entry data. Simple rules determine the pivot cells, and hence the propagated signal value, by defining the horizontal or vertical jump to the appropriate next pivot cell, depending upon whether the signal value 0 or 1 or 2 is present at the pivot cell under consideration.

However, if such arrays are to be used as universal arrays, they must by definition be capable of producing any one of the 3^{3^n} ternary functions of n variables. Miller and Muzio's published work extends to the consideration of a universal array for $n = 2$. This requires a 4 × 4 × 4 cubic array of cells, with input signals X_1, $\overline{X}_1$, $\overline{\overline{X}}_1$, X_2, $\overline{X}_2$, and $\overline{\overline{X}}_2$, together with the truth values 0, 1, and 2. These inputs are connected to the e input of the 64 cells as detailed in Table 6.6, where (*a*) each 4 × 4 matrix of numbers defines the e inputs to a 4 × 4 horizontal layer of cells (see Fig. 6.16), the left-hand matrix being the topmost layer inputs; (b) each column within each 4 × 4 matrix defines the e inputs of a front-to-back row of cells, the left-hand matrix column being the left-hand row of cell inputs; (c) each row within each 4 = 4 matrix defines the e inputs of a side-to-side row of cells, the top matrix row being the backmost row of cell inputs.

Table 6.6. The matrix of input connections required on a 4 × 4 × 4 universal ternary array in order to give output $f(X) = f(0)$ on input minterm 00, f_1 on minterm 01, f_2 on minterm 02, ..., f_8 on minterm 22.

$$\begin{bmatrix} X_1 & X_1 & X_2 & f_4 & | & X_1 & X_1 & f_5 & \overline{X_2} & | & X_2 & f_7 & \overline{X_1} & \overline{X_1} & | & f_8 & \overline{X_2} & \overline{X_1} & \overline{X_1} \\ X_1 & X_1 & f_3 & \overline{X_2} & | & X_1 & X_1 & \overline{X_2} & \overline{X_2} & | & f_6 & \overline{X_2} & \overline{X_1} & \overline{X_1} & | & \overline{X_2} & \overline{X_2} & \overline{X_1} & \overline{X_1} \\ X_2 & f_1 & \overline{X_1} & \overline{X_1} & | & f_2 & \overline{X_2} & \overline{X_1} & \overline{X_1} & | & \overline{X_1} & \overline{X_1} & \overline{X_1} & \overline{X_1} & | & \overline{X_1} & \overline{X_1} & \overline{X_1} & \overline{X_1} \\ f_0 & \overline{X_2} & \overline{X_1} & \overline{X_1} & | & \overline{X_2} & \overline{X_2} & \overline{X_1} & \overline{X_1} & | & \overline{X_1} & \overline{X_1} & \overline{X_1} & \overline{X_1} & | & \overline{X_1} & \overline{X_1} & \overline{X_1} & \overline{X_1} \end{bmatrix}$$

Pivot cell jumps from 0-valued cells one row downwards, from 1-valued cells one column to the right, and from 2-valued cells four columns to the right.

In order to use this array to synthesize any specific function $f(X)$ of the two variables X_1, X_2, the appropriate 0, 1, or 2 signal values must be applied to the nine cell inputs $f_0, f_1, \ldots, f_8$, each of these input signals then forming the function output value $f_{\ldots}(X)$ on the appropriate input minterm.

It is left as an interesting exercise for the reader to substitute 0, 1, or 2 values for both X_1 and X_2 in Table 6.6, and then to plot the moves of the pivot cells from the extreme top left-hand entry through the array. It will be found that the jumps in the matrix of numbers rapidly reach the appropriate f_j entry, $0 \leq j \leq 8$, which entry then propagates to the cell output. The grouping of the X_1, X_2 signals at a jump distance of one from each f_j entry will be found to be a significant factor in the progression of the pivot cell after the f_j selection.

Although the circuit action of this proposed ternary array is intriguing, it would not appear to be a very economical form of universal array when factors such as the number of cells, cell complexity, and required interconnections between the cells are taken into consideration. For example the universal $n = 2$ array detailed in Table 6.6 requires $2^{3n} = 64$ cells; this may be contrasted with a triangular array of T-gates (see Fig. 6.10 for example), which for the same duty would require only $\frac{1}{2}(3^n - 1) = 4$ gates. A comparison for $2 \leq n \leq 5$ is as follows.

	Cubic array 2^{3n} cells	Triangular T-gate array $\frac{1}{2}(3^n - 1)$ T-gates
$n = 2$	64	4
$n = 3$	512	13
$n = 4$	4096	40
$n = 5$	32768	121

The universal cellular array approach of Kodanapani and Setlur[33] employs an expansion similar to the generalized Reed–Muller expansion for a ternary function, originally discussed in Chapter 1, §1.8.2. It will be recalled that for a two-variable ternary function, for example, the generalized Reed–Muller expansion is

$$f(X) = \{a_{00} + a_{10}X_1 + a_{20}X_1^2 + a_{01}X_2 + a_{11}X_1X_2 + a_{21}X_1^2X_2 + a_{02}X_2^2 + a_{12}X_1X_2^2 + a_{22}X_1^2X_2^2\}_{\text{mod}3}$$

where the coefficients $a_0, a_1, \ldots, a_{22}$ are 0, 1, or 2 as required to realize $f(X)$. The expansion used by Kodanapani and Setlur, however, is

$$f(X) = \{a_0 + a_1 f_1 + a_2 f_2 + \ldots + a_{3^n-1} f_{3^n-1}\}_{\text{mod}3}$$

where $a_0, a_1, \ldots, a_{3^n-1}$ are coefficients taking the value 0, 1, or 2 as required and $f_1, f_2, \ldots, f_{3^n-1}$ are functions of the input variables X_i which take the values 0 or 1. This set, like the generalized Reed–Muller set, enables any three-valued function $f(X)$ to be realized, provided the two constants 1 and 2 are also available. Notice, however, that the values taken by the functions $f_1, f_2, \ldots$ are 0 or 1, and not the 0, 1, or 2 of the original input variables X_i.

The specification for the binary output functions f_1, f_2, onwards is that they progressively become 1 as the input minterms increase from $00\ldots 0$ through to $22\ldots 2$. For example, for $n = 2$ we have

Ternary inputs		Binary-valued f_p outputs							
X_1	X_2	f_1	f_2	f_3	f_4	f_5	f_6	f_7	f_8
0	0	0	0	0	0	0	0	0	0
0	1	1	0	0	0	0	0	0	0
0	2	1	1	0	0	0	0	0	0
1	0	1	1	1	0	0	0	0	0
1	1	1	1	1	1	0	0	0	0
1	2	1	1	1	1	1	0	0	0
2	0	1	1	1	1	1	1	0	0
2	1	1	1	1	1	1	1	1	0
2	2	1	1	1	1	1	1	1	1

which requirements may be realized by appropriate assemblies of modulo-3 carry functions. The assembly suggested by Kodanapani and Setlur is a regular array of mod_3 carry functions, as shown in Fig. 6.19. The b_{jk} inputs, $1 \leq j \leq n$, $1 \leq k \leq 3^n-1$, to the carry functions are 0,

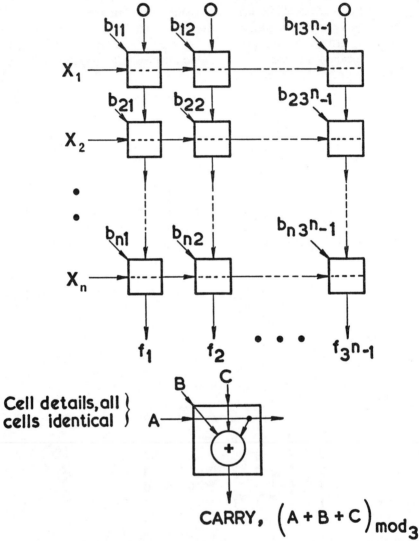

Fig. 6.19. The modulo-3 carry array of Kodanapani and Setlur to realize $3^n - 1$ binary-valued outputs f_1 to f_{3^n-1}.

The Design of Ternary Logic Networks

1, or 2 valued, as required to provide the binary-valued f_p output at the bottom of each column of the array. It is a simple exercise to show that for the first column f_1 the b_{jk} values require to be

2, 2, ..., 2

reading from top to bottom; for the second column f_2 they require to be

2, 2, ..., 1

and so on, to the final f_{3^n-1} column where they require to be

1, 0, ..., 0.

These f_p outputs now require to be multiplied by their appropriate a_p coefficient values, and finally summed by mod_3 addition. This is as illustrated in Fig. 6.20 The necessary a_p values to realize any given function $f(X)$ may be determined as follows.

For the 00...0 input minterm condition, all the f_p's will be zero valued. Hence coefficient a_0 must provide the function output value

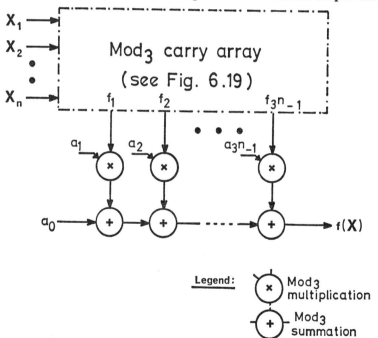

Fig. 6.20. The multiplication and summation modulo-3 of the array outputs of Fig. 6.19 in order to realize any $f(X)$.

$f(X)$ at this point. Coefficient a_0 is therefore set to the value of $f_{00\ldots0}(X)$. At minterm $00\ldots1$ $f_1 = 1$ is now present, and therefore in the final horizontal summation we have

$$\{a_0 + a_1 f_1\}_{\text{mod}3} = f_{00\ldots1}(X)$$
$$= \{a_0 + a_1\}_{\text{mod}3} = f_{00\ldots1}(X)$$

whence a_1 may be determined. This may be generalized to the requirement that

$$\{f_{p-1}(X) + a_p\}_{\text{mod}3} = f_p(X)$$

$1 \leq p \leq 3^n - 1$.

These calculations for the required a_p values may also be generated by a simple matrix evaluation[32], although in general they may be written down immediately from a consideration of the output truth table. Table 6.7 illustrates a simple two-variable function $f(X)$ and the required nine a_p coefficient values.

Table 6.7. The determination of the a_p coefficients required in a Kodanapani and Setlur ternary array for a two-variable function $f(X)$

Input X_1, X_2 Input minterm p	0,0 0	0,1 1	0,2 2	1,0 3	1,1 4	1,2 5	2,0 6	2,1 7	2,2 8
Output $f(X)$	0	1	2	2	1	0	2	0	0
Required value of $a_0, a_1, \ldots, a_8$	0	1	1	0	2	2	2	1	0

In comparison with Miller and Muzio's cubical array considered earlier in this section, this latter type of rectangular array is more compact. The cell count is $n(3^n - 1)$ carry cells plus $3^n - 1$ pairs of multiplication and addition functions, whilst the circuit complexity per cell is likely to be noticeably less than the cubic cell circuitry of the former array. Nevertheless the triangular T-gate array still appears to be a more practical solution to the quest for a universal ternary array assembly than either of these alternative universal proposals.

6.7. Ternary Threshold-Logic Considerations

Chapter 4 concerned itself exclusively with two-valued (binary) threshold-logic considerations. However, in the closing part of §1.9 of Chapter 1 a very brief mention was made of the possibility of higher-

The Design of Ternary Logic Networks

valued threshold-logic working. Ternary threshold-logic gates are clearly the next step upwards from binary threshold-logic gates.

The properties of a multi-valued ($R > 2$) threshold-logic operators have been studied by several authorities, with the main emphasis on the ternary-valued case. A practical procedure for optimal network synthesis using multi-valued threshold-logic gates, however, is still not known. Indeed this is not surprising, as even in the two-valued case, as we have seen in Chapter 4, no universally acceptable design procedure is yet available to utilize fully the power and flexibility of binary threshold-logic relationships. The one area where ternary threshold-logic gates may possibly be applied without the need for complex design procedures is in the area of arithmetic functions, where the symmetry of the requirements effectively defines the gate input and output threshold values.

The fundamental properties of binary linearly separable functions, which we have covered in this book, have been generalized to the multi-valued case[35-37]. The basic definition of an R-valued threshold-logic function is as follows:

$$
\begin{aligned}
f(X) &= 0 && \text{if } \sum_{i=1}^{n} A_i X_i < T_1 \\
&= 1 && \text{if } \sum_{i=1}^{n} A_i X_i \geq T_1, < T_2 \\
&\vdots \\
&= k && \text{if } \sum_{i=1}^{n} A_i X_i \geq T_k, < T_{k+1} \\
&= R-1 && \text{if } \sum_{i=1}^{n} A_i X_i \geq T_{R-1}
\end{aligned}
$$

where the inputs X_i may each take the value $0, 1, \ldots, R-1$, and the gate parameters A_i, T are (in general) positive integer values.

For the ternary case, the definition reduces to the obvious output definition:

$$
\begin{aligned}
f(X) &= 0 && \text{if } \sum_{i=1}^{n} A_i X_i < T_1 \\
&= 1 && \text{if } \sum_{i=1}^{n} A_i X_i \geq T_1, < T_2 \\
&= 2 && \text{if } \sum_{i=1}^{n} A_i X_i \geq T_2.
\end{aligned}
$$

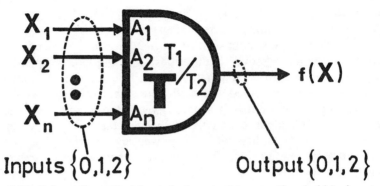

Fig. 6.21. Schematic symbol for a single-output ternary threshold-logic gate.

This is shown schematically in Fig. 6.21. Notice that the two threshold values T_1 and T_2 do not in this case denote a multi-output threshold-logic gate, but merely the different defining levels for the single 0-, 1-, or 2-valued gate output.

The basic properties of ternary threshold-logic functions follow closely from binary-valued single-threshold and multi-threshold considerations. For example, any given ternary function $f(X)$ is linearly separable and hence realizable by one ternary threshold-logic gate, iff the 3^n vertices (minterms) of the hypercube construction can be separated by two *parallel* separating planes, such that 0 minterms, 1 minterms, and 2 minterms lie completely on either side of the separating planes in this order. Further, all ternary functions which are realizable by one ternary threshold-logic gate are unate. Single threshold-logic gate realizability is maintained under permutations and negations of the input variables, and negation of the output, i.e. the normal NPN invariance operations associated with binary logic functions. Finally decomposition of a ternary threshold-logic function into two (or more) threshold-logic functions with fewer inputs per function is possible, linear-separability holding for all such decompositions. Further details and formal proofs of these features may be found published[36].

When we turn to consider how possibly to use such gates for ternary network synthesis, we find that three principal avenues have so far been considered. They are as follows.

(i) Ternary arithmetic requirements, which we have already suggested as being likely candidates for threshold realizations.

(ii) The realization of any random function $f(X)$ by the use of canonic look-up tables, that is the ternary equivalent of the canonic Chow-parameter tables.

The Design of Ternary Logic Networks

(iii) The specific case of the ternary threshold-logic gate being a ternary majority gate, giving rise to multi-valued majority-logic design considerations.

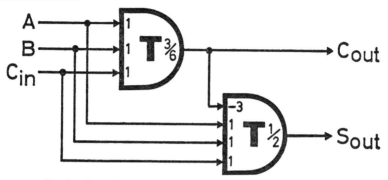

Fig. 6.22. Realization of the ternary full-adder requirements plotted in Fig. 6.8 using ternary threshold-logic gates.

The first of these three categories is amenable to intuitive design procedures, although noticeable differences of course result depending upon whether the arithmetic system is the asymmetrical 0, 1, 2 system or the symmetrical -1, 0, $+1$ system. Details relating to both may be found published[25,27,38]. The design for a symmetrical-system ternary full-adder, with the same input–output requirements as given in Figs. 6.8a and b, is illustrated in Fig. 6.22. Notice that, in the realization shown, a negative input weight is given to the carry input on the second-level summation gate; if only positive input weights are allowed, as would possibly be the case, then an appropriate 1 to 0 "inversion" of this carry signal input to the second gate would be necessary. The apparent simplicity of this realization for the ternary full-adder in comparison with, say, Figs. 6.8, 6.9, and 6.10, however, may be misleading, as the internal circuit complexity or even viability of the ternary threshold-logic gate should not be overlooked.

The possibility of canonic look-up tables for linearly separable ternary functions was mentioned in Chapter 2, §2.5.1. The main publication to date in this area is that of Moraga and Nazarala[39], who publish the minimum-integer weight and threshold values for all positive canonic linearly separable functions of $n \leq 3$.[5] The list for $n \leq 3$ has 528 entries, covering the total of 85 639 different ternary threshold functions of $n \leq 3$.

[5] Notice in §2.5.1 that we employed Moraga and Nazarala's choice of the symmetrical ternary values $\{-1,+1\}$ rather than $\{0,1,2\}$. Their classification table definitions may be translated into the latter area if desired.

However, whilst these tables are useful if the given ternary function $f(X)$ is linearly separable, like the binary Chow-parameter tables they are of little direct help when $f(X)$ is not linearly separable. Also the linearly separable ternary functions are an even smaller percentage of the total number of possible functions than in the corresponding binary case, and hence such classification tables are likely to be of limited direct use for ternary function synthesis. Much more research into the synthesis of non-linearly separable ternary functions remains to be done.

In the restricted area of ternary majority logic, rather more progress has been reported, but still the difficulty of efficient network synthesis remains generally unsolved. Hanson has shown that functional completeness can be achieved with ternary majority gates, and has produced an expansion for the realization of $f(X)$ using such operators[36,40]. This expansion is similar to the Mühldorf ternary expansion given in Chapter 1, §1.8.4 and consists of a lengthy sum-of-products type of expression in which only one term is non-zero-valued at any one time. Other work in this area is that of Varshavsky et al.[41,42].

However, we must still conclude, as was the case in 1971[36], that so far we are still unable to design efficient ternary threshold realizations for random-logic functions. This therefore remains an open area of logic theory and design.

6.8. Sequential Circuits

Leaving higher-valued threshold-logic considerations to further research, let us return to possibly less frustrating areas, using non-threshold-logic elements. Let us finally look at ternary sequential circuits.

Research and developments in the area of ternary sequential circuits fall into two main parts, namely (i) a theoretical consideration of generalized asynchronous sequential system design for $R \geq 2$, of which the ternary-valued case is merely the next higher case to binary, and (ii) the design and subsequent simple applications of tristable storage (or "memory") circuits, without necessarily any prior algebraic or structural requirements. These arrangements are usually clocked (synchronous) circuits.

The consideration of generalized sequential synthesis for any R has been approached by a few authorities. Zavisca and Allen in 1971 introduce the structure of a generalized asynchronous sequential sys-

The Design of Ternary Logic Networks

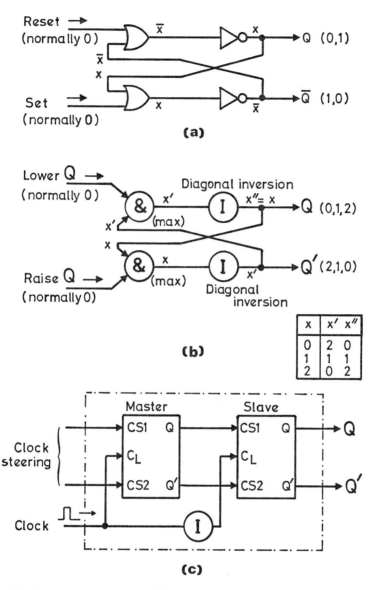

Fig. 6.23. The basic principles of R-valued storage elements, $R \geq 2$: (a) the bistable circuit, with cross-coupled OR/INVERTER functions; (b) the corresponding tristable circuit (note that the proposed INVERTER function does not alter the 1 input signal value); (c) basic master–slave configuration for clocked operation.

tem similar to that previously illustrated in Fig. 4.33 of Chapter 4, and continue to discuss the general realization of such a network using literal operators (see §6.5) and "zero-ordered-hold" storage functions[43]. The latter is a storage circuit which samples its single multi-valued input, signal values $0, 1, \ldots, R-1$, and provides the same output signal until the next sample is taken.

The work covered in this publication is extended by the publications of McDonald and Singh[44] and Wojcik[45]. Here concentration on state assignments and the difficult problem of formulating a minimal state-assignment table is considered. Both these publications highlight the difficulties of asynchronous sequential design for $R > 2$; the increasing choice of state assignments with $R > 2$ accentuates the difficulty of determining a "best" assignment for any given problem which together with the actual choice of circuit elements which may be used in the final realization makes this whole area a very difficult one to handle. The publications available to date must, as their authors point out, be considered as initial results and considerations in this area.

However, in the area of synchronous (clocked) circuits and applications, rather more practical progress may be found, as of course the problems here are not so intractable. A number of published papers consider ternary counters, ternary-coded-decimal ("TCD") counters, shift registers, and similar configurations which are commonly encountered in the binary field. In the ternary-coded-decimal area in particular a large number of possible TCD codes may be suggested, such as positive-weighted codes, positive-and-negative-weighted codes, "excess" codes comparable with the excess-three BCD code, and so on. Details of these may be found in the published literature[17,19,25,46]. Here we shall briefly look at some of the types of storage circuit elements which have been proposed for such duties, and indicate how they may be incorporated in simple ternary sequential networks.

A general extension from binary-valued storage elements to ternary- and higher-valued cases has been considered by Irving and Nagle[47]. Their work extended the basic principles of cross-coupled binary gates to form binary storage elements to that of cross-coupled higher-valued gates to form corresponding higher-valued storage elements. Figures 6.23a and b illustrate this concept, illustrating here the binary and the corresponding ternary cases. An extension of these principles to clocked master–slave assemblies, similar to the familiar binary master–slave JK circuit, was also covered, such higher-radix assemblies using two multi-valued storage elements (the master and the slave) to achieve unambiguous circuit action on receipt of the input clock. This is illustrated in Fig. 6.23c.

The Design of Ternary Logic Networks

The ternary equivalents of the familiar binary storage circuits, particularly the JK circuit, have been studied in their own right by a number of authorities, not necessarily considering such circuits as fundamentally cross-coupled radix-R unary functions. Unfortunately there is little agreement to date between precise terminologies and state tables for such assemblies, and each must be separately studied. As representatives of this work we shall look at the proposals of Porat[19,25], Moraga[46], and Duncan[29]. For a more comprehensive bibliography and general review reference may be made to Vranesic and Smith[48].

Porat's tristable circuit proposals cover both single-input–single-output and triple-input–triple-output circuits. Here we shall briefly mention the type D single-output circuit, and the type JKL triple-output circuit. Their input–output state tables are as detailed in Table 6.8.

It will be noted that in the type JKL tristable circuit the provision of three circuit outputs does not provide more than three different combinations of states, as the Y_0, Y_1, and Y_2 outputs are not independent of each other. There is also a high measure of redundancy in the JKL clock-steering inputs. Its relaxation to a normal binary type JK bistable circuit is straightforward, by applying 0 to L and using the J, K, Y_1, and Y_2 terminals only.

The design of synchronous sequential circuits using, say, the JKL circuit follows normal binary practice, the various steps being as follows.

(i) Formulation of the required output sequence tabulation.

(ii) Tabulation of the required next-state outputs against the present-state outputs.

(iii) Tabulation of the necessary clock-steering input signals to produce the required next-state outputs.

(iv) Minimization of each individual clock-steering input signal requirement.

(v) Realization of these requirements with the available combinatorial logic gates.

It is left as an exercise for the reader to try a simple synchronous sequential design using this tristable circuit. Notice that the clock input signal in this and all subsequent clocked tristable circuits may be considered to be a simple rectangular clock pulse, as in normal binary circuits.

Moraga's tristable circuit proposal[46] is a single-output circuit, with two clock-steering inputs S and C. The internal circuitry consists

Table 6.8. State tables for Porat's type D and type JKL tristable circuits

(a)

Present-state output Q_n	Required next-state output Q_{n-1}	Necessary clock-steering input D
State "0" $\begin{cases} 0 \\ 0 \\ 0 \end{cases}$	0 1 2	0 1 2
State "1" $\begin{cases} 1 \\ 1 \\ 1 \end{cases}$	0 1 2	0 1 2
State "2" $\begin{cases} 2 \\ 2 \\ 2 \end{cases}$	0 1 2	0 1 2

(b)

	Present-state outputs Q_n			Required next-state outputs Q_{n+1}			Necessary clock-steering inputs		
Outputs:	Y_0	Y_1	Y_2	Y_0	Y_1	Y_2	J	K	L
State "0" $\begin{cases} \\ \\ \end{cases}$	2 2 2	0 0 0	0 0 0	2 0 0	0 2 0	0 0 2	0 0 1/2	0 1/2 0	d.c. d.c. d.c.
State "1" $\begin{cases} \\ \\ \end{cases}$	0 0 0	2 2 2	0 0 0	2 0 0	0 2 0	0 0 2	0 0 1/2	d.c. d.c. d.c.	1/2 0 0
State "2" $\begin{cases} \\ \\ \end{cases}$	0 0 0	0 0 0	2 2 2	2 0 0	0 2 0	0 0 2	d.c. d.c. d.c.	0 1/2 0	1/2 0 0

(a) The single-output "delay" circuit.

(b) The triple-output circuit. Note, the three outputs Y_0, Y_1, and Y_2 are binary valued; also the clock-steering input 1/2 indicates that either a 1 or a 2 signal voltage is acceptable.

The Design of Ternary Logic Networks

basically of a pair of binary JK circuits with one of the four possible combinations of output states suppressed. The number of clock-steering instructions is severely limited, being summarized by the four-line state table given in Table 6.9. Notice that the clock-steering input C effectively acts as an inhibit input when set to 0, whilst when it is set to 2 effectively allows type D tristable circuit action with the second clock-steering input S. The use of this circuit in clocked sequential circuits is, like Porat's circuit proposals, basically quite straightforward.

Table 6.9. State table for Moraga's type SC tristable circuit

Present-state output Q_n	Required next-state output Q_{n+1}	Necessary clock-steering inputs	
		S	C
Q_n	Q_n	d.c.	0
d.c.	0	0	2
d.c.	1	1	2
d.c.	2	2	2

The final clocked tristable circuit which we shall review is more complex and hence more versatile than either of the previous circuits, although like Moraga's proposal it was conceived on the basis of twin coupled binary JK bistable elements. The state table for this circuit, termed by Duncan a type LMN circuit, is given in Table 6.10. Note that the circuit has one ternary output Q and three ternary clock-steering inputs L, M, and N. The state-transition diagram for this tristable circuit and also the two previous ones are shown in Fig. 6.24.

Table 6.10. The state table for Duncan's type LMN tristable circuit.

Present-state output Q_n	Clock-steering inputs			Resulting next-state output Q_{n+1}
	L	M	N	
0	l	d.c.	d.c.	l
1	d.c.	m	d.c.	m
2	d.c.	d.c.	n	n

Note: $l = \{0, 1, 2\}$; $m = \{0, 1, 2\}$; $n = \{0, 1, 2\}$; d.c. = don't care (immaterial).

Now there is one very important difference between the LMN tristable circuit and the two previous ones, and this concerns its action on receipt of a chain of clock pulses, with no change of its clock-

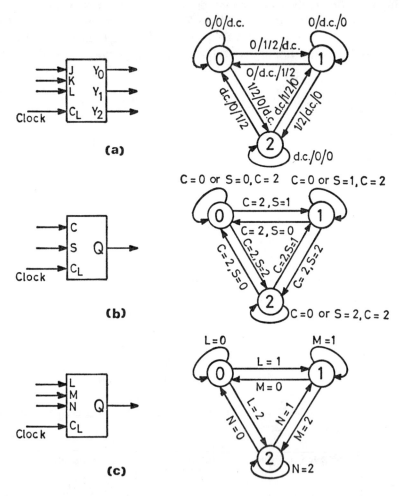

Fig. 6.24. The schematic symbol and state-transition diagram for the three tristable circuits considered in the text: (*a*) Porat's type JKL circuit; the three circuit states cited—state 0, state 1, and state 2—are defined in Table 6.8*b*. (*b*) Moraga's type SC tristable circuit; state 0 = output Q = 0, state 1 = output Q = 1, state 2 = output Q = 2. (*c*) Duncan's type LMN tristable circuit; outputs in states 0, 1, and 2 the same as (*b*).

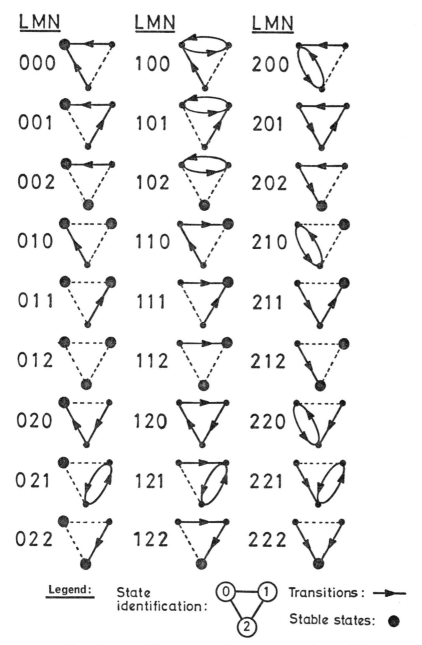

Fig. 6.25. The full range of 27 state transitions for Duncan's type LMN tristable circuit.

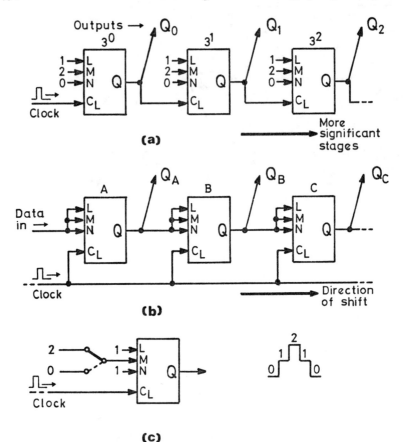

Fig. 6.26. Simple sequential designs using the type LMN tristable circuit with a single output Q and without any additional combinatorial gates for the clock-steering logic signals: (a) a ripple-through ternary counter assuming the falling edge of a clock input is the effective edge for the tristable circuits; (b) a ternary shift register (note that this may be formed into various ring counters by completing different loops from the last to the first circuit); (c) an up–down staircase generator, provided the clock-steering input M is changed from 2 to 0 either on the output $Q = 2$ or following $Q = 1$ state.

steering input data. It will be noticed that Porat's circuit can oscillate between two of its three stable states on receipt of consecutive clock pulses for certain fixed clock-steering inputs, for example between states 0 and 1 on clock-steering inputs $JKL = 011, 012, 021$, or 022, but no fixed clock-steering data will cause a circulation between all three possible states. Muraga's circuit is completely unaffected by sub-

The Design of Ternary Logic Networks 465

sequent clock pulses until the clock steering data are revised. However, the more versatile LMN circuit may be made to oscillate in many modes with fixed clock-steering data. In particular $LMN = 120$ will cause a clockwise circulation of states, whilst 201 will cause an anticlockwise circulation. The full range of these possibilities is shown in Fig. 6.25.[6]

Clearly in the type LMN tristable circuit we have a potentially very powerful function available; the debatable point is possibly whether all 27 natural modes shown in Fig. 6.25 would ever be required! Also, although only one circuit output Q is shown in our previous considerations, much greater circuit flexibility is available if two outputs, say Q and the diagonal inversion Q', are provided as follows (cf. the Q and $\bar{Q}$ outputs of a normal binary bistable circuit).

Q	Q'
0	2
1	1
2	0

The use of this type LMN circuit in simple sequential applications is entirely straightforward. Figure 6.26 illustrates some typical possibilities, obtainable without the use of a second output Q' and without the use of any additional clock-steering logic.

Various other tristable circuit proposals may be found in the literature[17,48]. Certainly many of these proposals may be readily fabricated by present-day integrated circuit technologies, although the best configuration to decide upon for possible monolithic realization is still a matter of debate and continuing research. The type LMN tristable circuit proposed by Duncan, however, is a very interesting possibility if a fast and efficient integrated circuit configuration could be engineered, particularly so if Q and Q' outputs are provided.

6.9. Chapter Summary

The synthesis of ternary networks which we have considered in this chapter is seen to be one of considerably greater complexity that that of the binary case where optimal network design, that is, for example,

---

[6] The other two tristable circuits may of course be arranged to produce any such sequence of states by the addition of appropriate logic networks on their clock-steering inputs.

using the fewest number of gates or interconnections, is required. This difficulty is compounded by the range of possible ternary functions which may be proposed to constitute the system "building blocks."

As far as choice of gates is concerned, an entirely academic approach would be to specify the types which give a minimum gate count per network, or which may be handled by a mathematically satisfying and easily manipulated algebra. This, however, entirely ignores the practicality or otherwise of the circuit realization of such gates. The other possible extreme is to see what range of functionally complete ternary gates can be designed with simple internal circuit configurations, and then use this functionally complete set in network realizations, with possibly a difficult algebra or minimization procedure in the synthesis of minimal networks. Somewhere between these two extremes should lie the optimum overall system realization.

We shall consider in the following chapter some of the ternary circuit configurations which have been proposed, but again we shall encounter a fundamental difficulty. This is that the majority of proposed circuits have binary signal paths within their structure, and hence the higher information content of ternary signal paths is not maintained within the gate structure. The circuits therefore are inherently more complex in terms of component or interconnection count than would be the case if exclusively ternary signals could be present. Such inefficiency of course applies also to the system synthesis, if quasi-binary networks such as we have considered in previous pages are used.

What we still require is an efficient truly three-state switching device, like the effectively obsolete three-state relay but in modern solid-state form. In some respects this want is a sad reflection of the complete swamping of the digital field by the present-day solid-state binary devices and systems. The 1957 comments of Shestakov[49] in discussing the development of ternary computers at that time, that "... another pressing development in computer technology is the development of ten-position (devices) for computing directly in the decimal system without translation of numbers into some other number system..." seems as far from fruition now as then. Indeed, even the ternary computer has slipped into present oblivion in the face of its binary competitor, but hopefully future resurgence of practical developments in the area of $R > 2$ will arise.

References

1. Karnaugh, M. A map method for the synthesis of combinational logic circuits. *Trans. AIEE* **72** (1), 593–99, 1953.

2. Quine, W.V. The problem of simplifying truth functions. *Am. Math. Mon.* **59**, 521–31, Oct. 1952.

3. McClusky, E.J. Minimisation of Boolean functions. *B.S.T J.* **35**, 1417–44, Nov. 1956.

4. Lewin, D. *Logical Design of Switching Circuits*. Elsevier/North-Holland, New York; Nelson, London, 1974.

5. Friedman, A.D. *Logical Design of Digital Systems*. Computer Science Press, Calif., 1975.

6. Hurst, S.L. An extension of binary minimisation techniques to ternary equations. *Computer J.* **11**, 277–86, Nov. 1968.

7. Hurst, S.L. Boolean minimisation techniques, *Electron. Lett.* **2**, 291, Aug. 1966.

8. Webb, D.L. Generation of an N-valued logic by one binary operation. *Proc. Natn. Acad. Sci.* **21**, 252–4, May 1935.

9. Martin, N.M. The Sheffer functions of 3-valued logic. *J. Symbol. Logic* **19**, 45–51, March 1954.

10. Rohleder, H. Three valued calculus of theoretical logics and its application to the description of switching circuits which consist of elements of two states. *Z. Angew. Math. Mech.* **34**, 308–11, 1954.

11. Goto, M. *Theory and Structure of the Automatic Relay Computer ETL Mk II*. Industrial Science Research of the Electrotechnical Lab., Japan, No. 556, 1956.

12. Vacca, R. *A Three-Value System of Logic and its Application to Base Three Digital Circuits*. UNESCO/NS/ICIP, G.2.14, 1957.

13. Yoeli, M., and Rosenfield, G. Logical design of ternary switching circuits. *Trans. IEEE* **EC14**, 19–29, Feb. 1965.

14. Pugh, A. Application of binary devices and Boolean algebra to the realisation of 3-valued logic circuits, *Proc. IEE* **114**, 335–8, March 1967.

15. Mühldorf, F. Ternare Schaltalgebra, *Arch. Elekt. Urbertrag.* **12**, 138–48, March 1958.

16. Lowenschuss, O. Non-binary switching theory. *IRE Natn. Conv. Rec.* **6** (4), 305–17, 1958.
17. Hurst, S.L. Semiconductor circuits for 3-state logic applications. *Electron. Eng* **40**, 197–202, April 1968; 256–9, May 1968.
18. Herrman, R.L. Selection and implementation of a ternary switching algebra. *Proc. Spring Joint Computer Conf. AFIPS* **32**, 283–90, 1968.
19. Porat, D.I. Three-valued digital systems. *Proc. IEE* **166**, 947-54, June 1969.
20. Birk, J.E., and Farmer, D.E. Design of multivalued switching circuits using principally binary components. *Proc. Int. Symp. on Multiple-Valued Logic, May 1974*, pp. 115–33.
21. Mouftah, H.T., and Jordan, I.B. Integrated circuits for ternary logic *Proc. Int. Symp. on Multiple-Valued Logic, May 1974*, pp. 285–303.
22. Su, S.Y.H., and Cheung, P.T. Computer minimization of multi-valued switching functions. *Trans. IEEE* **C21**, 995–1003, Sept. 1972.
23. Ostapako, D.L., Cain, R.G., and Hong, S.J. A practical approach to two-level minimisation of multivalued logic. *Proc. Int. Symp. on Multiple-Valued Logic, May 1974*, pp. 168–82.
24. Nutter, R.S. The algebraic simplification of a ternary full adder. *Conf. Rec. Symp. on the Theory and Application of Multiple-Valued Logic Design, Buffalo, N.Y., May 1972*, pp. 75–81.
25. Porat, D.I. *Three-Valued Combinational and Sequential Logic Networks*. Ph.D. Thesis, University of Manchester, U.K., 1967.
26. Smith, R.W. Minimization of multivalued functions. *Proc. Int. Symp. on Multiple-Valued Logic, May 1974*, pp. 27–43.
27. Sebastian, P., and Vranesic, Z.G. Ternary logic in arithmetic units. *Conf. Rec. Symp. on the Theory and Application of Multiple-Valued Logic Design, Buffalo, N.Y., May 1972*, pp. 153–62.
28. Hurst, S.L. *An Investigation into the Realization and Algebra of Electronic Ternary Switching Functions*. M.Sc. Thesis, University of London, 1966.
29. Duncan, F.G. A Three-Valued Logic; Description Implementation and Application. Unpublished Rep., Dept. of Computer Science, University of Bristol, U.K., Sept. 1974.

30. Su, S.Y.H., and Sarris, A.A. The relationship between multi-valued switching algebra and Boolean algebra under different definitions of complement. *Trans IEEE* **C21**, 479–85, May 1972.

31. Yoeli, M. Ternary cellular cascades. *Trans. IEEE* **C17**, pp. 66–7, Jan. 1968.

32. Miller, D.M., and Muzio, J.C. A ternary cellular array. *Proc. Int. Symp. on Multiple-Valued Logic, May 1974*, pp. 469–82.

33. Kodanapani, K.L., and Setlur, R.V. A cellular array for multivalued functions. *Proc. Int. Symp. on Multiple-Valued Logic, May 1974*, pp. 529–44.

34. Akers, S.B. A rectangular logic array. *Trans IEEE* **C21**, 848–57, Aug. 1972.

35. Carnevale, A. *Circuit Realization of Multi-Threshold Elements*. M.S. Thesis, Massachusetts Institute of Technology, Sept. 1965.

36. Ying, C., and Susskind, A.K. Building blocks and synthesis techniques for the realization of M-ary combinational switching functions. *Conf. Rec. Symp. on the Theory and Application of Multiple-Valued Logic Design, Buffalo, N.Y., May 1971*, pp. 183–205.

37. Druzeta, A., Vranesic, Z.G., and Sedra, A.S. Application of multithreshold elements in the realization of many-valued logic networks. *Trans. IEEE* **C23**, 1194–8, Nov. 1974.

38. Kookkenen, H., and Ojala, L. Some applications of many-valued threshold logic in digital arithmetic. *Conf. Rec. Symp. on Theory and Application of Multiple-Valued Logic Design, Buffalo, May 1972*, pp. 47–64.

39. Moraga, C.R., and Nazarala, J. Minimum realization of ternary threshold functions. *Proc. Int. Symp. on Multiple-Valued Logic, May 1974*, pp. 347–59.

40. Hanson, W.H. Ternary threshold logic. *Trans. IEEE* **EC12**, 191–97, June 1963.

41. Varshavsky, V.I. Ternary majority logic. *Avtom. Telemekh.* **25** (5), 673–84, 1964 (in Russian).

42. Varshavsky, V.I., and Ovsievich, B. Networks composed of ternary majority elements. *Trans. IEEE* **EC14**, 730–3, Oct. 1965.

43. Zavisca, E.G., and Allen, C.M. An approach to multiple-valued sequential circuit synthesis. *Conf. Rec. Symp. on the Theory and Application of Multiple-Valued Logic Design, Buffalo, N.Y., May 1971*, pp. 206–18.

44. McDonald, J.F., and Singh, I. Extensions of the Weiner–Smith algorithm for m-ary synchronous sequential circuit design. *Proc. Int. Symp. on Multiple-Valued Logic, May 1974*, pp. 135–54.

45. Wojcik, A.S. Multi-valued asynchronous sequential circuits. *Proc. Int. Symp. on Multiple-Valued Logic, May 1974*, pp. 155–67.

46. Moraga, C.R. A tristable switching circuit for synchronised sequential machines. *Conf. Rec. Symp. on the Theory and Application of Multiple-Valued Logic Design, Buffalo, N.Y., May 1971*, pp. 103–11.

47. Irving, T.A., and Nagle, H.T. An approach to multi-valued sequential logic. *Conf. Rec. Symp. on the Theory and Application of Multiple-Valued Logic Design, Toronto, Canada, May 1973*, pp. 89–105.

48. Vranesic, Z.G., and Smith, K.C. Engineering aspects of multi-valued logic systems. *IEEE Computer* **7**, 34–41, Sept. 1974.

49. Shestakov, V.I. Punched card method of synthesizing sequential systems of multi-position relays. *Avtom. Telemekh.* **19** (5), 1958 (in Russian). Transl. in *Autom. Remote Control, Instrum. Soc. Am.* **20**, 1457–66, 1959.

Chapter 7

Circuit Designs for Digital Logic Functions

Introduction

In all the preceding chapters of this book we have been dealing with appropriate algebraic expressions to define our logic system and using appropriate symbols to represent our logic building blocks. However, in this final chapter we shall at last discuss circuit configurations which can be proposed to realize some of these logic functions. We shall of course confine our discussions to solid-state circuit realizations.

Emphasis, however, will remain in the following pages on newer or less familiar circuits and functions. We shall assume the reader is already familiar with the principles and range of present-day widely used logic circuits, the present scope of which is generally as shown in Fig. 7.1.

In the bipolar digital field the TTL 7400 series and subsequent variants dominate the present marketplace. The ECL area comes into its own for computer and other spheres where very high speed of operation is required. The unipolar area, which includes all the many variations on MOS technology, is particularly important in the field of very high circuit complexity but not very high speeds and for the smaller production quantity custom-designed product. These MOS advantages, however, are being continuously challenged by bipolar developments, such as the collector-diffusion-isolated ("CDI") bipolar process. Detailed considerations of many of these aspects and others may be found in Hnateck and elsewhere[1-6].

Some further monolithic circuit concepts which appear particularly applicable to the design of logic circuits are, however, being developed. Let us in the following section briefly introduce some of these newer possibilities, although not all have currently reached the stage of commercial exploitation.

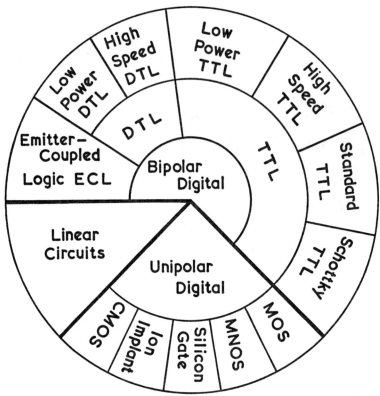

Fig. 7.1. The general range of present-day monolithic integrated circuits. Note that we are not considering the linear area at all in this book.

7.1. Newer Microelectronic Concepts

Three new silicon technologies for large scale integration have recently been publicized. Two are bipolar processes, and one an extension of MOS unipolar techniques. The three processes are (i) integrated-injection logic ("I²L"), a bipolar process; (ii) current-hogging logic ("CHL"), the further bipolar process; (iii) charge-coupled-device ("CCD") techniques, the unipolar process.

7.1.1. Integrated-Injection Logic

Integrated-injection logic, sometimes alternatively termed merged-transistor logic ("MTL"), employs multi-collector n.p.n. transistors as

Circuit Designs for Digital Logic Functions

active devices and p.n.p. transistors as collector loads. However, the emitter and base of the active n.p.n. transistor also act as the collector and base of the p.n.p. load transistor, and hence both active device and load are "merged" into one single transistor pair occupying about the space normally used for one transistor. Very high circuit densities rivaling MOS, but with the speed capability of bipolar technology, are achieved, up to 200 gates per mm^2 and a speed–power product of 0.1 pJ currently being quoted. The d.c. supply voltage can lie in the range of 1–15 V, the current per gate in the range of 1 nA to 1 mA, with an output logic voltage swing of about 0.6 V. The low output-voltage swing is inherent in this particular construction. The gate propagation delay may be as low as 10–20 ns[7-11].

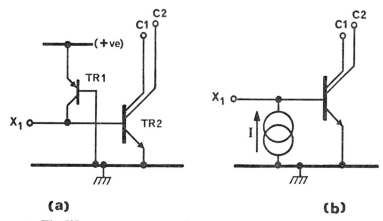

Fig. 7.2. The I^{2}L gate structure: (*a*) the "unmerged" structure, where the gate is regarded as two separate transistors TR1 and TR2 (note: more than two collectors C1 and C2 may be present); (*b*) the equivalent circuit, where transistor TR1 has been replaced by its equivalent constant-current source.

The basic structure, illustrating it as two "unmerged" transistors TR1 and TR2, is shown in Fig. 7.2*a*. Transistor TR1 acts as the load for a preceding stage and also as the base current source for TR2, which is reminiscent of an open-collector DTL logic configuration. With no connection at all at input x_1, TR2 will be on, and both collectors C1 and C2 will independently be acting as sinks to any following loads. When x_1 is pulled towards 0 V, then both C1 and C2 cease acting as sinks to following loads. Transistor TR1 may therefore be regarded as a current source which either supplies TR2 or which is sunk by a collector of the preceding stage. This is illustrated in Fig. 7.2*b*. The

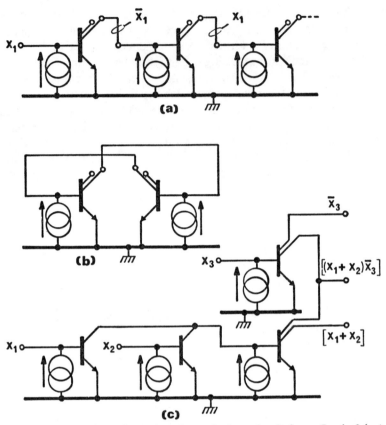

Fig. 7.3. Simple I²L configurations in equivalent circuit form. Logic 0 is (approximately) 0 V, whilst logic 1 is one base–emitter volt drop higher: (*a*) a simple cascade of inverters; (*b*) a simple cross-coupled bistable pair; (*c*) a three-output complex function.

value of this current is controlled without the need for any ohmic resistance by the doping ratio and silicon area of TR1. All gates on the chip share one common substrate for the emitter of the TR2 transistors.

If the logical operation of a simple cascade of such gates is considered, as illustrated in Fig. 7.3*a*, then simple inversion is performed at each stage. Two such circuits may be cross-connected to form a bistable circuit, as shown in Fig. 7.3*b*. Other logic configurations with series and parallel arrangements of gates such as shown in Fig. 7.3*c*

Circuit Designs for Digital Logic Functions

may readily be proposed. In many of these assemblies a great deal of merging of chip area between the parts of the circuit shown as separate parts in these diagrams may be made. Thus families of compact but logically complex functions can be proposed.

It will be noted that there is no fan-in facility available on the basic I²L gate such as we have illustrated. Instead the multiple collector outputs per gate provide isolation between the separate logic signals of a network. However, the addition of Schottky diodes on inputs to provide isolated fan-in as well as the previous isolated fan-out has been proposed[10], which will increase still more the flexibility and advantages of this particular monolithic technology.

7.1.2. Current-Hogging Logic

The second newer bipolar technology, CHL[12–14], is somewhat similar to I²L in that multiple-collector transistors are employed, but the circuit action is entirely different. Like I²L technology, it allows logic assemblies to be fabricated within one common isolation region on the silicon chip and thus provides very high packing density.

The basic CHL element is a p.n.p. transistor, but with an additional collector C_1, the "control" collector, provided between the emitter and the normal output collector C_0, as illustrated in Fig. 7.4a. With C_1 not in circuit, or otherwise not carrying any current, the E, B, C_0 structure operates as a normal p.n.p. transistor and C_0 collects all the resultant collector current. However, if C_1 is given a negative bias with respect to its emitter it captures ("hogs") all the carriers which were previously collected by the normal collector output C_0. Collector C_0 current therefore falls to a leakage-current value. The geometry of the structure is arranged such that when C_1 is collecting, no effective path for the emitter-to-collector carriers is present to the normal collector output C_0.

If a final output n.p.n. transistor is driven by the C_0 collector current, then this output transistor conducts or ceases to conduct in accordance with the presence or absence of C_0 collector current. Hence the final output current I_0 is present when I_1 is zero, and *vice versa*, as captioned for Fig. 7.4b.

More than one control collector C_1 may be provided on the CHL element. These additional collectors may be arranged geometrically in series or in parallel, as illustrated in Figs. 7.4c or d, or in combinations of series/parallel if desired. The schematic geometry of the element shown in Figs. 7.4c and d is a useful means of considering the CHL

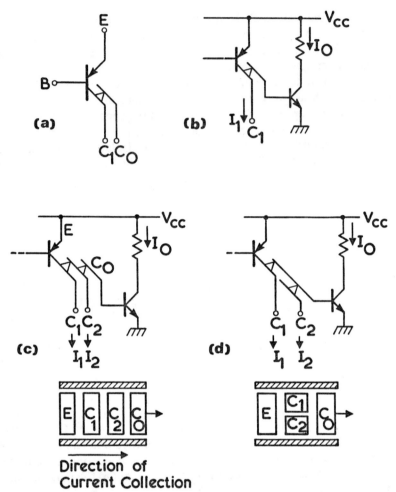

Fig. 7.4. Basic CHL logic configurations: (*a*) the symbol for a simple CHL element; (*b*) the simple CHL element used to drive an n.p.n. output transistor, giving $I_0 = \bar{I_1}$; (*c*) two control collectors in series on the CHL element, either of which can capture the CHL output current, giving $I_0 = \overline{[I_1 + I_2]}$; (*d*) two control collectors in parallel on the CHL element, neither of which by itself can capture the CHL output current, giving $I_0 = \overline{[I_1 \cdot I_2]}$.

action, as well as being the general device layout. The base region of the element is not shown, being basically a buried layer under the emitter and collector regions; the shaded areas in these latter figures represent appropriate doping areas to maintain the carriers within the desired boundaries.

Circuit Designs for Digital Logic Functions

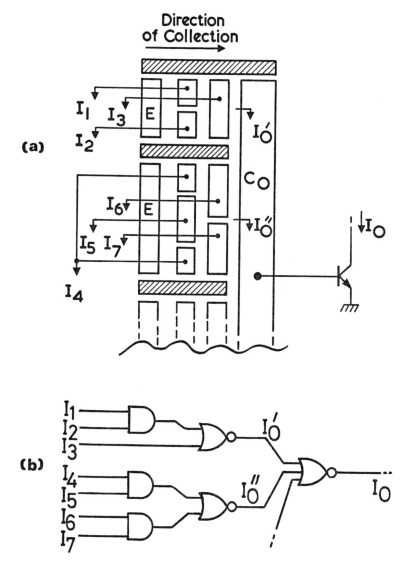

Fig. 7.5. A complex CHL function: (*a*) the geometric layout; (*b*) the equivalent logic circuit (note: the I_0 output current is a sink current, and all signal currents I_1 to I_7 are also sunk by the preceding-level circuits.)

For normal logic action the base–emitter region of all CHL elements is permanently forward biased. The binary logic signals are the presence and absence of currents in the control collectors and final output transistor. The action of complex logic configurations is readily comprehended by considering such currents, which may be likened very closely to the concept of transmission which characterizes the analysis of relay switching circuits. As an example of a CHL logic network which may be considered as one complex logic function, consider the geometric arrangement illustrated in Fig. 7.5a. The logical action of this layout is given in Fig. 7.5b. The ready translation of the one to the other should be apparent.

A comparison between CHL and I²L technologies seems to favor I²L, certainly in terms of speed–power product, being some two orders of magnitude smaller than CHL figures. However, CHL circuits appear to have higher voltage and current capability, and therefore may be very suitable for lower-speed higher-power industrial logic applications.

7.1.3. Charge-Coupled Devices

Charge-coupled devices are the ultimate extension of the concept of using packets of electrical charge as logic signals, for example as is employed in two-phase and four-phase dynamic MOS circuits. However, the latter contain active transistors in their configurations to destroy or create charge packets, whereas the CCD is an entirely passive device[1,15–17].

Basically a charge-coupled device consists of a conducting path through a silicon substrate, into which packets of charge can be introduced. These packets of charge are then held in place or moved through the substrate in accordance with patterns of electric fields produced by signals applied to the controlling electrodes of the device. The essential requirement for this concept to be feasible is for extremely high insulation resistance in the conducting channel and negligible charge recombination in order that the charge packets shall retain their identity and not rapidly decay. Even so the CCD is fundamentally a dynamic device and can never be used in a static mode, as eventually the packets of charge must decay to an unacceptable level.

Figure 7.6 illustrates in a schematic manner the basic CCD concept. Interleaved clock waveforms applied to the shifting electrodes move any charge packet by one clock pitch per clock cycle. The presence or absence of a charge package at different points along the channel may be detected, or alternatively the serial output from the end of the channel may be the required output. A three-phase device is illustrated, although with more complex double-layer electrode layouts two-phase CCD's are also possible.

Circuit Designs for Digital Logic Functions

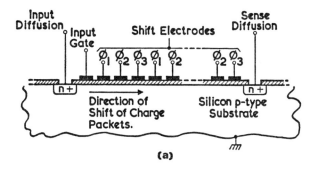

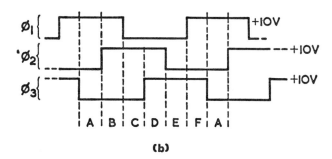

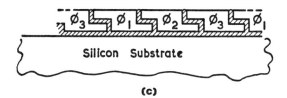

Fig. 7.6. The basic concept of a charge-coupled device ("CCD"): (a) diagramatic representation of one channel of a three-phase CCD; (b) the three clock phases, where

period A = charge packets held by electrode ϕ_1
period B = charge packets under electrodes ϕ_1 and ϕ_2
period C = charge packets held by ϕ_2
period D = charge packets under ϕ_2 and ϕ_3
period E = charge packets held by ϕ_3
period F = charge packets under ϕ_3 and ϕ_1;

(c) a cross-section of the possible ϕ_1, ϕ_2, and ϕ_3 electrode spacings.

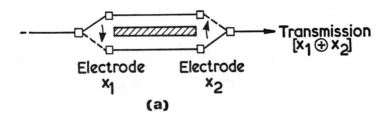

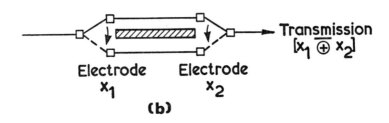

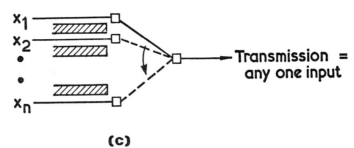

Fig. 7.7. Switching possibilities if routing of charge packets in CCD's could be done: (a) an exclusive-OR transmission circuit; (b) an exclusive-NOR transmission circuit; (c) a multiplexing circuit.

At present it is difficult to visualize this device being other than an extremely simple and compact shift-register type of storage device, unless possibly complex electrode geometries can be suggested to steer the charge packets into alternative paths. If it was possible to route a package into one or other of two paths, then we have the equivalent of a change-over mechanical contact, from which all the many very powerful switching circuits familiar to relay engineers could be built up. Figure 7.7 illustrates this idea.

Circuit Designs for Digital Logic Functions 481

However, the extreme simplicity of the CCD concept and its high packing density must make it a device of future commercial use, although whether its future lies outside the shift-register and storage area is debatable. Possibly I²L will be a better contender for the non-storage area.

7.2. Binary Exclusive-OR Gates and their Variants

The newer technologies briefly introduced in the previous section have the possibility of realizing more complex functions which may then be regarded as basic building blocks for the logic designer to employ in his network syntheses. At this stage, however, it is not clear what most useful family of complex non-vertex functions should be proposed, or the best technology with which they may be realized. Indeed the latter largely dictates the evolution and finalization of the former.

On the other hand, the exclusive-OR gate and its variants have already figured prominently in preceding chapters of this book. The idea of considering an exclusive-OR gate as being an assembly of separate vertex gates, realizing for the two-input case the expression $f(x) = [\bar{x}_1 x_2 + x_1 \bar{x}_2]$, has been mentioned as being an entirely wrong concept. Instead the exclusive gates should be considered as unique circuit designs in their own right.

Several possible circuit realizations for such gates using present-day bipolar and MOS technologies have been published and we shall therefore consider certain of them here to illustrate the concepts of some non-vertex function realizations.

7.2.1. Bipolar Circuits

A simple form of exclusive-OR configuration is shown in Fig. 7.8a[18]. Clearly no power gain is available from the circuit shown, but as most common vertex gate configurations have comprehensive output buffering as an integral part of their monolithic realization, such provision could be added to this basic two-transistor circuit without embarrassment. A three-input exclusive-OR configuration using the same circuit concepts is shown in Fig. 7.8b.

An alternative circuit concept is shown in Fig. 7.9a[19,20], the action of which is as follows. Suppose inputs x_1 and x_2 are both at 0 V, then no current flows through the resistors R_B and no current flows into the transistor base. The transistor is therefore non-conducting and the gate output is at logic 1 ($= V_{CC}$). If both x_1 and x_2 are at V_{CC}, then clearly

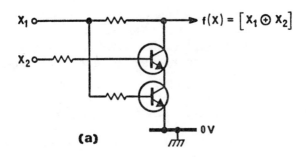

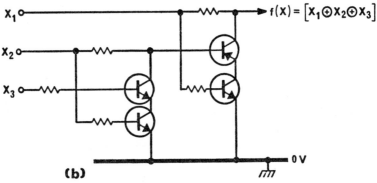

Fig. 7.8. Bipolar exclusive-OR signal-powered gates: (a) two-input exclusive-OR; (b) three-input exclusive-OR.

the whole configuration is at this voltage, and hence the gate output remains at this logic 1 level. However, if x_1 or x_2 is at V_{CC} (logic 1) with the second input at 0 V (logic 0), current will now flow through the two resistors R_B and a potential will exist between the transistor base and the emitter connected to 0 V. Base–emitter current will flow in this path, sufficient to cause the transistor to conduct. The gate output will now fall to the logic 0 level. Hence the gate output will be 0 if one but not both inputs are at 1, which is the two-input exclusive-NOR relationship.

When one but not both inputs of this circuit is at V_{CC}, not a very large current flows through the R_B resistor which is connected to the 0 V input, as this resistor is effectively shunted by the transistor base–emitter diode. However, negligible direct cross-coupling between the

Circuit Designs for Digital Logic Functions

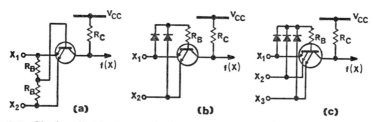

Fig. 7.9. Single-rail bipolar exclusive gates: (*a*) two-input exclusive-NOR; (*b*) modified two-input exclusive-NOR; (*c*) a three-input exclusive-type circuit.

inputs may be achieved by the modified circuit of Fig. 7.9*b*, the basic action of which remains unchanged. Additional inputs and emitters such as shown in Fig. 7.9*c* realize further exclusive-type functions, the example shown in Fig. 7.9*c* realizing

$$f(x) = [(x_1 \overline{\oplus} x_2)(x_2 \overline{\oplus} x_3)] = [\bar{x}_1\bar{x}_2\bar{x}_3 + x_1x_2x_3].$$

7.2.2. MOS Circuits

Exclusive-NOR MOS circuit realizations using a single voltage rail are similarly possible. Figure 7.10*a* illustrates a simple exclusive-NOR configuration, one or other transistor being on when one but not both inputs are energized. It may be noted that this circuit configuration is also possible and is currently used in bipolar form, but suffers from a disadvantage in that a single energized input is clamped to the deenergized input via a conducting base–emitter diode. Hence it is not feasible to attempt to share one input signal between different gates as no guarantee of equal base–emitter currents can be given[20]. This problem of input current hogging is not of course present in the alternative bipolar circuits of Fig. 7.9, or in any of the MOS circuits where no input current has to be provided.

An alternative configuration providing the exclusive-OR function is shown in Fig. 7.10*b*[19]. When both inputs are at 0 V the whole circuit including the output is at 0 V; when both inputs are at a high voltage then no voltage exists across either transistor gate source, and therefore both still remain off; when, however, one input is high but the second remains low, then one of the two transistors conducts and the high input voltage (logic 1) appears at the output. Hence the output is high if one and only one input is energized, giving the exclusive-OR relationship $f(x) = [x_1 \oplus x_2]$.

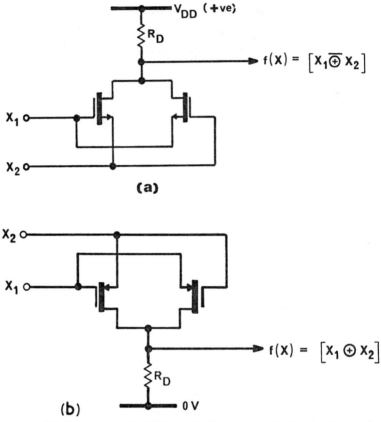

Fig. 7.10. Single-voltage-rail MOS exclusive gates: (*a*) the basic two-input exclusive-NOR gate (n-channel enhancement); (*b*) an exclusive-OR gate (p-channel enhancement). (Note: The resistors shown may be replaced with MOS transistors acting as loads in the usual manner.)

7.2.3. More Complex Functions

Many additional complex functions based upon the single-voltage-rail configurations of Figs. 7.9 and 7.10 may be proposed. All are characterized by potentially very small silicon areas for their logical power, this criterion being a vital consideration for monolithic circuits.

Figure 7.11 gives some representative examples of more complex functions which may readily be built around an exclusive-type gate, together with their Karnaugh map coverage. Notice that permutation

Circuit Designs for Digital Logic Functions 485

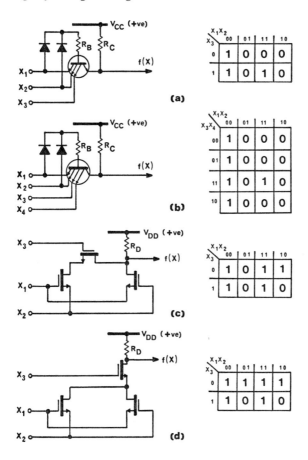

Fig. 7.11. Complex function realizations based upon exclusive configurations:

(a) Bipolar function, $f(x) = [(x_1 \overline{\oplus} x_2)(\bar{x}_1 + x_3)]$,
$$= [\bar{x}_1\bar{x}_2 + x_1x_2x_3]$$

(b) Bipolar function, $f(x) = [(x_1 \overline{\oplus} x_2)(\bar{x}_1 + x_3x_4)]$,
$$= [\bar{x}_1\bar{x}_2 + x_1x_2x_3x_4]$$

(c) n-channel enhancement
MOS function, $f(x) = [(x_1 \overline{\oplus} x_2) + x_1\bar{x}_3]$,
$$= [\bar{x}_1\bar{x}_2 + x_1x_2 + x_1\bar{x}_3]$$

(d) n-channel enhancement
MOS function, $f(x) = [(x_1 \overline{\oplus} x_2) + \bar{x}_3]$,
$$= [\bar{x}_1\bar{x}_2 + x_1x_2 + \bar{x}_3].$$

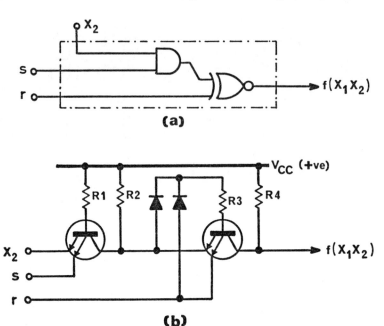

Fig. 7.12. Murugesan's universal logic gate: (a) circuit schematic, $r,s \in \{0,1,x_1,\bar{x}_1\}$; (b) bipolar realization.

and negation of the input variables together with negation of the output enables a family of functions to be realized with each circuit, the one illustrated being the positive canonic representative of each such NPN group of functions.

Finally, the idea of universal logic modules which can be used to realize any combinatorial function $f(x)$ of n variables has been covered in Chapter 5. To produce all the 16 possible functions of $n = 2$ the very simple circuit configuration similar to that shown in Fig. 7.12 has been suggested[1]. The 16 functions f_0 through to f_{15} of $f(x_1,x_2)$, where f_0 to f_{15} are as defined in Table 1.2 of Chapter 1, are realized as detailed in Table 7.1. For $n > 2$ a triangular array of such modules is proposed. Further developments along these lines may be expected.

---

[1] This module appears to have first been published in a paper dealing with *sequential* machine design, entitled "Uniform modular realization of sequential machines", by E. P. Hsieh, C. J. Tan, and M. N. Newborn, published in *Proc. ACM Conf.* pp. 613–21, 1968. Possibly because of the title of the paper this disclosure is not widely cited as an $n = 2$ universal combinatorial module. The circuit seems to have been subsequently re-invented by Murugesan[21], based on a general evolution of universal logic modules.

Circuit Designs for Digital Logic Functions

Table 7.1. Input signals to r and s to generate all 16 functions of x_1, x_2 using the universal logic gate shown in Fig. 7.12

Control inputs required	Output function															
	f_0	f_1	f_2	f_3	f_4	f_5	f_6	f_7	f_8	f_9	f_{10}	f_{11}	f_{12}	f_{13}	f_{14}	f_{15}
r	1	1	$\bar{x}_1$	$\bar{x}_1$	1	1	$\bar{x}_1$	$\bar{x}_1$	x_1	x_1	0	0	x_1	x_1	0	0
s	0	x_1	x_1	0	$\bar{x}_1$	1	1	$\bar{x}_1$	$\bar{x}_1$	1	1	$\bar{x}_1$	0	x_1	x_1	0

From all these examples it is beginning to be feasible to consider basic building blocks for logic synthesis which are fundamentally more powerful, though not necessarily more costly in silicon area, than simple vertex gates. Clearly, however, more sophisticated synthesis techniques to exploit these new functions fully become necessary, and this is where the concepts considered in Chapter 5 must mature hand-in-hand with the design and fabrication of these and other newer functions.

7.3. Binary Threshold-Logic gates

In the majority of textbooks on threshold logic currently available, no mention is made of possible circuits by which threshold-logic gates may be made. This is equally true of the papers published by learned societies on this subject. Nevertheless, a considerable amount of work has been done on possible circuits, although regrettably no monolithic threshold-logic gates are yet commercially available.

One very early development was the magnetic-core logic element, which we have already illustrated and commented upon in Chapter 3. At least three binary computers using magnetic cores were manufactured,[2] all of which used the magnetic cores as majority decision functions.

The first non-magnetic threshold-logic gates were simple potential-divider RTL circuits built up from discrete components, as illustrated in Fig. 7.13. The action of this p.n.p. circuit is that until a certain number of inputs are energized with a negative input signal (logic 1) the base potential-divider network holds TR1 reverse-biased, and hence output $f(x) = 0$. When sufficient inputs are energized, however, the base input is no longer maintained reverse-biased and TR1 therefore conducts, giving output $f(x) = 1$. Different input weightings

[2] The Japanese MUSASINO 1 computer and the Ferranti UK SIRIUS and ORION 1 computers, the former using sinusoidal input signals, the latter using input pulses[22-24].

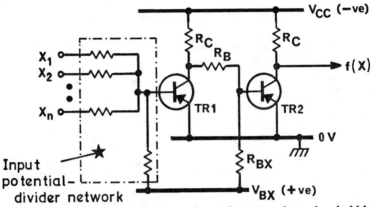

Fig. 7.13. An early potential-divider action resistor–transistor threshold-logic gate employing p.n.p. bipolar transistors.

and gate threshold may be obtained varying the value of the various resistors in the input potential-divider network.

This type of circuit was employed in a small evaluation computer, the General Electric DONUT computer[23,25]. However, the concept of employing *voltage summations* at gate inputs to give the threshold action is one in which tolerance on the input signals becomes very troublesome and which led rapidly on to the development of alternative *current-summation* techniques. The concept of current summation in fact characterizes the majority of threshold-logic gates subsequently developed after these early magnetic-core and voltage-summation proposals.

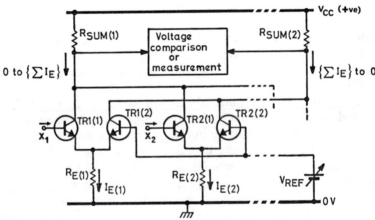

Fig. 7.14. The principle of current-summation threshold-logic gates.

7.3.1. Current-Summation Gates

The fundamentally more satisfactory method of establishing the input summation part of a threshold-logic gate is by establishing precise units of current from each of the input signals. This is readily done by using emitter-coupled transistor input pairs, as illustrated in Fig. 7.14.

The action of each input pair is similar to that in normal ECL logic circuits. As an input signal voltage rises from the logic 0 level to logic 1 level, so the current in the emitter resistor is switched from the right-hand to the left-hand transistor of the pair, the value of this current, which is a measure of the input weighting at this point, being controlled by the emitter resistor and not directly by the value of the logic 1 input signal voltage. Hence preset units of current may be switched by the several input signals, these units of current being summed by the common summing resistors.

The threshold detection part of the circuit may either be an absolute measurement of the resultant voltage across one or other summing resistor R_{SUM} or a comparison of the voltages across the two, in the latter case detecting when, say, the voltage across $R_{SUM(1)}$ rises above the voltage across $R_{SUM(2)}$. The latter is the technique employed in earlier circuits of this form, including the first fully integrated threshold-logic gate produced[26,27]. The circuit of this monolithic gate is shown in Fig. 7.15.

Considerable development of this basic circuit concept has taken place, including the possibilities of single-sided summation using one summing resistor $R_{SUM(1)}$ only and entirely deleting the opposite resistor, and other circuit embellishments to enhance the overall performance. Applications and fabrications have also been considered in detail[27-29]. The advantages which inherently are present in monolithic circuit realizations, such as matching of transistor and other circuit parameters, are used to full advantage in these circuits.

However, there are two problems associated with these ECL-type circuits. Firstly internal tolerances must ultimately limit the fan-in and/or threshold discrimination which can reliably be produced by such gate structures, and secondly the input and output signal levels with such gates are characteristically ECL-type swings.

The problem of quantifying the permissible parameter tolerances of current-summation threshold-logic gates is a peculiar problem. Intuitively it is clear that a simple gate, say a three-input majority gate $\langle 1,1,1 \rangle_2$, will be more tolerant of internal parameter variations than, say, a five-input gate of specification $\langle 3,2,2,1,1 \rangle_5$, but just exactly how

490 The Logical Processing of Digital Signals

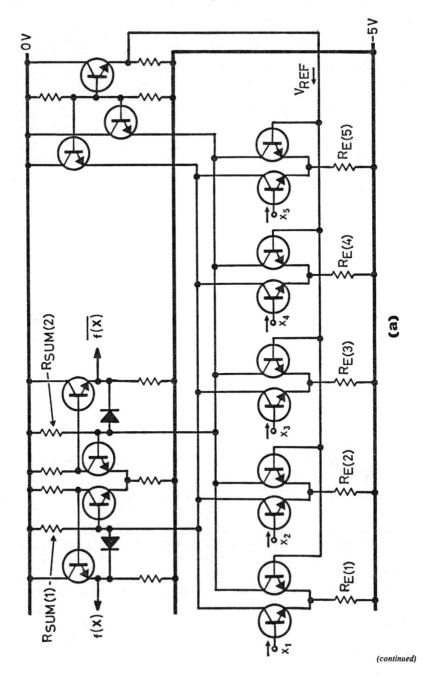

(a)

(continued)

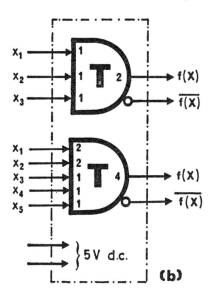

Fig. 7.15. The RCA monolithic integrated circuit threshold-logic gate of 1966: (a) circuit of the five-input gate, specification $\langle 2,2,1,1,1 \rangle_4$, resistors $R_{E(1)}$ to $R_{E(5)}$ having resistor ratios of $\frac{1}{2},\frac{1}{2},1,1,1$, respectively; (b) the two gates provided per 14-pin package.

input fan-in and gate threshold fit together in this context of permissible tolerance is not immediately clear.

The problem has been considered in several ways[30-34]. One analysis for a balanced current–summation realization is given in Appendix F, from which the rather surprising result emerges that the allowable tolerances on internal parameters, assuming they all have the same tolerance value, is inversely proportional to the sum of the $n + 1$ Chow parameters of the gate. In particular it is suggested that a value for the maximum allowable tolerance δ_{max} is given by

$$\delta_{max} = \pm \left\{ \frac{0.5}{\sum_{i=0}^{n} a_i} \right\} \times 100\%$$

It will be recalled that the full $n + 1$ Chow parameters embrace both gate-input and gate-threshold information, and hence this δ_{max} expression is correctly bringing all the necessary gate parameters into the tolerance equation. This is yet another example of the concentration of information which is present in these remarkable parameters.

Values for δ_{max} given by the above expression will be found in Appendix F. These results have been applied to indicate which range of gate specifications are feasible and which are not feasible, if one assumes a value for δ_{max} of $\pm 5\%$. This consideration shows that all threshold-logic gates covered by the $n \leq 4$ Chow-parameter listing are feasible, but for $n \leq 5$ 10 of the 21 canonic entries require a tighter

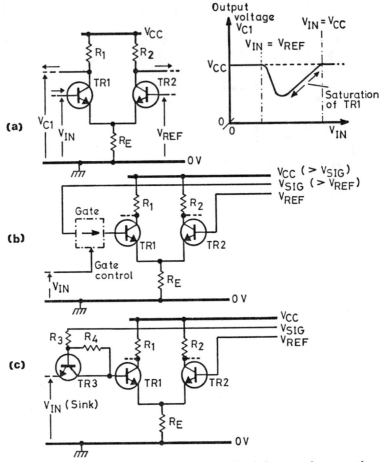

Fig. 7.16. Modifications to the emitter-coupled-pair input to improve the permitted input voltage swing: (*a*) the normal input characteristics; (*b*) the principle of not using the applied input signal to control TR1 directly; (*c*) one possible circuit realization for (*b*).

tolerance than ± 5% for their successful realization[34]. Whether these conclusions are precisely accurate for any given monolithic technology of course requires more exact data, but it is likely that they correctly indicate the general limit of current-summation threshold-logic realizations.

On the further question of the ECL-type input and output signal-voltage swings, which are typically 800 mV between the logic 0 and 1 signal values, various circuit improvements can be proposed which provide TTL-compatible input and output voltages.

Figure 7.16a illustrates the normal input voltage characteristic of a simple emitter-coupled transistor pair. The input voltage required to switch current from the right-hand side of the circuit to the left-hand side must be greater than V_{REF}, but must not be sufficient to cause saturation of TR1. However, if the actual input signal is used merely to gate an appropriate preset voltage, whose value is such as to cause the correct transfer of current from TR2 to TR1 without saturation, then the actual input signal voltage can be allowed to swing through much wider limits, for example from 0 to V_{CC} as in TTL circuits. This possibility is illustrated in Figs. 7.16b and c. The internal gate voltage V_{SIG} may conveniently be generated by the same circuit as generates V_{REF}, V_{SIG} being one base–emitter volt drop higher than V_{REF}. Such a circuit is illustrated in Fig. 7.17a. Analysis of the output voltage V_{REF} of this particular circuit will show that it is $\frac{1}{3}V_{CC}$, provided the base–emitter characteristics of the transistors are matched[28,35], as of course they would be within close limits in a monolithic realization[3]. Finally, a possible voltage-comparison circuit which provides (approximately) 0 to V_{CC} output voltage swing is shown in Fig. 7.17b. Again this simple circuit depends for its remarkable voltage discrimination upon close matching of the transistors in the voltage comparison part of the circuit, its difference-mode hysteresis between fully on and fully off being only about ± 150 mV.

The possibility of multi-output threshold-logic gates using current-summation techniques has also been considered[36]. This requires a number of summing resistors in series for each summing side instead of the single resistor R_{SUM} per side that we have previously illustrated. Very close component tolerances would seem to be necessary for this particular form of realization, and also the chip area taken up by the chain of summing resistors in addition to the individual emitter resistors may prove to be an embarrassment.

7.3.2. Digital-Summation Gates

In contrast to the previous circuits, which all employ some form of analog summation, the principle of digital summation to provide the threshold action has been investigated. The immediate advantage of

---

[3] This circuit is one of a series of very interesting voltage-divider circuits ("d.c. transformers") with low output impedances which generate an output voltage equal to $1/x$ of the supply rail voltage, where x is an integer ≥ 2. Each increase in x requires an additional transistor in the circuit configuration to cancel out internal V_{BE} volt drops, the total number of transistors per circuit always totaling x.

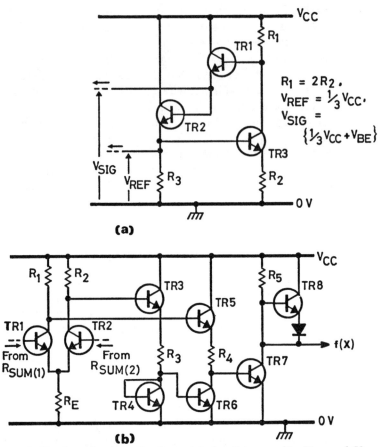

Fig. 7.17. Further circuit realizations: (*a*) circuit to provide V_{REF} and V_{SIG}, as shown in Fig. 7.16c; (*b*) output circuit to give an output voltage swing of (approximately) 0 to V_{CC}.

this alternative approach is that it entirely removes the cumulative tolerance problems but in exchange brings in certain disadvantages of its own, as we shall shortly see.

The principle of a digital-summation threshold-logic ("DSTL") gate is shown in Fig. 7.18. However, whilst the possibility of sequential summation is entirely feasible, it has been shown that combinatorial networks to provide the summation are generally preferable[37]. The basic combinatorial network proposed for a DSTL gate is a regular array structure, as illustrated in Fig. 7.19a. All the cells in the com-

Circuit Designs for Digital Logic Functions

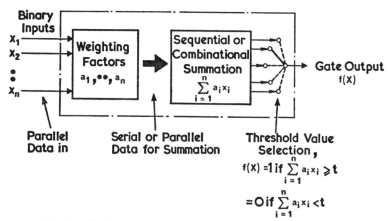

Fig. 7.18. Basic principle of the digital-summation type of threshold-logic gate.

plete array are identical. The action of each cell is such that a logic 1 input at Q is transmitted diagonally across the cell to output R, unless there is also a simultaneous 1 input at cell input P. In the latter case the P input signal is routed to output R, whilst the Q input signal is deflected across to output S. The logic for this action is an AND/OR pair of gates, as shown in Fig. 7.19b.

The action of the complete array of cells is therefore such that logic 1 signals on the cell input lines y_1 to y_m cascade diagonally to the left through the array to energize the m output lines z_1 to z_m consistently from left to right. An interesting feature is that no matter in what order the 1 input signals are applied and removed, the outputs on lines z_1 to z_m always "fill up" regularly from left to right, and "empty" regularly from right to left. There is no ripple-through action on the output lines with transistory wrong output signals during changing input data.

For input weightings of greater than unity, two (or more) of the cell input lines y_i, $i = 1 \ldots m$, are hard-wired together, as illustrated in Fig. 7.19 for input x_n. Notice that the number of cell inputs m has to be made equal to the total input summation. Also if the full triangular array is present, then all threshold detection outputs from threshold $t = 1$ (OR) through to $t = m$ (AND) are available. Thus the DSTL gate inherently possesses multi-threshold and variable-threshold capability, unlike all the simple analog summation types of threshold-logic gate. Inversion on the z_1 to z_m output lines will of course provide the various outputs from NOR through to NAND[38].

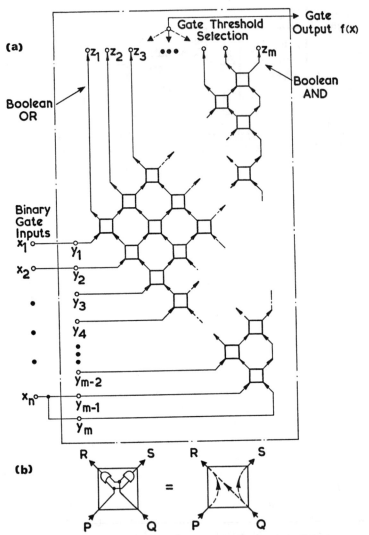

Fig. 7.19. Schematic arrangement of the combinatorial digital-summation threshold-logic gate: (*a*) complete cell assembly; (*b*) cell details, all cells identical.

For specific applications the full triangular array may be truncated in several ways to eliminate unnecessary cells[37,39]. For example, if only one fixed threshold output is required, it is unnecessary to provide any of the cells which feed the higher-valued threshold outputs and a truncation such as shown in Fig. 7.20a becomes possible. Similarly for

Circuit Designs for Digital Logic Functions 497

input weightings of greater than unity, certain cells may be dispensed with at the bottom of the array, as suggested in Fig. 7.20b.

It will be noticed that the basic triangular array, and any simple truncations such as shown in Fig. 7.20, has the valuable property that the cell interconnections are direct, and contain no cross-over paths.

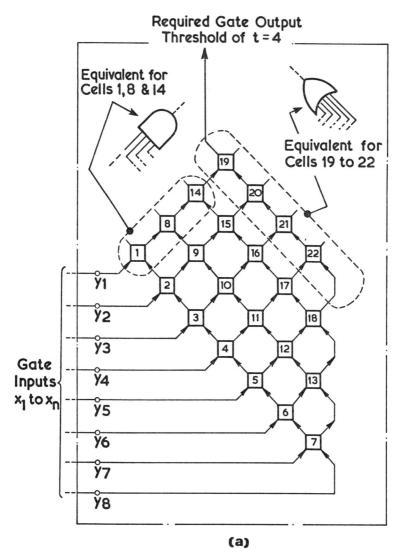

(a)

(continued)

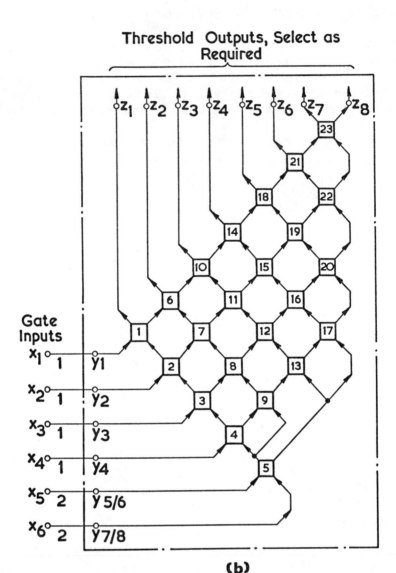

Fig. 7.20. Truncation of the DSTL gate array: (*a*) output truncation of an $m = 8$ array, where the required gate threshold is $t = 4$ only; (*b*) input truncations on an $m = 8$ array, with input weightings of 2 on x_5 and x_6 inputs.

Hence easy monolithic layout of such structures becomes possible, with peripheral metalization connecting input and output points. However, against these layout advantages must be set two adverse factors, namely (i) slow gate-response times[37], owing to the appreciable number of cells which may be present in series between inputs and outputs, and (ii) more cells are used than are theoretically necessary (see the following developments and Fig. 7.23).

One method which has been suggested to reduce the overall maximum gate-response time is to divide the array into two halves, and then to combine the reduced-array outputs by suitable additional gates[39]. For example an $m = 8$ array could be "bifurcated" into two $m = 4$ arrays, each of which could then be further split into two halves if desired. Figure 7.21 illustrates this concept applied to a simple $m = 4$

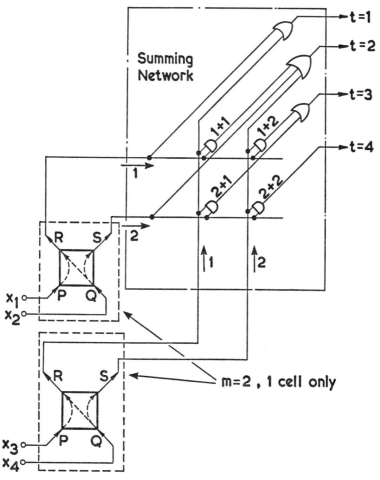

Fig. 7.21. Method of bifurcating an $m = 4$ DSTL array into two $m = 2$ arrays.

case. The ultimate extension of this approach is shown in Figs. 7.22a and b. However, this method leads to a very large number of crossovers in the final gate structure and a negligible saving in the total number of vertex gates per array. The original regular simplicity of the DSTL gate structure is of course completely lost in the extreme case shown in Fig. 7.22a.

A more sophisticated redesign of the original DSTL structure has been made by Edwards[40] using symmetry techniques as introduced in Chapter 5. Figure 7.23 illustrates the form of these array realizations, which minimize the number of AND/OR cells required for any m but

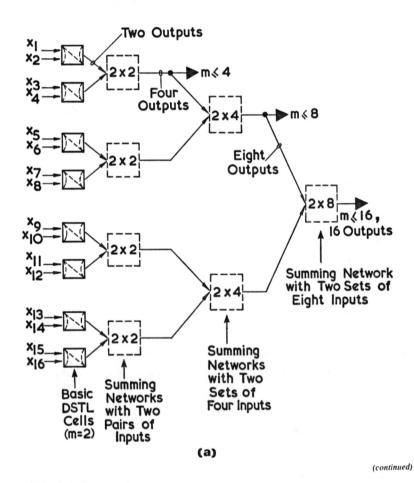

(a)

(continued)

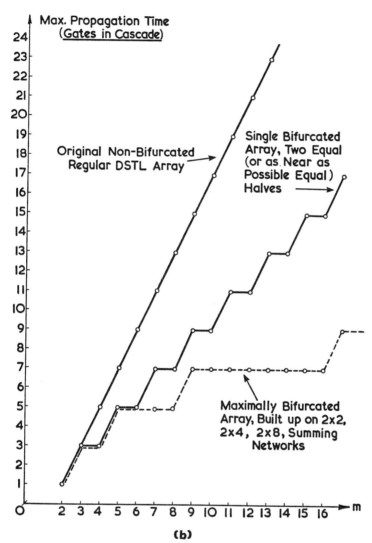

Fig. 7.22. The extreme possibility of bifurcating a DSTL array: (*a*) the general structure for $m \leq 16$; (*b*) maximum propagation time per array, considering the longest input–output path.

at the expense of a certain number of cross-overs in the cell interconnections. Table 7.2 below summarizes the performance of these designs in comparison with earlier arrangements.

Table 7.2. A comparison of the complexity and speed of the three types of DSTL array of Figs. 7.19, 7.21, and 7.23 repectively

	Fan-in m	Original DSTL gate	Single-bifurcated gate	Fig. 7.23 DSTL designs
Number of AND/OR cells	2 4 6 8	1 6 15 28	* * * *	1 5 12 19
Maximum gate delay (cells in series)	2 4 6 8	1 5 9 13	1 3 5 7	1 3 6 7
Number of signal path crossovers	2 4 6 8	0 0 0 0	0 7 22 56	0 1 6 9

Asterisks indicate approximately the same total complexity as the original DSTL gate.

An entirely different approach to a digital threshold-logic gate is proposed in a paper by Winn[41], being based upon Boolean product-of-sums expansions. Further work is necessary possibly to refine this approach, particularly to attempt to provide multi-threshold capability from one network assembly.

The whole philosophy of employing vertex gates in assemblies to realize threshold-logic functions may possibly be questioned. Clearly if one considers the number of vertex functions in a DSTL assembly as *entirely separate* vertex gates, then any given function $f(x)$ realized using such assemblies will never employ a *fewer* total number of vertex gates than designs based directly on Boolean algebra and individual vertex gates. However, as shown in Fig. 7.24, it may not be entirely realistic to consider each AND/OR cell of a DSTL array as two entirely separate vertex gates. Also the input and output inversions of this circuit may not always be necessary in a complete array, as they may be arranged in series to cancel with each other.

Circuit Designs for Digital Logic Functions 503

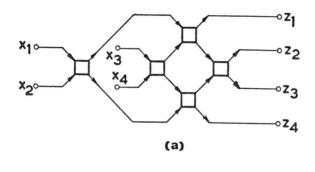

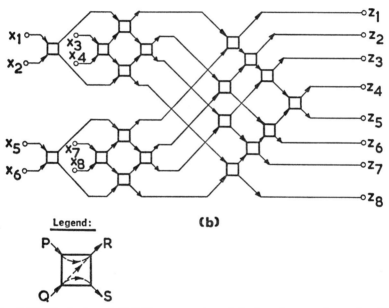

Fig. 7.23. Redesigned DSTL arrays to minimize the number of cells: (a) $m = 4$ array; (b) $m = 8$ array.

What the DSTL concept may provide is a compact way of fabricating a multi-threshold assembly, possibly using a smaller total silicon area than current-summation ECL solutions, and which with exclusive OR/NOR gates may provide a very powerful commercial package for the random logic market[42]. What it cannot match is the higher-speed capability of the non-saturating ECL solutions.

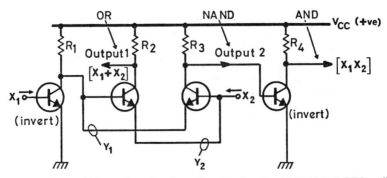

Fig. 7.24. A possible outline circuit realization for the AND/OR DSTL cell:
output 1 = $\overline{[y_1\bar{y}_2]} = \overline{[\bar{x}_1\bar{x}_2]} = [x_1 + x_2]$
output 2 = $\overline{[\bar{y}_1 y_2]} = \overline{[x_1 x_2]}$.

7.4. Ternary Circuits

7.4.1. Multi-State Electronic Devices

We have already seen in Chapter 6 how ternary network synthesis can be investigated using ternary functions which satisfy certain requirements, such as realizing a given algebra or otherwise providing functional completeness, without undertaking any detailed consideration of the possible circuits of such functions. Indeed a large percentage of the published papers in the ternary field come under this category.

On the other hand, development of actual ternary circuit configurations has been hampered by the general non-availability of electronic devices which can naturally assume three (or more) stable states. The bipolar or MOS transistors are, through development, excellent and superbly reliable two-state switching devices, but any intermediate state between the two extremes of fully conducting (saturated) and non-conducting (cut-off) is in general inconvenient to define and stabilize. Nevertheless, because of their availability, the majority of published circuits in the ternary field have been forced to adopt the binary transistor switch as the active element in their circuit configurations.

There have, however, been attempts to develop devices which inherently provide three stable states. For interest and completeness we shall here very briefly review the main protagonists which have

Circuit Designs for Digital Logic Functions

appeared, but regretfully all are now obsolete with no immediate successors in the present era of binary domination.

The first solid-state device which exhibited three stable states was possibly the Rutz commutating transistor[43]. This was a point-contact transistor having two collectors and one emitter, as illustrated in Fig. 7.25. The characteristics of this transistor are such that for zero-input emitter current, there is no output current in either collector C1 or C2; for an increasing emitter input current, however, first collector C1, then collector C2, and finally both C1 and C2 will collect current. These characteristics are shown in Fig. 7.25b. By appropriate choice of collector-resistor values the collectors C1 and C2 may each be made to swing between the collector supply rail voltage V_{CC} and 0 V (approximately) when non-conducting and conducting. A configuration as shown in Fig. 7.25c will therefore realize the three j_k-type functions following.

X_i	$j_0'(X_i)$	$j_1'(X_i)$	$j_2'(X_i)$
0	0	2	2
1	2	0	2
2	2	2	0

An entirely different approach was that of attempting to use different sinusoidal frequencies, or different phases of a single frequency, to represent the ternary values. Edson suggested the use of a multi-frequency oscillator for logic purposes, but no practical advantages accrued[44]. The single-frequency, variable-phase concept, using three phases separated by 120 electrical degrees to represent the ternary signals, had somewhat greater development; here a single frequency source f_p was subdivided by the circuit elements to $\frac{1}{3}f_p$, three phase-locked conditions ϕ_1, ϕ_2, and ϕ_3 thus being available. Such phase-locked parametric oscillator devices reached the stage of circuit implementation before they were overtaken by binary solid-state developments[45–47].

A third early approach to ternary circuit realization was the employment of square-loop ferrite cores[48–51]. The possibility of two directions of saturation, giving $+B_{max}$ and $-B_{max}$ residual states as well as the demagnetized state, provided the total of three possibilities. Again, however, in spite of considerable work including the use of such devices in at least one computer[4], such developments were made nonviable by subsequent binary developments.

[4]The SETUN computer of c. 1960[48,52].

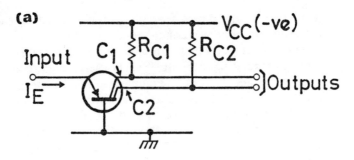

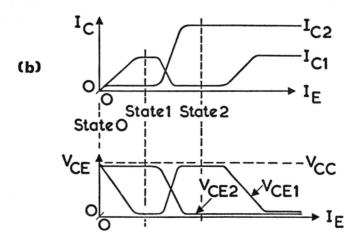

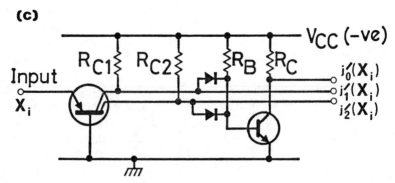

Fig. 7.25. The Rutz two-collector transistor: (*a*) basic circuit configuration; (*b*) I_C/I_E and V_{CE}/I_E characteristics; (*c*) realization of j'_k functions using one Rutz transistor plus a following NAND gate.

Circuit Designs for Digital Logic Functions

These and other early ideas, therefore, did not attain the status of commercial viability, and the pursuit of possibilities at the fundamental device stage has very largely lapsed. Yet with all the expertise which is now available on monolithic circuit design and fabrication maybe it is time to look again at this problem. We shall return to comment on this point in the closing part of this section.

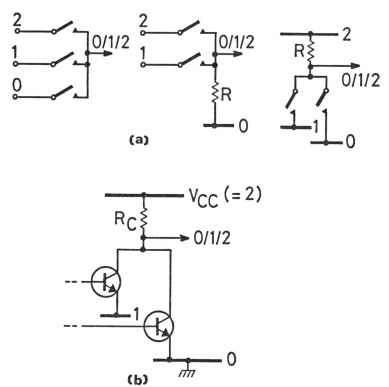

Fig. 7.26. Basic concepts for generating three logic voltage levels, assuming only one switch is closed at any one time: (*a*) possible mechanical contact realizations; (*b*) a bipolar switching transistor equivalent realization.

7.4.2. Bipolar Gate Circuits

The p.n.p. and n.p.n. bipolar transistors were the first efficient solid-state switching devices available to circuit designers, and it is therefore natural that they were applied to the problem of realizing ternary switching circuits. Let us look first at voltage-mode circuits.

Some basic ways of producing three-level output voltages are shown in Fig. 7.26*a*. As the bipolar transistor is a reasonably efficient on/off switch, the substitution of transistors for switches can be proposed, for example as shown in Fig. 7.26b. Working between two levels only, say between 0 and 2, is entirely straightforward, being a normal single-voltage-rail binary circuit configuration, which is a reason why many proposed ternary circuit realizations largely use two levels, say 0 and 2, in preference to the equal use of all three logic levels.

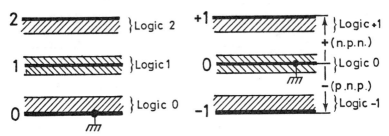

Fig. 7.27. The ternary logic levels considered as asymmetrical 0, 1, 2, values, and as symmetrical $-1, 0, +1$ values.

The adoption of complementary forms of transistor enables the 1 to 0 voltage band to be treated as one polarity and the 1 to 2 voltage band as the opposite polarity; in effect logic 1 becomes the common voltage rail of the system and hence is directly related to the symmetrical $\{-1, 0, +1\}$ ternary set of values. Figure 7.27 illustrates this principle. Notice that we may still refer to such a system as a $\{0, 1, 2\}$ set without difficulty, if preferred.

The majority of proposed transistor cirucits thus use two voltage supplies (two voltage rails plus a common or "earthy" rail) in order to obtain the ternary signal voltages. The use of one voltage rail only with some non-linear circuit elements to define an intermediate circuit condition (logic 1) in addition to the on/off circuit conditions (logic 0 and 2) has been tried, but in general is not economical[53]. However, a large variety of ternary gate circuits using two (or more) voltage rails and binary swtiches may be found in the published literature[54-62]; Fig. 7.28 illustrates some of these possibilities. Without exception all these circuits were developed using discrete devices, and hence the more sophisticated circuit techniques which are viable in fully integrated circuit designs are not employed.

It should be noticed in Mühldorf's p.n.p. circuits that a 1- or a 2-signal voltage input will fully switch on any transistor with a single

Circuit Designs for Digital Logic Functions

input resistor R_B, but signal 1 is inadequate when an input resistor-divider R_B, R'_B is present. In Hurst's circuits, this input resistor-divider to discriminate between 1 and 2 input signals is replaced by a complementary common-base input transistor, the base voltage of which is biased so as to make the circuit input appear as an (ideally) infinite impedance until the input voltage rises above the base voltage[63].

The final-level maximum-of or minimum-of output functions for many of these proposed circuit families are frequently simple diode-AND and diode-OR circuits, as shown in Fig. 7.29. The loss of signal voltage level owing to diode forward volt drop is the principal disadvantage of these simple circuits. Voltage-level restoring circuits are not widely publicized, although some circuits have been mentioned[60-64].

For quasi-binary systems, such as Pugh, the combinatorial logic gates can be conventional binary gates, one half being of p.n.p. construction, the other half being n.p.n. The necessary "translating" gates between the two polarity halves are typically as shown in Figs. 7.30a and b, whilst the final level "combining" circuit is typically as shown in Fig. 7.30c. One common disadvantage of the output circuit of many of these complementary realizations is that the output impedance at the center logic level is poor, often being merely a passive clamp to the center voltage rail.

In general it may be concluded that no great sophistication is revealed by the majority of voltage-mode ternary circuits so far published. All generally employ fully saturating common-emitter transistor configurations, exploiting the normal fully-on/fully-off characteristics of this configuration.

The non-saturating transistor in current-mode working, however, has been previously covered in this chapter, particularly in connection with binary-output threshold-logic gates. Clearly we have in this mode the possiblity of three (or more) levels of current working, which may be applied to ternary gate realization. Initial work into these possibilities has been started, and preliminary considerations published[65-67].

The basic circuit configuration in these proposals remains the emitter-coupled transistor pair, in which a constant emitter current is switched from one side of the pair to the other as the input signal voltage changes. By suitable choices of internal base voltages and emitter resistor values, so units of current representing the ternary values can be switched and summed.

Fig. 7.28. Basic circuit arrangements of some typical bipolar ternary circuits developed using discrete components: (*a*) Mühldorf's individual j_k functions using p.n.p. transistors; Porat's circuits for these functions are basically the same but with catching diodes etc. to improve the circuit performance. Note the three separate j_k functions may be combined to form one composite circuit assembly, as in Porat. (*b*) A selection of Hurst's unary and multi-input functions; the unary function input–output relationships are given by the 0/1/2 numbering, whilst the multi-input function shown is a 1-level AND function, output $f(X) = 0$ when all inputs are 1, otherwise 2 (see Table 1.13 of Chapter 1). (*c*) Thelliez's T-gate proposal: $0 = -6V$, $1 = 0V$, $2 = +6V$, TR1 and TR2 on when $X_4 = 0$, TR1 off TR2 on when $X_4 = 1$, TR1 and TR2 off when $X_4 = 2$.

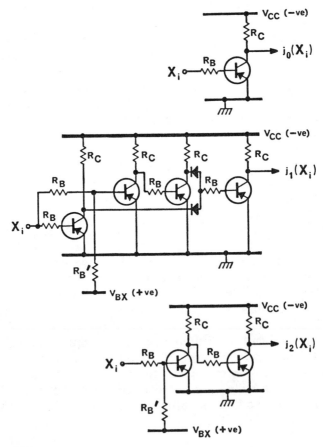

(a) *(continued)*

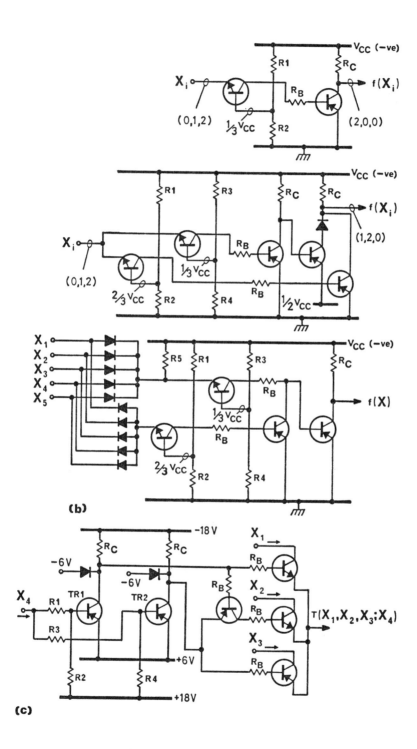

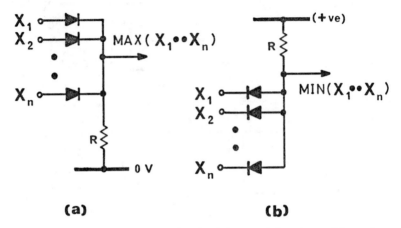

Fig. 7.29. Final-level maximum-of and minimum-of functions: (*a*) maximum-of function, assuming positive logic voltages; (*b*) minimum-of function, assuming positive logic voltages. (Note: emitter followers may be added to provide buffering, but additional loss of signal level then results.)

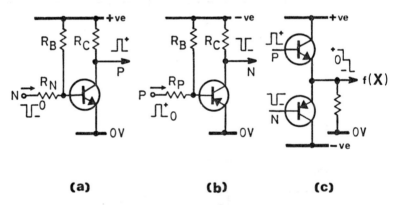

Fig. 7.30. Basic "translating" and "combining" circuits for complementary quasi-binary systems: (*a*) negative input-signal to positive output-signal translation; (*b*) positive input-signal to negative output-signal translation; (*c*) final-level combining circuit.

Figure 7.31a illustrates the j_k circuit configurations proposed by Dunderdale[66,67] (cf. Fig. 7.28a, where the three separate j_k circuits there shown may be amalgamated into one composite assembly). Dunderdale's original diagrams also provided two additional outputs, namely the diagonal inversions of $j_0(X_i)$ and $j_2(X_i)$, as indeed was pro-

Circuit Designs for Digital Logic Functions

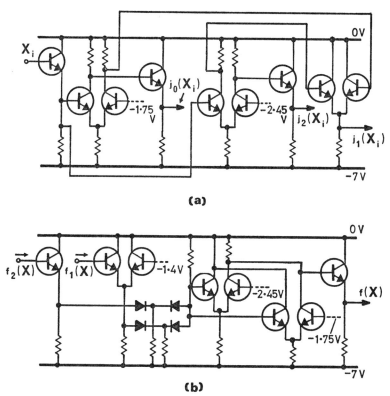

Fig. 7.31. The basic arrangement of Dunderdale's proposed non-saturating emitter-coupled ternary circuits, where logic $0 = -0.7$ V, logic $1 = -1.4$ V, and logic $2 = -2.1$ V: (a) the three j_k functions; (b) the output circuit to provide

$$f(X) = [f_2(X) \vee \{1 \& f_1(X)\}]$$

where $f_2(X)$ is a 0,2 binary signal, value 2 when $f(X) = 2$, and $f_1(X)$ is also a 0,2 binary signal, value 2 when $f(X) = 1$.

vided by Porat in his saturating circuit arrangements[59]. Figure. 7.31b illustrates a proposed output configuration to provide the final output

$$f(X) = [f_2(X) \vee \{1 \& f_1(X)\}].$$

Circuits to realize the binary-output $f_1(X)$ and $f_2(X)$ functions are also disclosed in these contributions. For successful realization of all these emitter-coupled circuits, matching of the transistor characteristics and resistor values is very desirable. Therefore they are likely to be much more satisfactory when produced in monolithic form than when breadboard realizations in non-monolithic form are attempted.

The emitter-coupled circuits of Druzeta et al.[65] employ similar basic concepts, but are directed specifically towards multi-valued threshold-logic functions, of which the ternary case may be considered as merely one possibility. Very close matching of both active and passive device parameters are likely to be necessary in such circuits, although in principle they provide an extremely fast and logically very powerful gate configuration. There is clearly much work to be continued in this area, although basic circuit concepts may not drastically change.

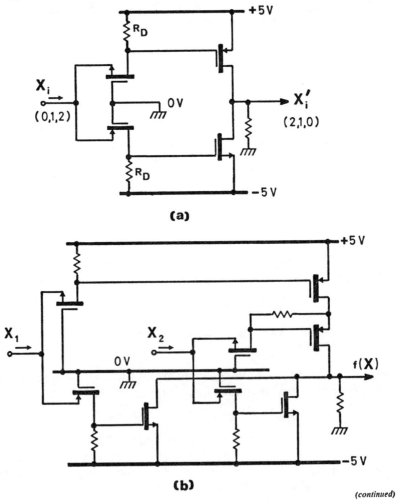

(a)

(b)

(continued)

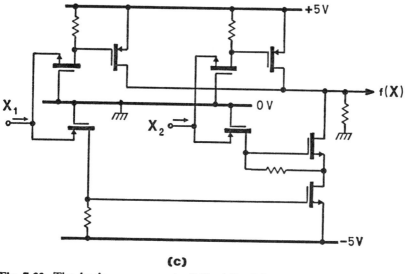

(c)

Fig. 7.32. The basic arrangement of Kaniel's COSMOS ternary functions: (a) the diagonal inversion function, as previously defined in Fig. 6.23; (b) a two-input maximum-of gate with diagonal inversion of the output; (c) a two-input minimum-of gate, with diagonal inversion of the output.

7.4.3. MOS Gate Circuits

MOS circuit realizations of ternary functions have not received such wide attention as bipolar circuits. Of the main categories of possible MOS digital circuits, namely static logic, two-phase dynamic logic, and four-phase dynamic logic[1–5], it is the static-logic category using complementary n-channel and p-channel devices ("COSMOS") which has received detailed attention[68–70]. Dynamic logic realizations do not yet seem to have been pursued for the ternary field.

The complementary circuits proposed by Kaniel[68] may be considered as typical of the COSMOS type of circuit realizations. Figs. 7.32a to c illustrate three typical gate configurations, in which the signal voltage values for the three logic levels are typically as follows

Logic level	Circuit voltage (V)
0 (or −1)	−5
1 (or 0)	0
2 (or +1)	+5

As both of the multi-input gates shown in Fig. 7.32 possess output inversion, then it is possible directly to cross-couple pairs of each type

516   The Logical Processing of Digital Signals

to form tristable assemblies. This is exactly as illustrated in Fig. 6.23*b* of Chapter 6, where maximum-of plus diagonal inversion functions were shown forming a tristable circuit configuration, with outputs Q and Q'.

However, there would still appear room for more circuit sophisti-

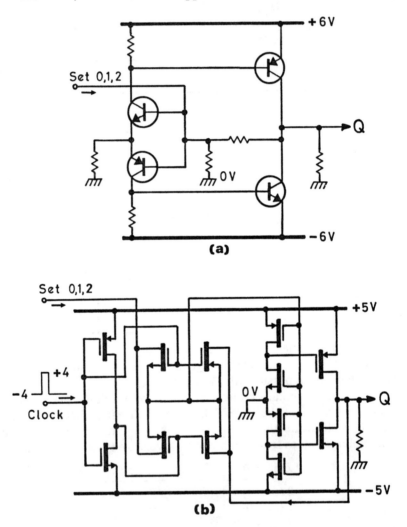

Fig. 7.33. Tristable circuit configurations based upon complementary active devices: (*a*) an unclocked tristable bipolar circuit of Smith[70]; (*b*) a clocked tristable COSMOS circuit discussed by Vranesic *et al.*[73].

Circuit Designs for Digital Logic Functions

cation to be applied not only to bipolar ternary function circuits, but also to this newer MOS field of ternary circuit design.

7.4.4. Tristable Circuits

The circuit configuration of tristable circuits may be based upon several possible lines of development.

(1) Cross-coupled unary gates, as illustrated in Fig. 6.23.

(2) Special designs, often employing non-linear circuit elements to achieve the three stable states.

(3) Cascade or ring-of-three circuit configurations.

(4) Interleaved bistable circuit configurations, with one of the four possible combinations of states being unused or suppressed.

The first of the above categories requires no further mention as its concepts have been adequately covered in previous pages. The second category, however, is one in which great ingenuity has often been applied, although frequently the proposed circuits would not be amenable to monolithic realization owing to the differing non-linear characteristics required from the component parts. Examples of such circuits are those of Mine *et al.*, Taub, and also others [56,57,71,72]. In general two cross-coupled transistors are used, with the non-linear components arranged to impose a third stable state in the complete configuration.

Special tristable designs using more than two active devices in complementary configurations have also been proposed, examples of which are illustrated in Figs. 7.33*a* and *b*. A problem with many of these realizations is a poorly defined center signal-voltage output, as only a passive pull-down to the center voltage rail is directly available. The other two signal-voltage outputs are generally actively connected to the outer supply-voltage rails, and are hence well defined.

The cascaded or ring-type tristable assemblies are in theory extendable to any higher-radix storage requirement. In principle, for R storage states, $R \geq 2$, there will be R circuit elements in cascade, only one of which has a specific output state at any one time. Generally, there will be R separate output points available, one from each circuit element, but these separate output signals are not of course independent of each other. Many ternary circuit proposals based upon such concepts have been published [55,57,59,72,73]. Should each circuit element be capable of providing all three output signal levels, then the total tristable output signals may be as shown in Table 7.3*a* below; if only two signals, say 0 or 2, are available then output states as in Table 7.3*b* may be present.

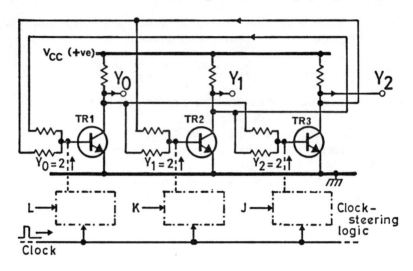
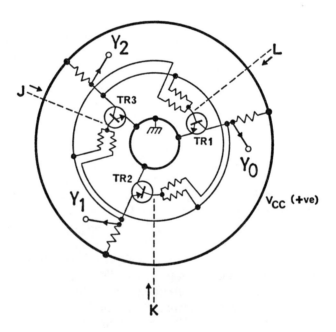

Fig. 7.34. Porat's bipolar ring-of-three type JKL tristable circuit, drawn in conventional form and in circular form to emphasize its symmetry.

Circuit Designs for Digital Logic Functions 519

Table 7.3. Possible outputs from a cascade of three circuit elements assembled to form a tristable assembly

(a)

Tristable circuit states	Tristable circuit outputs		
	Circuit 1	Circuit 2	Circuit 3
State 0	2	0	1
State 1	1	2	0
State 2	0	1	2

(b)

Tristable circuit states	Tristable circuit outputs		
	Circuit 1	Circuit 2	Circuit 3
State 0	2	0	0
State 1	0	2	0
State 2	0	0	2

(a) 0, 1, 2 outputs available from each circuit.
(b) 0, 2 outputs only available from each output.

Porat's type JKL tristable circuit is representative of many of these proposals. This bipolar circuit is shown in Fig. 7.34 being drawn both in conventional form and in circular form to emphasize its ring-of-three construction.

Finally, let us look at tristable circuits which are formed from some assembly of basically conventional bistable circuits. Two variations on this final theme may be found, the first where two bistable circuits are intimately interleaved with each other to give the tristable condition, and the second where the two bistable circuits are effectively in parallel with each other, with appropriate combining circuits on the bistable output points to provide the final tristable output[58,60,74,75].

An example of an interleaved bistable circuit design providing tristable action is illustrated in Fig. 7.35. This configuration provides output-signal values of (approximately) V_{CC}, $\frac{1}{2}V_{CC}$, and 0 V in accordance with the following circuit states.

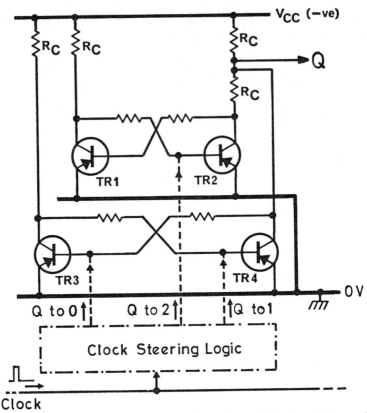

Fig. 7.35. The interleaved bistable configuration of Hurst which provides tristable circuit action.

Output signal at Q	State of TR1/TR2 bistable	State of TR3/TR4 bistable
$\approx V_{CC}$	TR1 on/TR2 off	TR3 on/TR4 off
$\approx \frac{1}{2}V_{CC}$	TR1 off/TR2 on	TR3 on/TR4 off
≈ 0 V	TR1 off/TR2 on	TR3 off/TR4 on

The fourth condition of TR1 on/TR2 off, TR3 off/TR4 on is not a stable state of the circuit, as no voltage feedback is available to the base of TR1 to hold it conducting under such conditions. Hence the circuit as a whole has three stable states only. The rather poor output impedance given by this particular circuit at output Q may be corrected by means

Circuit Designs for Digital Logic Functions

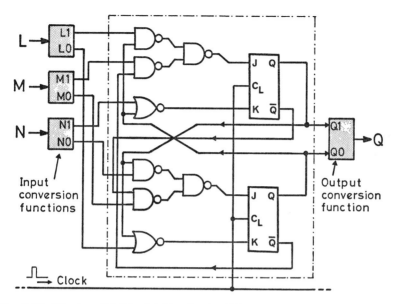

Fig. 7.36. The type LMN tristable circuit of Duncan using quasi-binary techniques for the ternary states.

of a following buffer amplifier, if necessary. Various clock-steering arrangements can be proposed, including a simple shift-register interconnection between stages which can conveniently use the various collector voltages TR1, TR2, TR3, and TR4 for this duty in place of the single ternary output signal Q[60].

The tristable circuit proposals formed by employing conventional binary circuits in parallel have largely been suggested because of the ready availability of (in particular) type JK bistable circuit packages. Thus although it is very convenient to produce prototype tristable circuit assemblies using this technique, it is very debatable whether such an arrangement would be optimum if a single monolithic integrated circuit was being designed from scratch to perform the same duties. Nevertheless the logic operation may remain unchanged, even though the total circuit configuration may be optimized.

As an example of this final work, consider the circuit illustrated in Fig. 7.36, this being the type LMN circuit of Duncan[75]. The complete circuit is a quasi-binary network, with ternary to quasi-binary relationships as follows:

Ternary value T	Quasi-binary values T_0	T_1
0	1	0
1	0	0
2	0	1

Note that the quasi-binary values of both-simultaneously-1 are not allowed.

On every input of the LMN circuit except the clock input C there must be a separate single-wire ternary/two-wire quasi-binary input converter circuit, and similarly on the quasi-binary output Q_0, Q_1 there must be a two-wire/single-wire output converter circuit, assuming single-wire ternary signals are required at inputs and output. The clock input C of this circuit is a normal 0, 1 binary signal. The whole circuit of Fig. 7.36 is therefore an appreciable assembly in terms of total number of active devices involved. However, this factor is not necessarily a serious restriction on the commercial viability of a circuit if a final realization in monolithic integrated circuit form is envisaged.

7.4.5. Summary and Future of Higher-Radix Circuits

In summarizing the present state of ternary logic, it is apparent that the lack of circuit realizations which can economically compete with binary circuits is the major problem in the introduction of ternary digital systems. The circuits which many researchers have introduced to date, typical representatives of which we have considered in the preceding pages, have been fundamentally inefficient, as two-valued rather than three-valued signals on the interconnecting wires predominate, whereas for maximum efficiency each wire should be capable of transmitting the full digital information capacity of the higher-valued radix.

To some extent this situation is inevitable. The capital expenditure required to introduce a new logic family in monolithic integrated circuit form is beyond the financial resources of most research teams, and therefore hardware reliance by such teams on available binary devices and circuits becomes necessary. Integrated circuit manufacturers on the other hand are clearly cautious about investing in wider fields when the binary market remains an expanding and still profitable area.

Nevertheless with the vast knowledge of integrated circuit technology and circuit design now available, it is surely only time before an efficient three-state device is developed from which families of monolithic ternary logic circuits may be produced. A start along this path may be found[76,77], but clearly great resources have to be engaged before the present binary dominance of the digital field is challenged.

As far as higher radices than ternary are concerned, the problems both of efficient system synthesis and also circuit realizations become increasingly difficult, although of course one has a greater information capacity per wire if the system can be made. Possibly further ternary research will show whether $R > 3$ is a feasible future proposition for the digital world.

7.5. Final Conclusions

In the past chapters and pages of this book we have been looking at some of the ideas in the world of digital logic which have not yet reached the stage of familiarity that Boolean logic and vertex gates currently possess. Some of the topics chosen may never reach maturity, and yet each adds something, however little, to our knowledge and appreciation of the whole subject matter.

As indicated in the preface and now hopefully shown, if there is any one underlying theme to the material of this book it is that we must be increasingly critical of the efficiency of present-day digital system designs based upon binary vertex gates and Boolean algebra. We have there three areas of potential inefficiency, namely (i) the use of simple binary gates with limited logical discrimination, (ii) inefficient design methods owing to the fundamental limitations of Boolean algebra, and (iii) the fundamentally low information content of a binary signal, giving only one out of two possible bits of information at any instant of time.

We have seen how more efficient binary gates than the AND, OR, NAND, NOR vertex family can be proposed, particularly the exclusive-OR and threshold-logic gates, but the design techniques to exploit their increased logical power successfully are more difficult. Certainly techniques based upon Boolean algebra will not cope in this area, and the main hope would seem to be in the newer spectral and symmetry techniques discussed in Chapter 5.

Radices higher than the two-valued binary case, whilst theoretically attractive, have the fundamental problem of efficient circuit realization still to solve. The three-valued digital system possesses many theoretical advantages for control systems and other industrial applications, owing to the three-state nature of many industrial control instructions, whilst at a much higher radix, a 10-state (denary) digital system would be ideal for many mathematical computers and arithmetic machines, rather than the present-day binary-based digital computers.

In all the design concepts covered in this book the size of the problem being considered, that is the number of input variables or circuit states, has had to be limited to that which can be conveniently written and displayed. Therefore four binary-input variables, for example, has in general constituted the biggest problem we have considered. Real-life problems of course may not have this fortunate limit! However, it is hoped that in considering a limited size of problem, the underlying theories and methods of handling the digital data may be clear and that application of these principles to larger problems with digital computer aid may be straightforward. In this context it may be remarked that programs to handle and manipulate digital data in the spectral domain seem to be particularly attractive and straightforward, much more so than the manipulation of binary data by Boolean-based algorithms and equations.

To conclude, there still remains a vast field of research, development, and exploitation ahead in the digital world. The years ahead are likely to be as full of new concepts and their application as the exciting and revolutionary years we have expreienced during the past three decades.

References

1. Hnatek, E.R. *A User's Handbook of Integrated Circuits* John Wiley, New York, 1973.
2. Lynn, D.L., Meyer, C.S., and Hamilton, D.J. (eds) *Analysis and Design of Integrated Circuits*. McGraw-Hill, New York, 1968.
3. Mavor, J. *MOST Integrated-Circuit Engineering*. IEE Peter Peregrinus, Stevenage, U.K., 1973.
4. Hibberd, R.G. *Integrated Circuit Pocket Book*. Newnes–Butterworth, London, 1972.
5. Kostopoulos, G.K. *Digital Engineering*. John Wiley Interscience, New York, 1975.
6. Murphy, B.T., Glinski, V.J., Gary, P.A., and Pederson, R.A. Collector diffusion isolation integrated circuits. *Proc. IEEE* **57**, 1523–27, Sept. 1969.

7. Altman, L. Logic's leap ahead creates new design tools for old and new applications. *Electronics* 83–96, 21 Feb. 1974.

8. Horton, R.L. I²L takes bipolar integration a significant step forward. *Electronics* 83–90, 6 Feb. 1975.

9. Hart, C.M., and Slob, A. Integrated injection logic (I²L). *Phillips Tech. Rev.* **33**, (3) 76–85, 1973.

10. Blatt, V., Kennedy, L.W., Walsh, P.S., and Ashford, R.C.A. Substrate-fed logic—an improved form of injection logic. *Tech. Dig. IEEE Int. Electronic Devices Meeting, Washington D.C., December 1974*, pp. 511–20.

11. Berger, H.H., and Wiedman, S.K. Merged transistor logic (MTL)—a new low-cost bipolar logic concept. *Trans. IEEE* **SC7**, 340–6, Oct. 1972.

12. Lehning, H. Current hogging logic (CHL)—a new bipolar logic for l.s.i. *IEEE J. Solid-St. Circuits* **SC9**, 228–33, Oct. 1974.

13. Lehning, H. Current hogging logic—a new logic for l.s.i. with noise immunity. *IEEE Int. Solid-State Circuits Conf. Dig. 1974*, pp. 18–19.

14. Muller, R. Current hogging injection logic: new functionally integrated circuits. *IEEE Int. Solid-State Circuits Conf. Dig., 1975*, pp. 174–5.

15. Boyle, W.S., and Smith, G.E., Charge-coupled semiconductor devices. *Bell Syst. Tech. J.* **49**, 587–93, April 1970.

16. Walden, R.H., Krambeck, R.H., Strain, R.J., McKenna, J., Schryer, N.L., and Smith, G.E. The buried channel charge-coupled device. *Bell Syst. Tech. J.* **51**, 1635–40, 1972.

17. Beynon, J.D.E. Charge-coupled devices—concepts, technologies and applications, *Radio Electron. Eng.* **45**, 647–56, Nov. 1975.

18. Dighe, K.D. Low-cost exclusive-OR needs no power supply. *Electronics* 56, April 1971.

19. Edwards, C.R. Some novel Exclusive-OR/NOR circuits. *Electron. Lett.* **11**, 3–4, Jan. 1975.

20. Edwards, C.R. Novel digital-integrated circuit configurations based upon spectral techniques. *Proc. 1st European Solid-State Circuits Conf., Canterbury, U.K., Sept. 1975*, pp. 82–83.

21. Murugesan, S. Universal logic gate and its applications. *Int. J. Electron.* **42**(1), 55–63, 1977.

22. Muroga, S., and Takashima, K. The parametron digital computer MUSASINO 1. *Proc. IRE* **EC8**, 308–16, Sept. 1959.

23. Lewis, P.M., and Coates, C.L. *Threshold Logic*. John Wiley, New York, 1967.

24. Hurst, S.L *Threshold Logic—An Engineering Survey* Mills & Boon, London, 1971.

25. Coates, C.L., and Lewis, P.M. DONUT: a threshold gate computer. *Trans. IRE* **EC13**, 240–7, June 1964.

26. Amodei, J.J., Hampel, D., Mayhew, T.R., and Winder, R.O. An integrated threshold gate. *Dig. Int. Circuits Conf. 1967. New York*, pp. 114–5.

27. Beinart, J.H., Hampel, D., Micheel, L.J., Prost, K.J., and Winder, R.O. Threshold logic for l.s.i. *Proc. IEEE National Aerospace Electronics Conf., 1969*, pp. 453–9

28. Beinart, J.H., Hampel, D., Prost, K.J., and Winder, R.O. *Integrated Threshold Logic*. Air Force Avionics Laboratory Rep. AFAL-TR-69-195, 1969.

29. Hampel, D., Beinart, J.H., and Prost, K.J. *Threshold Logic Implementations of a Modular Computer System Design* NASA Rep. NASA-CR-1668, Oct. 1970

30. Dertouzos, M.L. *Threshold-Logic–A Synthesis Approach*. MIT Press, Cambridge, Mass, 1965

31. Haring, D.H. A technique for improving the reliability of certain classes of threshold elements. *Trans IEEE*, **C17**, 997–9, Oct 1968

32. Sheng, C.L. *Threshold Logic*. Academic Press, London, 1969

33. Hurst, S.L. Sensitivity performance of threshold-logic gates using Chow parameters. *Electron. Lett.* **7**, 371–3, July 1971.

34. Hurst, S.L. Threshold logic network synthesis with specific threshold-gate sensitivities, *Radio Electron. Eng.* **42**, 295–300, June 1972.

35. Hurst, S.L. Improvements in circuit realisation of threshold-logic gates. *Electron. Lett.* **9**, pp. 123–5, March 1973.

36. Hampel, D. Multifunction threshold gates. *Trans. IEEE* **C22**, 197–203, Feb. 1973.

37. Hurst, S.L. Digital-summation threshold-logic gates: a new circuit element. *Proc. IEE* **120**, 1301–7, Nov. 1973.

38. Hurst, S.L. Threshold logic arrays. *Electron* No. 80, 34–43, Sept. 1975.

39. Reddy, V.C.V.P., and Swamy, P.S.N. Note on digital-summation threshold-logic gates. *Proc. IEE* **121**, 1085–6, Oct. 1974.

40. Edwards, C.R. *Some Improved Designs for the Digital-Summation Threshold-Logic (DSTL) Gate. Computer J.* **21**, 73–78, Feb. 1978.

41. Winn, G.C.E. A digital approach to the efficient synthesis of threshold gates. *Computer J.* **18**, 239–42, Aug. 1975.

42. Hurst, S.L. Improvements for general-purpose s.s.i. logic packages, and m.s.i./l.s.i. logic sub-systems. *Electron. Lett.* **11**, 78–9, Feb. 1975.

43. Rutz, R.F. Two-collector transistor for binary full-addition. *IBM J. Res. Dev.* **1**, 212–22, July 1957.

44. Edson, W.A. *Frequency Memory in Multi-Mode Oscillators.* Stanford University Tech. Rep. No. 16, July 1954.

45. Wigington, R.L. A new concept in computing. *Proc. IRE* **47**, 516–23, April 1959.

46. Sobornikov, Y.P. Three-valued parametron logic in Post's algebra. *Avtomatika* **2**, 45–53, 1965 (in Ukranian).

47. Komolov, V.P., and Roshal, A.S. Logic circuits with ternary parametrons. *Radiofizika* **1**, 177–82, 1965 (in Russian).

48. Brusentzov, N.P. Development of the ternary computer. *Vest. Moscow Univ.* No. 2, 39–48, 1965 (in Russian).

49. Alexander, W. The ternary computer. *IEE Electron. Power,* **10**, 36–9, Feb. 1964.

50. Anderson, D.J., and Deitmeyer, D.L A magnetic ternary device. *Trans. IEEE* **EC12**, 911–4, Dec. 1963

51. Santos, J., Arango, H., and Pascual, M. A ternary storage element using a conventional ferrite core. *Trans. IEEE* **EC14**, 248, Apr. 1965.

52. Brusentzov, N.P., Zhogolev, Y.A., Verigin, V.V., Maslov, S.P., and Tishulina, A.M. The SETUN small automatic digital computer. *Vest. Moscow Univ.* No. 4, 3–12, 1962.

53. Henle, R.A. A multistable transistor circuit. *Trans. AIEE.* **74**, 568–71, Nov. 1955.

54. Mühldorf, E. Schaltungen für trenäre Schaltvariable. *Arch. Elekt. Übertrag.* **12**, 176–82, April 1958 (in German).

55. Benetazzo, L., and Debiasi, G.B. Criteri de progetto di un circuito tristabile. *Elettrotecnia* **52**, 223–30, March 1965 (in Italian).

56. Mine, H., Hasegawa, T., Ikeda, M., and Shintani, T. A construction of ternary logic circuits. *Electron. Commun. Japan* **51C** (12) 133–40, 1968.

57. Hallworth, R.P. and Heath, F.G. Semiconductor circuits for ternary logic. *Proc. IEE* **109C,** 219–25, March 1962.

58. Pugh, A. Application of binary devices and Boolean algebra to the realisation of 3-valued logic circuits. *Proc. IEE* **114,** 335–8, March 1967.

59. Porat, D.I. Three-valued digital systems. *Proc. IEE* **116,** 947–54, June 1969.

60. Hurst, S.L. Semiconductor circuits for 3-state logic applications. *Electron. Eng.* **40,** 197–202 and 256–9, April/May 1968.

61. Thelliez, S. Note on the synthesis of ternary combinatorial networks using T gate operator. *Electron. Lett.* **3,** 204–5, May 1967, (in French).

62. Higuchi, T., and Kameyama, M. Ternary logic systems based on the T gate, *Proc. Int. Symp. on Multiple-Valued Logic, May 1975,* pp. 290–304.

63. Hurst, S.L. D.C. voltage level detection. *Electron. Eng.* **37,** 668, Oct. 1965.

64. Smith, K.C., Vranesic, Z.G., and Janczewski, L. Circuit implementations of multivalued logic. *Conf. Rec. Symp. on Theory and Applications of Multiple-Valued Logic, Buffalo, N.Y., May 1971,* pp. 133–9.

65. Druzetta, A., Vranesic, Z.G., and Sedra, A.S. Application of multi-threshold elements in the realization of many-valued logic networks. *Trans. IEEE* **C23,** 1194–8, Nov. 1974.

66. Dunderdale, H. Current-mode circuits for ternary-logic realisation. *Electron. Lett.* **5,** 575–7, May 1969,

67. Dunderdale, H. Current-mode circuits for the unary functions of a ternary variable. *Electron. Lett.* **6,** 15–16, Jan. 1970.

68. Kaniel, A. Trilogic, a three-level logic system provides greater memory density. *EDN* **18,** 80–3, April 1973.

69. Mouftah, H.O., and Jordan, I.B. Integrated circuits for ternary logic. *Proc. Int. Symp. on Multiple-Valued Logic, May 1974,* pp. 285–302.

70. Vranesic, Z.G., and Smith, K.C. Engineering aspects of multi-valued logic systems. *IEEE Comput.* **7,** 34–41, Sept. 1974.

71. Taub, D.M. Tristable circuit with 4ns switching time. *Electron. Lett.* **2,** 61–62, Feb. 1966.

72. Braddock, R.C., Epstein, G., and Yamanaka, H. Multiple-valued logic design and applications in binary computers. *Conf. Rec. Symp. on the Theory and Applications of Multiple-Valued Logic Design, May 1971*, Buffalo, N.Y., pp. 13–25.

73. Vranesic, Z.G., Smith, K.C., and Druzeta, A. Electronic implementation of multi-valued logic networks. *Proc. Int. Symp on Multiple-Valued Logic, May 1974*, pp. 59–78.

74. Moraga, C.R. A tristable circuit for synchronized sequential machines, *Conf. Rec. Symp. of Theory and Application of Multiple-Valued Logic, Buffalo, N.Y.,* May 1971, 103–11.

75. Duncan, F.G. *A Three-Valued Logic; Description, Implementation and Application*. Unpublished Rep., Dept. of Computer Science, University of Bristol, U.K., Sept. 1974.

76. Abu-Zeid, M.M., and Kerssen, C. Multi-stable circuit using complementary field-effect transistors. *Electron. Lett.* **11,** 613–4, Dec. 1975.

77. Abraham, G. Variable radix multistable integrated circuits. *IEEE Computer* **7,** 42–59, Sept. 1974.

Appendix A

The Canonic Characteristic Weight-Threshold Vectors, or Chow-Parameter Classifications, for All Linearly Separable Binary Functions of $n \leq 6$

Notes

(1) For any n-binary function $f(y)$ with binary inputs y_i, $i = 1 \ldots n$, $f(y)$, $y_i \in \{-1, +1\}$, the Chow parameters are defined as

$$b_i, i = 0 \ldots n, = \sum_{p=0}^{2^n-1} \{f(y) \cdot y_i\}$$

where $y_0 \triangleq +1$.

This may be translated into the definitions following:

$b_0 = \{$(number of true minterms in $f(x)$) − (number of false minterms in $f(x)$)$\}$
$\quad = \{2$ (number of true minterms) − $2^n\}$

$b_i, i = 1 \ldots n, = \{$(number of agreements between the input variable x_i and the function output $f(x)$) − (the number of disagreements between the input variable x_i and the function output $f(x)$)$\}$
$\quad = 2 \{$(number of minterms for which both $f(x)$ and x_i are 1) − (number of minterms where $f(x)$ is 1 but x_i is 0)$\}$

(2) The canonic tables list the $|b_i|$ values for all the linearly-separable functions in descending magnitude order. Each entry uniquely defines one and only one standard (or "representative") function. If the Chow parameters for any non-linearly-separable function are computed, the resultant parameter values will not be found in these standard tabulations.

(3) The minimum-integer realizing weight/threshold values $|a_i|$, $i = 0 \ldots n$, are tabulated against each $|b_i|$ classification entry. Although the maximum, co-equal, and minimum values of the $|a_i|$'s reflect the maximum, co-equal, and minimum values of the $|b_i|$'s, there is no simple arithmetic relationship between them.

(4) For any chosen set of gate input weights a_1 to a_n, the resultant gate-threshold value is given by taking the remaining tabulated a_i value ($= a_0$) and evaluating

$$t = \tfrac{1}{2}\left\{\left(\sum_{i=1}^{n} a_i\right) - a_0 + 1\right\}.$$

(5) Notice that all the entries for any n appear in the subsequent tabulation for $n + 1$, but with all values multiplied by 2 in the latter and with a further zero-valued component. The multiplication by 2 is because there are twice the number of minterms present in the $n + 1$ case compared with the n-valued case.

(6) Finally, for the 2470 entries that consitute the Chow parameters for $n \leq 7$, reference may be made to R. O. Winder: *Threshold Functions Through $n = 7$*, Sci. Rep. No. 7, A.F.C.R.L. Contract AF19 (604)–8423, Oct. 1969.

n		$\lvert b_i \rvert$					$\lvert a_i \rvert$					
$n \leq 3$												
	1	8	0	0	0		1	0	0	0		
	2	6	2	2	2		2	1	1	1		
	3	4	4	4	0		1	1	1	0		
$n \leq 4$												
	1	16	0	0	0	0	1	0	0	0	0	
	2	14	2	2	2	2	3	1	1	1	1	
	3	12	4	4	4	0	2	1	1	1	0	
	4	10	6	6	2	2	3	2	2	1	1	
	5	8	8	8	0	0	1	1	1	0	0	
	6	8	8	4	4	4	2	2	1	1	1	
	7	6	6	6	6	6	1	1	1	1	1	
$n \leq 5$												
	1	32	0	0	0	0	0	1	0	0	0	0	0
	2	30	2	2	2	2	2	4	1	1	1	1	1
	3	28	4	4	4	4	0	3	1	1	1	1	0
	4	26	6	6	6	2	2	5	2	2	2	1	1
	5	24	8	8	4	4	4	4	2	2	1	1	1
	6	24	8	8	8	0	0	2	1	1	1	0	0
	7	22	10	10	6	2	2	5	3	3	2	1	1
	8	22	10	6	6	6	6	3	2	1	1	1	1
	9	20	12	12	4	4	0	3	2	2	1	1	0
	10	20	12	8	8	4	4	4	3	2	2	1	1
	11	20	8	8	8	8	8	2	1	1	1	1	1
	12	18	14	14	2	2	2	4	3	3	1	1	1

Appendix

n			$\|b_i\|$						$\|a_i\|$				
	13	18	14	10	6	6	2		5	4	3	2	2	1
	14	18	10	10	10	6	6		3	2	2	2	1	1
	15	16	16	16	0	0	0		1	1	1	0	0	0
	16	16	16	12	4	4	4		3	3	2	1	1	1
	17	16	16	8	8	8	0		2	2	1	1	1	0
	18	16	12	12	8	8	4		4	3	3	2	2	1
	19	14	14	14	6	6	6		2	2	2	1	1	1
	20	14	14	10	10	10	2		3	3	2	2	2	1
	21	12	12	12	12	12	0		1	1	1	1	1	0
$n \leq 6$														
	1	64	0	0	0	0	0	0	1	0	0	0	0	0	0
	2	62	2	2	2	2	2	2	5	1	1	1	1	1	1
	3	60	4	4	4	4	4	0	4	1	1	1	1	1	0
	4	58	6	6	6	6	2	2	7	2	2	2	2	1	1
	5	56	8	8	8	8	0	0	3	1	1	1	1	0	0
	6	56	8	8	8	4	4	4	6	2	2	2	1	1	1
	7	54	10	10	10	6	2	2	8	3	3	3	2	1	1
	8	54	10	10	6	6	6	6	5	2	2	1	1	1	1
	9	52	12	12	12	4	4	0	5	2	2	2	1	1	0
	10	52	12	12	8	8	4	4	7	3	3	2	2	1	1
	11	52	12	8	8	8	8	8	4	2	1	1	1	1	1
	12	50	14	14	14	2	2	2	7	3	3	3	1	1	1
	13	50	14	14	10	6	6	2	9	4	4	3	2	2	1
	14	50	14	10	10	10	6	6	6	3	2	2	2	1	1
	15	50	10	10	10	10	10	10	3	1	1	1	1	1	1
	16	48	16	16	16	0	0	0	2	1	1	1	0	0	0
	17	48	16	16	12	4	4	4	6	3	3	2	1	1	1
	18	48	16	16	8	8	8	0	4	2	2	1	1	1	0
	19	48	16	12	12	8	8	4	8	4	3	3	2	2	1
	20	48	12	12	12	12	8	8	5	2	2	2	2	1	1
	21	46	18	18	14	2	2	2	7	4	4	3	1	1	1
	22	46	18	18	10	6	6	2	9	5	5	3	2	2	1
	23	46	18	14	14	6	6	6	5	3	2	2	1	1	1
	24	46	18	14	10	10	10	2	7	4	3	2	2	2	1
	25	46	14	14	14	10	10	6	7	3	3	3	2	2	1
	26	44	20	20	12	4	4	0	5	3	3	2	1	1	0
	27	44	20	20	8	8	4	4	7	4	4	2	2	1	1
	28	44	20	16	16	4	4	4	6	4	3	3	1	1	1
	29	44	20	16	12	8	8	4	8	5	4	3	2	2	1
	30	44	20	12	12	12	12	0	3	2	1	1	1	1	0
	31	44	16	16	16	8	8	8	4	2	2	2	1	1	1
	32	44	16	16	12	12	12	4	6	3	3	2	2	2	1
	33	42	22	22	10	6	2	2	8	5	5	3	2	1	1
	34	42	22	22	6	6	6	6	5	3	3	1	1	1	1
	35	42	22	18	14	6	6	2	9	6	5	4	2	2	1
	36	42	22	18	10	10	6	6	6	4	3	2	2	1	1

n				$\|b_i\|$						$\|a_i\|$			
$n \leq 6$ (cont'd)	37	42	22	14	14	10	10	2	7	5	3	3	2	2	1
	38	42	18	18	18	6	6	6	5	3	3	3	1	1	1
	39	42	18	18	14	10	10	6	7	4	4	3	2	2	1
	40	42	18	14	14	14	14	2	5	3	2	2	2	2	1
	41	40	24	24	8	8	0	0	3	2	2	1	1	0	0
	42	40	24	24	8	4	4	4	6	4	4	2	1	1	1
	43	40	24	20	12	8	4	4	7	5	4	3	2	1	1
	44	40	24	20	8	8	8	8	4	3	2	1	1	1	1
	45	40	24	16	16	8	8	0	4	3	2	2	1	1	0
	46	40	24	16	12	12	8	4	8	6	4	3	3	2	1
	47	40	20	20	16	8	8	4	8	5	5	4	2	2	1
	48	40	20	20	12	12	8	8	5	3	3	2	2	1	1
	49	40	20	16	16	12	12	4	6	4	3	3	2	2	1
	50	40	16	16	16	16	16	0	2	1	1	1	1	1	0
	51	38	26	26	6	6	2	2	7	5	5	2	2	1	1
	52	38	26	22	10	10	2	2	8	6	5	3	3	1	1
	53	38	26	22	10	6	6	6	5	4	3	2	1	1	1
	54	38	26	18	14	10	6	2	9	7	5	4	3	2	1
	55	38	26	18	10	10	10	6	6	5	3	2	2	2	1
	56	38	26	14	14	14	6	6	5	4	2	2	2	1	1
	57	38	22	22	14	10	6	6	6	4	4	3	2	1	1
	58	38	22	22	10	10	10	10	3	2	2	1	1	1	1
	59	38	22	18	18	10	10	2	7	5	4	4	2	2	1
	60	38	22	18	14	14	10	6	7	5	4	3	3	2	1
	61	38	18	18	18	14	14	2	5	3	3	3	2	2	1
	62	36	28	28	4	4	4	0	4	3	3	1	1	1	0
	63	36	28	24	8	8	4	4	6	5	4	2	2	1	1
	64	36	28	20	12	12	4	0	5	4	3	2	2	1	0
	65	36	28	20	12	8	8	4	7	6	4	3	2	2	1
	66	36	28	16	16	12	4	4	6	5	3	3	2	1	1
	67	36	28	16	12	12	8	8	8	7	4	3	3	2	2
	68	36	24	24	12	12	4	4	7	5	5	3	3	1	1
	69	36	24	24	12	8	8	8	4	3	3	2	1	1	1
	70	36	24	20	16	12	8	4	8	6	5	4	3	2	1
	71	36	24	20	12	12	12	8	5	4	3	2	2	2	1
	72	36	24	16	16	16	8	8	4	3	2	2	2	1	1
	73	36	20	20	20	12	12	0	3	2	2	2	1	1	0
	74	36	20	20	16	16	12	4	6	4	4	3	3	2	1
	75	34	30	30	2	2	2	2	5	4	4	1	1	1	1
	76	34	30	26	6	6	6	2	7	6	5	2	2	2	1
	77	34	30	22	10	10	6	2	8	7	5	3	3	2	1
	78	34	30	18	14	14	2	2	7	6	4	3	3	1	1
	79	34	30	18	14	10	6	6	9	8	5	4	3	2	2
	80	34	30	14	14	10	10	10	7	6	3	3	2	2	2
	81	34	26	26	10	10	6	6	5	4	4	2	2	1	1
	82	34	26	22	14	14	6	2	9	7	6	4	4	2	1

Appendix

n				$\|b_i\|$						$\|a_i\|$			
$n \leq 6$ (cont'd)	83	34	26	22	14	10	10	6	6	5	4	3	2	2	1
	84	34	26	18	18	14	6	6	5	4	3	3	2	1	1
	85	34	26	18	14	14	10	10	6	5	4	3	3	2	2
	86	34	22	22	18	14	10	2	7	5	5	4	3	2	1
	87	34	22	22	14	14	14	6	4	3	3	2	2	2	1
	88	34	22	18	18	18	10	6	5	4	3	3	3	2	1
	89	32	32	32	0	0	0	0	1	1	1	0	0	0	0
	90	32	32	28	4	4	4	4	4	4	3	1	1	1	1
	91	32	32	24	8	8	8	0	3	3	2	1	1	1	0
	92	32	32	20	12	12	4	4	5	5	3	2	2	1	1
	93	32	32	16	16	16	0	0	2	2	1	1	1	0	0
	94	32	32	16	16	8	8	8	4	4	2	2	1	1	1
	95	32	32	12	12	12	12	12	3	3	1	1	1	1	1
	96	32	28	28	8	8	8	4	6	5	5	2	2	2	1
	97	32	28	24	12	12	8	4	7	6	5	3	3	2	1
	98	32	28	20	16	16	4	4	6	5	4	3	3	1	1
	99	32	28	20	16	12	8	8	7	6	5	4	3	2	2
	100	32	28	16	16	12	12	12	5	4	3	3	2	2	2
	101	32	24	24	16	16	8	0	4	3	3	2	2	1	0
	102	32	24	24	16	12	12	4	5	4	4	3	2	2	1
	103	32	24	20	20	16	8	4	6	5	4	4	3	2	1
	104	32	24	20	16	16	12	8	7	6	5	4	4	3	2
	105	32	20	20	20	20	8	8	3	2	2	2	2	1	1
	106	30	30	30	6	6	6	6	3	3	3	1	1	1	1
	107	30	30	26	10	10	10	2	5	5	4	2	2	2	1
	108	30	30	22	14	14	6	6	4	4	3	2	2	1	1
	109	30	30	18	18	18	2	2	5	5	3	3	3	1	1
	110	30	30	18	18	10	10	10	3	3	2	2	1	1	1
	111	30	30	14	14	14	14	14	2	2	1	1	1	1	1
	112	30	26	26	14	14	10	2	6	5	5	3	3	2	1
	113	30	26	22	18	18	6	2	7	6	5	4	4	2	1
	114	30	26	22	18	14	10	6	8	7	6	5	4	3	2
	115	30	26	18	18	14	14	10	6	5	4	4	3	3	2
	116	30	22	22	22	18	6	6	4	3	3	3	2	1	1
	117	30	22	22	18	18	10	10	5	4	4	3	3	2	2
	118	28	28	28	12	12	12	0	2	2	2	1	1	1	0
	119	28	28	24	16	16	8	4	5	5	4	3	3	2	1
	120	28	28	20	20	20	4	0	3	3	2	2	2	1	0
	121	28	28	20	20	12	12	8	4	4	3	3	2	2	1
	122	28	28	16	16	16	16	12	3	3	2	2	2	2	1
	123	28	24	24	20	20	4	4	5	4	4	3	3	1	1
	124	28	24	24	20	16	8	8	6	5	5	4	3	2	2
	125	28	24	20	20	16	12	12	7	6	5	5	4	3	3
	126	26	26	26	18	18	6	6	3	3	3	2	2	1	1
	127	26	26	22	22	22	2	2	4	4	3	3	3	1	1
	128	26	26	22	22	14	10	10	5	5	4	4	3	2	2

n		$\|b_i\|$							$\|a_i\|$						
$n \leq 6$	129	26	26	18	18	18	14	14	4	4	3	3	3	2	2
(cont'd)	130	26	22	22	22	14	14	14	4	3	3	3	2	2	2
	131	24	24	24	24	24	0	0	1	1	1	1	1	0	0
	132	24	24	24	24	12	12	12	2	2	2	2	1	1	1
	133	24	24	20	20	16	16	16	5	5	4	4	3	3	3
	134	22	22	22	18	18	18	18	3	3	3	2	2	2	2
	135	20	20	20	20	20	20	20	1	1	1	1	1	1	1

Appendix B

Summary Rademacher–Walsh, or Spectral, Canonic Classifications for All Binary Functions of $n \leq 5$ Enumerated Under Spectral Translation Methods

Notes

(1) The spectral coefficients of a binary function are the coefficient values of S] which result from the transform

$$[T] \; F] = S]$$

where F] is the given binary function and [T] is the Rademacher–Walsh orthogonal transform matrix. The spectral classification is the subsequent re-ordering of all such sets of coefficient values in descending-magnitude order of the first $n + 1$ ("primary") coefficient values.

The operation of the above transform yields 2^n coefficients for an n-variable function $f(x)$ which are numerically equal to the following.

Primary Coefficients

$r_0 = \{(\text{number of false minterms in } f(x)) - (\text{number of true minterms in } f(x))\}$

$r_i, i = 1 \ldots n, = \{(\text{number of agreements between the input variable } x_i \text{ and the function output } f(x)) - (\text{the number of disagreements between the input variable } x_i \text{ and the function output } f(x))\}$

Secondary Coefficients

$r_i, i = 12, 13, \cdots, 12 \ldots n = \{(\text{number of agreements between the function output } f(x) \text{ and the appropriate exclusive-OR of the } x_i \text{ input variables}) - (\text{number of disagreements between the function output } f(x) \text{ and the appropriate exclusive-OR of the } x_i \text{ inputs})\}$

Note: the primary coefficients are identical to the Chow parameters b_0 to b_n except for the sign of r_0.

(2) The full 32 coefficient values for each classified entry have not been printed. Instead the primary coefficient values only are given, together with a summary which indicates the total number of different coefficient values in each entry. For example 6×10; 10×6; 16×2 indicates that there are six 10's, ten 6's, and sixteen 2's in the full complement of 32 coefficient values.

(3) The first 21 entries of $n \leq 5$ will be seen to be linearly separable functions, as their magnitudes are identical to the $n \leq 5$ Chow-parameter classification. This means that 21 of the $n \leq 5$ canonic functions are threshold-logic functions.

(4) The $n + 1$ primary coefficient values are sufficient to identify unambiguously each entry classification of $n \leq 5$, except for functions 35 and 36, and 45a and 45b.

(5) The sum of the squares of the primary coefficient values is sufficient to distinguish between the first 21 linearly separable functions, and the remaining non-linearly separable functions. The lowest sum of the squares in the linearly separable class is 688 whilst the highest in the non-linearly separable class is 672. It is not known whether this holds for $n \leq 6$, but it is thought to be unlikely.

(6) Finally, for a further reference to this $n \leq 5$ classification, see C.R. Edwards: Characterization of threshold functions under the Walsh transform and linear translation. *Electron. Lett.* **11**, 563–65, Nov. 1975.

Standard canonic function no.	Primary spectral coefficients	Complete spectral summary	Linearly separable
1	32 0 0 0 0 0	1×32; 31×0	Yes
2	30 2 2 2 2 2	1×30; 31×2	Yes
3	28 4 4 4 4 0	1×28; 15×4; 16×0	Yes
4	26 6 6 6 2 2	1×26; 7×6; 24×2	Yes
5	24 8 8 4 4 4	1×24; 3×8; 16×4; 12×0	Yes
6	24 8 8 8 0 0	1×24; 7×8; 24×0	Yes
7	22 10 10 6 2 2	1×22; 3×10; 4×6; 24×2	Yes

Appendix 539

Standard canonic function no.	Primary spectral coefficients	Complete spectral summary	Linearly separable
8	22 10 6 6 6 6	$1 \times 22; 1 \times 10; 10 \times 6; 20 \times 2$	Yes
9	20 12 12 4 4 0	$1 \times 20; 3 \times 12; 12 \times 4; 16 \times 0$	Yes
10	20 12 8 8 4 4	$1 \times 20; 1 \times 12; 4 \times 8; 14 \times 4; 12 \times 0$	Yes
11	20 8 8 8 8 8	$1 \times 20; 6 \times 8; 15 \times 4; 10 \times 0$	Yes
12	18 14 14 2 2 2	$1 \times 18; 3 \times 14; 28 \times 2$	Yes
13	18 14 10 6 6 2	$1 \times 18; 1 \times 14; 2 \times 10; 6 \times 6; 22 \times 2$	Yes
14	18 10 10 10 6 6	$1 \times 18; 3 \times 10; 9 \times 6; 19 \times 2$	Yes
15	16 16 16 0 0 0	$4 \times 16; 28 \times 0$	Yes
16	16 16 12 4 4 4	$2 \times 16; 2 \times 12; 14 \times 4; 14 \times 0$	Yes
17	16 16 8 8 8 0	$2 \times 16; 8 \times 8; 22 \times 0$	Yes
18	16 12 12 8 8 4	$1 \times 16; 2 \times 12; 4 \times 8; 14 \times 4; 11 \times 0$	Yes
19	14 14 14 6 6 6	$3 \times 14; 1 \times 10; 7 \times 6; 21 \times 2$	Yes
20	14 14 10 10 10 2	$2 \times 14; 4 \times 10; 4 \times 6; 22 \times 2$	Yes
21	12 12 12 12 12 0	$6 \times 12; 10 \times 4; 16 \times 0$	Yes
22	20 12 4 4 4 4	$1 \times 20; 1 \times 12; 30 \times 4$	No
23	18 14 6 6 6 6	$1 \times 18; 1 \times 14; 12 \times 6; 18 \times 2$	No
24	18 10 6 6 6 6	$1 \times 18; 1 \times 10; 15 \times 6; 15 \times 2$	No
25	16 16 8 8 4 4	$2 \times 16; 4 \times 8; 16 \times 4; 10 \times 0$	No
26	16 12 8 8 8 8	$1 \times 16; 1 \times 12; 6 \times 8; 15 \times 4; 9 \times 0$	No
27	16 8 8 8 8 8	$1 \times 16; 12 \times 8; 19 \times 0$	No
28	16 8 8 8 8 4	$1 \times 16; 8 \times 8; 16 \times 4; 7 \times 0$	No
29	14 14 10 10 6 6	$2 \times 14; 2 \times 10; 10 \times 6; 18 \times 2$	No
30	14 10 10 10 10 6	$1 \times 14; 5 \times 10; 7 \times 6; 19 \times 2$	No
31	14 10 10 10 6 6	$1 \times 14; 3 \times 10; 13 \times 6; 15 \times 2$	No
32	14 10 10 6 6 6	$1 \times 14; 3 \times 10; 13 \times 6; 15 \times 2$	No
33	12 12 12 12 8 4	$4 \times 12; 4 \times 8; 12 \times 4; 12 \times 0$	No
34	12 12 12 12 4 4	$4 \times 12; 28 \times 4$	No
35	12 12 12 8 8 8	$4 \times 12; 4 \times 8; 12 \times 4; 12 \times 0$	No
36	12 12 12 8 8 8	$3 \times 12; 6 \times 8; 13 \times 4; 10 \times 0$	No
37	12 12 12 4 4 4	$4 \times 12; 28 \times 4$	No
38	12 12 8 8 8 8	$2 \times 12; 8 \times 8; 14 \times 4; 8 \times 0$	No
39	12 12 8 8 8 4	$2 \times 12; 8 \times 8; 14 \times 4; 8 \times 0$	No
40	12 8 8 8 8 8	$1 \times 12; 10 \times 8; 15 \times 4; 6 \times 0$	No
41	10 10 10 10 10 10	$6 \times 10; 10 \times 6; 16 \times 2$	No
42	10 10 10 10 10 6	$6 \times 10; 10 \times 6; 16 \times 2$	No

Standard canonic function no.	Primary spectral coefficients	Complete spectral summary	Linearly separable
43	10 10 10 10 10 2	6 × 10; 10 × 6; 16 × 2	No
44	10 10 10 10 6 6	4 × 10; 16 × 6; 12 × 2	No
₁{ 45a	8 8 8 8 8 8	16 × 8; 16 × 0	No
45b	8 8 8 8 8 8	16 × 8; 16 × 0	No
46	8 8 8 8 8 4	12 × 8; 16 × 4; 4 × 0	No
47	8 8 8 8 8 0	16 × 8; 16 × 0	No

Finally notice that in Lechner's $n = 5$ classification tables (see reference 38 Chapter 2), compiled under group-theory procedures, his "split-classes" classifications correspond as follows to the above classification entries.

Classification above	Lechner's classification
Functions 31, 32	Classes 30A, 30B
Functions 33, 35	Classes 32A, 32B
Functions 34, 37	Classes 33A, 33B
Functions 38, 39	Classes 35A, 35B
Functions 41, 43	Classes 37A, 37B
Functions 45a, 45b, 47	Classes 39A, 39B
Functions 42	Not listed by Lechner, but should lie within his split classification 37

1. The two entries 45a and 45b are the only two entries for $n \leq 5$ which possess absolutely the same primary coefficient values when re-ordered under the spectral translation methods, and which also have exactly the same complete spectral summary. However, it proves impossible to map from a function in classification 45a to a function in classification 45b under the invariance operations used in the classification. These two canonic functions of $n \leq 5$ are illustrated in Fig. B.1.

Appendix

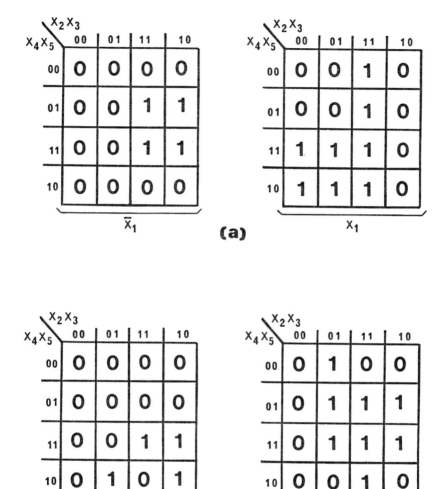

Fig. B.1. The two $n = 5$ canonic binary functions which possess the same canonic classification coefficient values, but which cannot be mapped from one to the other by any Rademacher–Walsh invariance operations: (*a*) function 45*a*. (*b*) function 45*b*.

Appendix C

Example of the Interchanging of Chow-Parameter Coefficient Values

At the end of §2.3.3 of Chapter 2 we illustrated the practical meaning of the various invariance operations which may be undertaken on the Chow-parameter classification values. Figure 2.4 illustrated on Karnaugh map layouts how interchanging coefficient values produced different functions within the same positive canonic classification entry.

A final point was made that an interchange of the b_0 value with a b_i value, $i \neq 0$, produced a function whose map pattern did not appear to have any immediate correlation with its predecessors. However, having now covered the subsequent theory of Rademacher–Walsh spectral coefficients and illustrated in Chapter 3 how Karnaugh maps show those invariance operations, we can now demonstrate the development of Fig. 2.4d from Fig. 2.4a.

Figure 2.4a was the linearly separable function

$$f(x) = [x_1(x_2 + x_3 + x_4) + x_2(x_3 + x_4)]$$

and had the following Chow-parameter values.

b_0	b_1	b_2	b_3	b_4
4	8	8	4	4

These values correspond to the following primary Rademacher–Walsh values.

r_0	r_1	r_2	r_3	r_4
−4	8	8	4	4

Now from the theory of the Rademacher–Walsh spectral transforms the interchange of the first coefficient value r_0, corresponding to the Chow coefficient b_0, with another coefficient value in the primary set will also produce a re-arrangement of all the remaining primary and secondary Rademacher–Walsh coefficients. Let us therefore evaluate the full Rademacher–Walsh set of coefficient values for this function and perform the necessary spectral coefficient interchanges.

The full Rademacher–Walsh coefficient values for the given function $f(x)$ are given in the first entry of the table below. Let us now interchange the r_0 and the r_1 coefficient values

Appendix

This spectral invariance operation corresponds to replacing the function output $f(x)$ by $[f(x) \oplus x_1]$ and requires the interchange of all the following coefficient values:

$r_0 \leftrightarrow r_1$
$r_2 \leftrightarrow r_{12}$
$r_3 \leftrightarrow r_{13}$
$r_4 \leftrightarrow r_{14}$
$r_{23} \leftrightarrow r_{123}$
$r_{24} \leftrightarrow r_{124}$
$r_{34} \leftrightarrow r_{134}$
$r_{234} \leftrightarrow r_{1234}$

This interchange is shown in the second entry on the table.

Now let us interchange r_1 with r_{12} so as to bring down a coefficient value 8 into the r_1 position. This corresponds to the circuit replacement of input x_1 by $[x_1 \oplus x_2]$, and requires the full interchange of

$r_1 \leftrightarrow r_{12}$
$r_{13} \leftrightarrow r_{123}$
$r_{14} \leftrightarrow r_{124}$
$r_{134} \leftrightarrow r_{1234}$

This interchange is shown in the third entry of the table.

Next let us interchange r_3 with r_{23} so as to bring down a 4 into the r_3 position. This corresponds to the circuit replacement of input x_3 by $[x_3 \oplus x_2]$ and requires the full interchange of

$r_3 \leftrightarrow r_{23}$
$r_{13} \leftrightarrow r_{123}$
$r_{34} \leftrightarrow r_{234}$
$r_{134} \leftrightarrow r_{1234}$

This interchange is shown in the fourth entry of the table.

Now interchange r_4 with r_{34} so as to bring the fourth 4 into the primary coefficient set. This corresponds to the circuit replacement of x_4 by $[x_4 \oplus x_3]$, and requires the full interchange of

$r_4 \leftrightarrow r_{34}$
$r_{14} \leftrightarrow r_{134}$
$r_{24} \leftrightarrow r_{234}$
$r_{124} \leftrightarrow r_{1234}$

This interchange is shown in the fifth entry of the table.

Next let us negate the complete function, which results in the sign change of every spectral coefficient value, as shown in the sixth entry. Next let us negate x_1, which will result in a sign change of all coefficients containing a 1 in their identification, and finally let us negate x_2, which will result in a sign change of all coefficients containing a 2 in their identification. These final two operations are shown in the last two entries, respectively, of the table.

Table C.1. Spectral coefficient values starting with the given function $f(x)$ and then subject to the seven invariance operations detailed above

	r_0	r_1	r_2	r_3	r_4	r_{12}	r_{13}	r_{14}	r_{23}	r_{24}	r_{34}	r_{123}	r_{124}	r_{134}	r_{234}	r_{1234}
(a)	−4	8	8	4	4	4	0	0	0	0	4	−4	−4	0	0	−4
(b)	8	−4	4	0	0	8	4	4	−4	−4	0	0	0	4	−4	0
(c)	8	8	4	0	0	−4	0	0	−4	−4	0	4	4	0	−4	4
(d)	8	8	4	−4	0	−4	4	0	0	−4	−4	0	4	4	0	0
(e)	8	8	4	−4	−4	−4	4	4	0	0	0	0	0	0	−4	4
(f)	−8	−8	−4	4	4	4	−4	−4	0	0	0	0	0	0	4	−4
(g)	−8	8	−4	4	4	−4	4	4	0	0	0	0	0	0	4	4
(h)	−8	8	4	4	4	4	4	4	0	0	0	0	0	0	−4	−4

Now the function we have finished up with in the eighth line of the tabulation is the positive canonic linearly separable function, whose Chow-parameter classification is

$$\begin{array}{ccccc} b_0 & b_1 & b_2 & b_3 & b_4 \\ 8 & 8 & 4 & 4 & 4 \end{array}$$

that is the function which results in the simple interchange of the original b_0 and b_2 Chow coefficient values.

Thus these detailed Rademacher–Walsh spectral operations have produced the final function illustrated in Fig. 2.4d. Hence the crucial links between the map pattern shown in Fig. 2.4a and the pattern of Fig. 2.4d are the exclusive-OR operations, operations which are not directly associated with the Chow linearly separable classification area. For final interest the seven individual steps in the above operation of translating from the first function of Fig. 2.4a to the final function of Fig. 2.4d are shown in Fig. C.1. Reference back to Figs. 3.17–3.19 of Chapter 3 will confirm these map evolutions.

Appendix

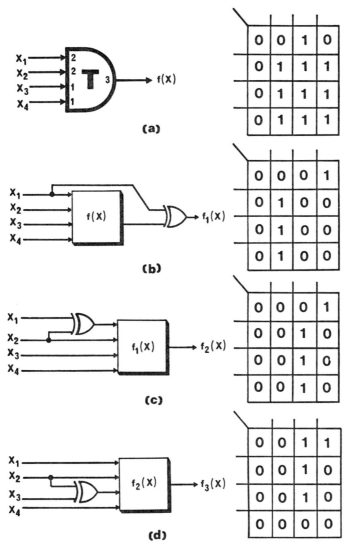

Fig. C.1. Mapping of the seven operations detailed above: (*a*) the given function, equal to Fig. 2.4*a* of Chapter 2; (*b*) first operation, compare with Fig. 3.19 of Chapter 3; (*c*) second operation, check with Fig. 3.18*b* of Chapter 3; (*d*) third operation, check with Fig. 3.18*j* of Chapter 3; (*e*) fourth operation, check with Fig. 3.18*n* of Chapter 3; (*f*) negation of (*e*); (*g*) input x_1 negated, check with Fig. 3.17*a* of Chapter 3; (*h*) input x_2 negated, check with Fig. 3.17*b* of Chapter 3.

546   The Logical Processing of Digital Signals

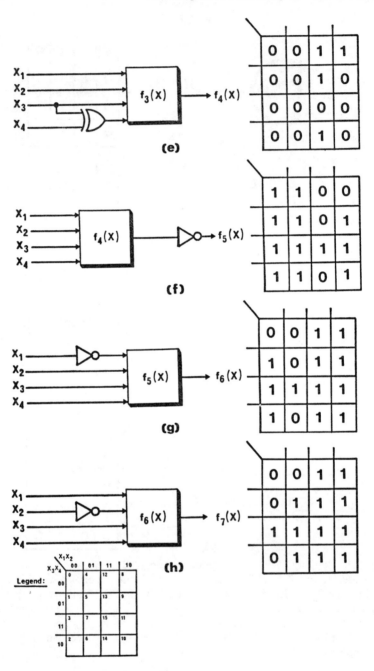

Appendix D

The Canonic Characteristic Karnaugh Map Patterns for All Linearly Separable Functions of $n \leq 4$

Notes

(i) The standard map patterns detailed here are based upon the positive canonic Chow-parameter classifications covering all linearly separable functions of $n \leq 4$. The seven entries in the $n \leq 4$ Chow table, corresponding to the SD classification for such functions, have first to be expanded into the 15 NPN classification entries. This is shown in Table D.1 below.

Table D.1. Expansion of the seven Chow SD classifications for $n \leq 4$ into the 15 NPN classifications

Chow classification		Possible input weights	Corresponding gate threshold[1]	Pattern density D_k	Pattern no. (see later)
$\|b_i\|$	$\|a_i\|$				
16,0,0,0	1,0,0,0	1,0,0,0	1	8/8	13
		0,0,0,0	0 or 1	16/0 or 0/16	1
14,2,2,2,2	3,1,1,1,1	3,1,1,1	3 or 4	9/7 or 7/9	11
		1,1,1,1	1 or 4	15/1 or 1/15	2
12,4,4,4,0	2,1,1,1,0	2,1,1,1	3	8/8	14
		2,1,1,0	2 or 3	10/6 or 6/10	9
		1,1,1,0	1 or 3	14/2 or 2/14	3
10,6,6,2,2	3,2,2,1,1	3,2,2,1	4 or 5	9/7 or 7/9	12
		3,2,1,1	3 or 5	11/5 or 5/11	7
		2,2,1,1	2 or 5	13/3 or 3/13	4
8,8,8,0,0	1,1,1,0,0	1,1,1,0	2	8/8	15
		1,1,0,0	1 or 2	12/4 or 4/12	5
8,8,4,4,4	2,2,1,1,1	2,2,1,1	3 or 4	10/6 or 6/10	10
		2,1,1,1	2 or 4	12/4 or 4/12	6
6,6,6,6,6	1,1,1,1,1	1,1,1,1	2 or 3	11/5 or 5/11	8

[1] The two complementary functions of each NPN classification are listed in these columns where dissimilar.

When using the standard map patterns for design purposes, as covered in Chapter 4, some parameter to provide easy reference to these patterns becomes necessary. A very convenient parameter is the number of true and number of false minterms in the Karnaugh map pattern. This parameter has been termed the "map pattern density," designation D_k. Thus $D_k = 7/9$ is a map pattern with seven true and nine false minterms in its complete build-up.

(ii) The 15 NPN canonic functions above produce the 15 characteristic Karnaugh map patterns which are detailed later. However, it proves useful to list and cross-reference these map patterns in order of their D_k value, which is done in Table D.2. It will be noticed that this table shows clearly the duality of the 15 NPN classifications when expanded into their complementary PN listing; the three self-dual linearly separable functions of $n \leq 4$ are the three middle $D_k = 8/8$ entries.

Table D.2. The Boolean and threshold realization of the 15 standard map patterns for $n \leq 4$, listing also the complement of the shaded map areas in positive canonic form

Pattern density D_k	Map pattern reference number	Function in positive canonic form Boolean	Function in positive canonic form Threshold
16/0	1	$[1]$ (Trivial case)	$\langle 0 \rangle_0$
15/1	2	$[x_1 + x_2 + x_3 + x_4]$	$\langle x_1 + x_2 + x_3 + x_4 \rangle_1$
14/2	3	$[x_1 + x_2 + x_3]$	$\langle x_1 + x_2 + x_3 \rangle_1$
13/3	4	$[x_1 + x_2 + x_3x_4]$	$\langle 2x_1 + 2x_2 + x_3 + x_4 \rangle_2$
12/4	5	$[x_1 + x_2]$	$\langle x_1 + x_2 \rangle_1$
	or 6	$[x_1 + x_2x_3 + x_2x_4 + x_3x_4]$	$\langle 2x_1 + x_2 + x_3 + x_4 \rangle_2$
11/5	7	$[x_1 + x_2(x_3 + x_4)]$	$\langle 3x_1 + 2x_2 + x_3 + x_4 \rangle_3$
	or 8	$[x_1(x_2 + x_3 + x_4) + x_2(x_3 + x_4) + x_3x_4]$	$\langle x_1 + x_2 + x_3 + x_4 \rangle_2$
10/6	9	$[x_1 + x_2x_3]$	$\langle 2x_1 + x_2 + x_3 \rangle_2$
	or 10	$[x_1(x_2 + x_3 + x_4) + x_2(x_3 + x_4)]$	$\langle 2x_1 + 2x_2 + x_3 + x_4 \rangle_3$
9/7	11	$[x_1 + x_2x_3x_4]$	$\langle 3x_1 + x_2 + x_3 + x_4 \rangle_3$
	or 12	$[x_1(x_2 + x_3 + x_4) + x_2x_3]$	$\langle 3x_1 + 2x_2 + 2x_3 + x_4 \rangle_4$
8/8	13	$[x_1]$	$\langle x_1 \rangle_1$
	or 14	$[x_1(x_2 + x_3 + x_4) + x_2x_3x_4]$	$\langle 2x_1 + x_2 + x_3 + x_4 \rangle_3$
	or 15	$[x_1x_2 + x_1x_3 + x_2x_3]$	$\langle x_1 + x_2 + x_3 \rangle_2$
7/9	11	$[x_1(x_2 + x_3 + x_4)]$	$\langle 3x_1 + x_2 + x_3 + x_4 \rangle_4$
	or 12	$[x_1(x_2 + x_3) + x_2x_3x_4]$	$\langle 3x_1 + 2x_2 + 2x_3 + x_4 \rangle_5$

Appendix

Pattern density D_k	Map pattern reference number	Function in positive canonic form	
		Boolean	Threshold
6/10	9	$[x_1(x_2 + x_3)]$	$\langle 2x_1 + x_2 + x_3 \rangle_3$
	or 10	$[x_1(x_2 + x_3x_4) + x_2x_3x_4]$	$\langle 2x_1 + 2x_2 + x_3 + x_4 \rangle_4$
5/11	7	$[x_1(x_2 + x_3x_4)]$	$\langle 3x_1 + 2x_2 + x_3 + x_4 \rangle_5$
	or 8	$[x_1(x_2x_3 + x_2x_4 + x_3x_4) + x_2x_3x_4]$	$\langle x_1 + x_2 + x_3 + x_4 \rangle_3$
4/12	5	$[x_1x_2]$	$\langle x_1 + x_2 \rangle_2$
	or 6	$[x_1(x_2x_3 + x_2x_4 + x_3x_4)]$	$\langle 2x_1 + x_2 + x_3 + x_4 \rangle_4$
3/13	4	$[x_1x_2(x_3 + x_4)]$	$\langle 2x_1 + 2x_2 + x_3 + x_4 \rangle_5$
2/14	3	$[x_1x_2x_3]$	$\langle x_1 + x_2 + x_3 \rangle_3$
1/15	2	$[x_1x_2x_3x_4]$	$\langle x_1 + x_2 + x_3 + x_4 \rangle_4$
0/16	1	$[0]$ (Trivial case)	$\langle 0 \rangle_1$

(iii) The map patterns for $n > 4$ are all appropriate combinations of these $n \leq 4$ map patterns. Chapter 4 covers this aspect; additional details will be found in reference 18, Chapter 3.

Map pattern No. 1, pattern density $D_k = 16/0$ or 0/16

= 1 position

= 2 possible degerate Boolean functions (trivial case, $n = 0$)

Chow parameters 16 0 0 0 0, weights 0 0 0 0

Thresholds 0 for pattern, 1 for complement of pattern

Map pattern No. 2, pattern density $D_k = 15/1$ or 1/15

= 16 possible positions

= 32 possible non-degenerate Boolean functions

Chow parameters 14 2 2 2 2, weights 1 1 1 1

Thresholds 1 for pattern, 4 for complement of pattern

Map pattern No. 3, pattern density $D_k = 14/2$ or 2/14

    = 32 possible positions

    = 64 possible degenerate Boolean functions ($n = 3$)

Chow parameters 12 4 4 4 0, weights 1 1 1 0

Thresholds 1 for pattern, 3 for complement of pattern

Map pattern No. 4, pattern density $D_k = 13/3$ or 3/13

    = 96 possible positions,

    = 192 possible non-degenerate Boolean functions

Chow parameters 10 6 6 2 2, weights 2 2 1 1

Thresholds 2 for pattern, 5 for complement of pattern

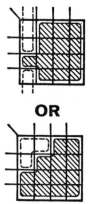

Map pattern No. 5, pattern density $D_k = 12/4$ or 4/12

    = 24 possible position

    = 48 possible degenerate Boolean functions ($n = 2$)

Chow parameters 8 8 0 0 0, weights 1 1 0 0

Thresholds 1 for pattern, 2 for complement of pattern

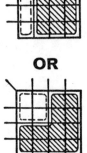

Appendix

Map pattern No. 6, pattern density $D_k = 12/4$ or $4/12$

    = 64 possible positions

    = 128 possible non-degenerate Boolean functions

Chow parameters 8 8 4 4 4, weights 2 1 1 1

Thresholds 2 for pattern, 4 for complement of pattern

Map pattern No. 7, pattern density $D_k = 11/5$ or $5/11$

    = 192 possible positions

    = 384 possible non-degenerate Boolean functions

Chow parameters 10 6 6 2 2, weights 3 2 1 1

Thresholds 3 for pattern, 5 for complement of pattern

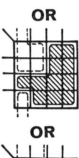

OR

OR

Map pattern No. 8, pattern density $D_k = 11/5$ or $5/11$

    = 16 possible positions

    = 32 possible non-degenerate Boolean functions

Chow parameters 6 6 6 6 6, weights 1 1 1 1

Thresholds 2 for pattern, 3 for complement of pattern

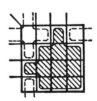

Map pattern No. 9, pattern density $D_k = 10/6$ or $6/10$

    = 96 possible positions

    = 192 possible degenerate Boolean functions ($n = 3$)

Chow parameters 12 4 4 4 0, weights 2 1 1 0

Thresholds 2 for pattern, 3 for complement of pattern

OR

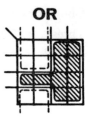

Map pattern No. 10, pattern density $D_k = 10/6$ or $6/10$

    = 96 possible positions

    = 192 possible non-degenerate Boolean functions

Chow parameters 8 8 4 4 4, weights 2 2 1 1

Thresholds 3 for pattern, 4 for complement of pattern

OR

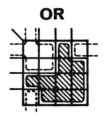

Map pattern No. 11, pattern density $D_k = 9/7$ or $7/9$

    = 64 possible positions

    = 128 possible non-degenerate Boolean functions

Chow parameters 14 2 2 2 2, weights 3 1 1 1

Thresholds 3 for pattern, 4 for complement of pattern

Appendix 553

Map pattern No. 12, pattern density $D_k = 9/7$ or 7/9

= 192 possible positions

= 384 possible non-degenerate Boolean functions

Chow parameters 10 6 6 2 2, weights 3 2 2 1

Thresholds 4 for pattern, 5 for complement of pattern

OR

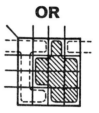

OR

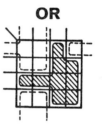

Map pattern No. 13, pattern density $D_k = 8/8$

= 8 possible positions

= 8 possible degenerate Boolean functions ($n = 1$)

Chow parameters 16 0 0 0 0, weights 1 0 0 0

Thresholds 1 for pattern, 1 for complement of pattern

Map pattern No. 14, pattern density $D_k = 8/8$

= 64 possible positions

= 64 possible non-degenerate Boolean functions

Chow parameters 12 4 4 4 0, weights 2 1 1 1

Thresholds 3 for pattern, 3 for complement of pattern

Map pattern No. 15, pattern density $D_k = 8/8$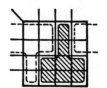
  = 32 possible positions
  = 32 possible degenerate Boolean functions ($n = 3$)

Chow parameters 8 8 8 0 0, weights 1 1 1 0

Thresholds 2 for pattern, 2 for complement of pattern

Appendix E

Standard Symmetry Mappings for Combinatorial Functions and the Associated Re-Mapping Functions

In §5.4 of Chapter 5 we considered various symmetries which may exist in the minterm patterns of given combinatorial functions $f(x_1,\ldots,x_i,x_j,\ldots,x_n)$, involving the input variables x_i, x_j, $i \neq j$, $1 \leq i, j \leq n$.

We summarize here the principal types of symmetry, together with possible combinatorial networks which map one group of minterms into the other identical group, thus generating "don't-care" areas in the next level of realization. Note that each square in the following summary maps is $(n-2)$-dimensional.

1. Non-Equivalence Symmetries
   $\text{NES}\{x_i, x_j\}$

(1.1) Symmetry Definition
   $f(x_1,\ldots,0,1,\ldots,x_n) = f(x_1,\ldots,1,0,\ldots,x_n)$
(1.2) Summary of the identical minterm groupings:

(1.3) Re-mapping functions to move the $\bar{x}_i x_j$ group of minterms into the $x_i \bar{x}_j$ position, thus leaving the former as a don't care area:

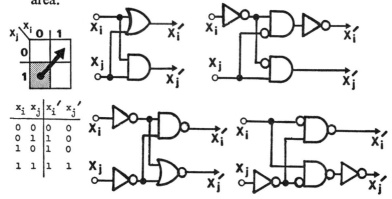

(1.4) Re-mapping functions to move the $x_i\bar{x}_j$ group of minterms:

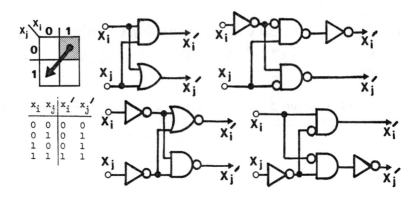

2. Equivalence Symmetries ES$\{x_i, x_j\}$

(2.1) Symmetry definition
$$f(x_1,\ldots,0,0,\ldots,x_n) = f(x_1,\ldots,1,1,\ldots,x_n)$$

(2.2) Summary of the identical minterm groupings:

(2.3) Re-mapping functions to move the $\bar{x}_i\bar{x}_j$ group of minterms:

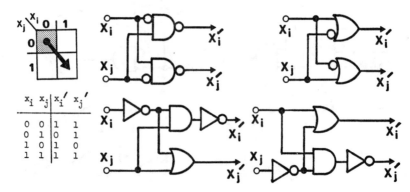

Appendix

(2.4) Re-mapping functions to move the $x_i x_j$ group of minterms:

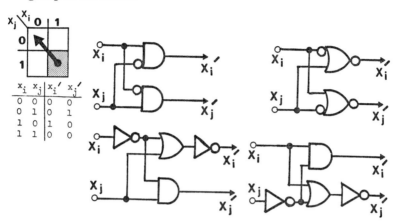

3. Multiform Symmetries MS$\{x_i, x_j\}$

(3.1) Symmetry definition
$$f(x_1,\ldots,0,0,\ldots,x_n) = f(x_1,\ldots,1,1,\ldots,x_n)$$
and
$$f(x_1,\ldots,0,1,\ldots,x_n) = f(x_1,\ldots,1,0,\ldots,x_n)$$

(3.2) Summary of the identical minterm groupings:

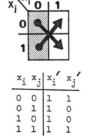

(3.3) Re-mapping function to move the $\bar{x}_i \bar{x}_j$ and $\bar{x}_i x_j$ groups of minterms:

(3.4) Re-mapping function to move the $x_i\bar{x}_j$ and $x_i x_j$ groups of minterms:

x_i	x_j	x_i'	x_j'
0	0	0	0
0	1	0	1
1	0	0	1
1	1	0	0

4. Single-variable Symmetries $\{SVSx_i\}\bar{x}_j$ or $\{SVSx_i\}x_j$

(4.1) Symmetry definition
$$f(x_1,\ldots,0,0,\ldots,x_n) = f(x_1,\ldots,1,0,\ldots,x_n)$$
or
$$f(x_1,\ldots,0,1,\ldots,x_n) = f(x_1,\ldots,1,1,\ldots,x_n).$$

(4.2) Summary of the identical minterm groupings:

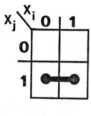 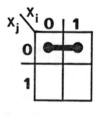

$\{SVSx_i\}\bar{x}_j$ $\qquad\qquad\qquad\qquad$ $\{SVSx_i\}x_j$

(4.3) Re-mapping functions to move the $\bar{x}_i\bar{x}_j$ group of minterms:

Appendix

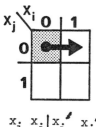
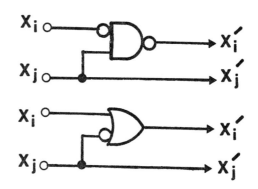

(4.4) Re-mapping functions to move the $x_i \bar{x}_j$ group of minterms:

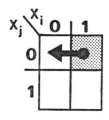
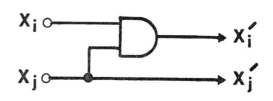
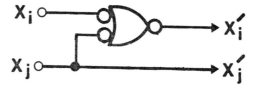

(4.5) Re-mapping functions to move the $\bar{x}_i x_j$ group of minterms:

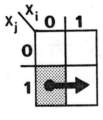
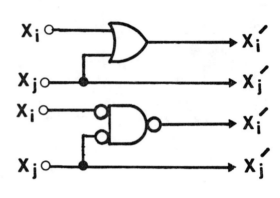

(4.6) Re-mapping functions to move the $x_i x_j$ group of minterms:

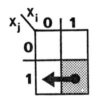
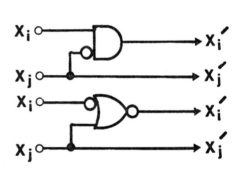

5. Multiple Single-Variable Symmetries

(5.1) Symmetry definition:
$$f(x_1,\ldots,x_i^*,x_j^*,\ldots,x_n) = f(x_1,\ldots,x_i^{**},x_j^{**},\ldots,x_n)$$
where $\{x_i^*,x_j^*\}, \{x_i^{**},x_j^{**}\} =$
$\{0,0\},\{0,1\}$ *and* $\{0,0\},\{1,0\}$

Appendix

or

$\{0,0\},\{0,1\}$ and $\{0,1\},\{1,1\}$

or

$\{0,0\},\{1,0\}$ and $\{1,0\},\{1,1\}$

or

$\{1,0\},\{1,1\}$ and $\{0,1\},\{1,1\}$.

(5.2) Summary of the identical minterm groupings:

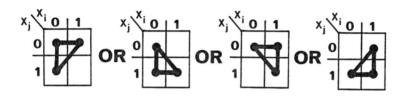

(5.3) Re-mapping functions to generate two don't-care minterm areas $x_i \bar{x}_j$ and $\bar{x}_i x_j$ (Note: it is generally immaterial which two identical minterm areas are made the don't-care areas):

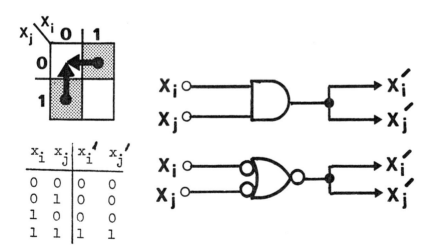

x_i	x_j	x_i'	x_j'
0	0	0	0
0	1	0	0
1	0	0	0
1	1	1	1

(5.4) Re-mapping functions to generate two don't-care minterm areas $\bar{x}_i x_j$ and $x_i x_j$:

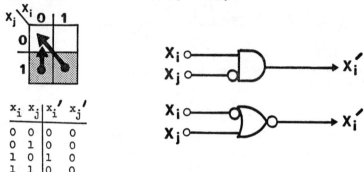

x_i	x_j	x_i'	x_j'
0	0	0	0
0	1	0	0
1	0	1	0
1	1	0	0

(5.5) Re-mapping functions to generate two don't-care minterm areas $x_i \bar{x}_j$ and $x_i x_j$:

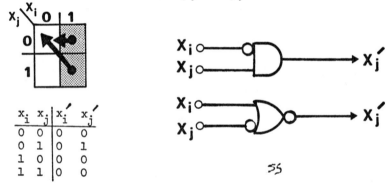

x_i	x_j	x_i'	x_j'
0	0	0	0
0	1	0	1
1	0	0	0
1	1	0	0

(5.6) Re-mapping functions to generate two don't-care minterm areas $\bar{x}_i x_j$ and $x_i x_j$:

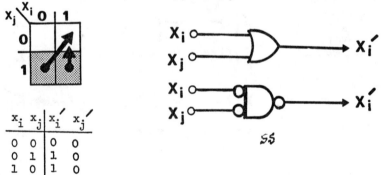

x_i	x_j	x_i'	x_j'
0	0	0	0
0	1	1	0
1	0	1	0
1	1	1	0

Appendix

The re-mapping functions detailed above are all simple functions associated with the standard symmetry mappings. Note that these symmetries between variables x_i, x_j hold over all the n-space of the function. More complex symmetries, for example symmetries between variables x_i, x_j but only over the n-space x_k, require more complex re-mapping functions to generate the don't-care minterm areas. The emphasis in this area, however, will tend to be as follows.

(*a*) What complex functions are available for use in our synthesis procedure?

(*b*) What symmetry re-mappings do these complex functions provide?

(*c*) Does the given function being synthesised contain any such complex symmetries?

(*d*) If it does, realize the function using these complex symmetry properties.

Appendix F

Sensitivity and Tolerance Considerations in Analog-Summation Threshold-Logic Gates

The question of the sensitivity of analog-summation types of threshold-logic gate to internal parameter variations was raised in Chapter 7, §7.3.1. Clearly if the gate action relies upon the switching of analog quantities within the gate structure, then tolerances on these values and their detection is of major significance. Such a consideration, however, is not present in the alternative digital-summation threshold-logic gate, as here normal digital signals only are involved.

The definition of the binary threshold-logic gate which we have adopted throughout this book has been

$$f(x) = 1 \text{ if } \sum_{i=1}^{n} a_i x_i \geq t$$
$$= 0 \text{ if } \sum_{i=1}^{n} a_i x_i < t.$$

This definition involving only one gate-threshold value t is adequate if we assume perfect gate action, and is completely unambiguous for the digital-summation threshold-logic gate construction. However, when analog-summation gates are involved, then we should more precisely specify two gate thresholds, an upper gate threshold t_u and a lower gate threshold t_ℓ, where

$$f(x) = 1 \text{ if } \sum_{i=1}^{n} a_i x_i \geq t_u$$
$$= 0 \text{ if } \sum_{i=1}^{n} a_i x_i < t_\ell$$

the value of $f(x)$ being indeterminate should the input summation lie between these two limits. These two thresholds may also be referred to as t_1 and t_2, respectively, and therefore gate specifications may be found written as

$$\langle a_1 x_1 + a_2 x_2 + \ldots + a_n x_n \rangle_{t_u : t_\ell}$$

or

$$\langle a_1 x_1 + a_2 x_2 + \ldots + a_n x_n \rangle_{t_1 : t_2}$$

Appendix

We shall use the latter designations of $t_1:t_2$ in the following developments.

In considering the sensitivity characteristics of an analog-summation gate, the action of the gate may be considered as comprising two parts, firstly the input summation part of the gate and secondly the threshold detection output part. In the summation part, variation of (a) the 0 and 1 input signal voltages, (b) the weighting factors $a_1, \ldots, a_n$, and (c) the efficiency of the summation may be present. These three factors all contribute to an overall variation in the resultant weighted sum $\sum_{i=1}^{n} a_i x_i$, and considering worst-case conditions may all be lumped together to give a maximum fractional deviation of $\pm \delta_S$ from the nominal value. Therefore δ_S may be defined as the maximum fractional perturbation of the weighted sum.

Similarly the threshold detection part of the gate is subject to variations. Ideally the detection half would possess a single threshold detection value T, an input summation greater than or equal to T giving an output of 1, less than T giving an output of 0. Clearly this is not practical in an analog-summation gate unless the detection circuit has infinite gain and zero hysteresis, and therefore in any practical circuit there must be a "dead space" where the output of the gate is indeterminate. This may be considered as a tolerance of $\pm \delta_T |T|$ on T, where δ_T may be defined as the maximum fractional perturbation of the ideal threshold T. It should be noted that the previously defined upper and lower nominal thresholds of t_1 and t_2 respectively and these parameters must be related by

$$t_2 \leq \{T - \delta_T |T|\} \leq T \leq \{T + \delta_T |T|\} \leq t_1.$$

The critical action of a complete gate may therefore be illustrated by the following model.

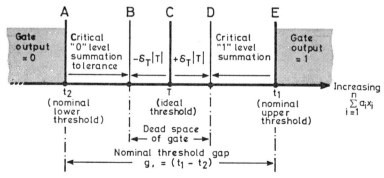

From this model we may state the following.

(i) Value A ($= t_2$) is the "nominal maximum-false input summation," that is the maximum value the nominal input summation can take with the gate output remaining zero. (Note: this limit may occur at more than one minterm of any given linearly separable function.)

(ii) Value E ($= t_1$) is the "nominal minimum-true input summation," that is the minimum value the nominal input summation can take with the gate output remaining 1. (Note: this limit may occur at more than one minterm.)

(iii) The actual input summation may increase from A to B without giving an incorrect or ambiguous output from the gate.

(iv) Similarly, the actual input summation may decrease from E to D without giving an incorrect or ambiguous output from the gate.

Sensitivity analyses using considerations such as illustrated above, with several variations and simplifications, have been made by Lewis and Coates, Dertouzos, Haring, Sheng, and others (see references 23, 30–32, Chapter 7). These approaches may be found compared in reference 30, Chapter 4. The approach we shall illustrate here is based upon the latter reference, and relates directly to the classic current-summation types of threshold-logic gate, as illustrated in Chapter 7, Fig. 7.14.

In these emitter-coupled circuits good noise immunity is inherently present on both logic 0 and logic 1 input signal voltages. Input voltage tolerances will not markedly influence the internal gate currents which constitute the analog signals of the gate. Hence provided the logic input voltages lie within their specified range we may disregard such variations of x_i input voltages, and consider that the variation of the gate input summation $\sum_{i=1}^{n} a_i x_i$ is dictated entirely by variation of the a_i weighting factors. Hence gate tolerance and sensitivity is now controlled entirely by internal gate parameters.

With a symmetrical gate construction as in Fig. 7.14, with its two summing resistors $R_{SUM(2)}$ and $R_{SUM(1)}$, the threshold detection circuit effectively sees a maximum current swing of $2 \times \sum_{i=1}^{n} a_i$, as the maximum current in $R_{SUM(2)}$ decreases from $\sum_{i=1}^{n} a_i$ to zero, whilst the current in $R_{SUM(1)}$ increases from zero to $\sum_{i=1}^{n} a_i$. Notice that for this analysis we need not specifically consider on which input minterms these maximum values occur; suffice to appreciate that they will each occur on appropriate input conditions. However, it is usually convenient to assume that one extreme is present with no inputs energized (all at logic 0), whilst the other extreme is present when all inputs are

Appendix

energized (all at logic 1), which implies that all the a_i weighting values are positive.

The threshold detection model for this symmetrical detection action is illustrated below:

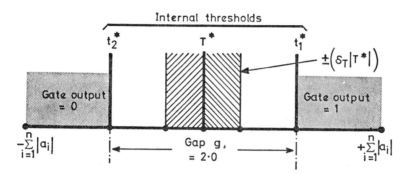

With no inputs energized, $-\sum_{i=1}^{n}|a_i|$ is present in the threshold detection part of the gate; with all inputs energized $+\sum_{i=1}^{n}|a_i|$ is present. For each a_i of unity weighting that is energized the resultant change in the balance of the summation presented to the detector is therefore 2.0. The ideal internal threshold detection level T^* may lie anywhere in the range $-\sum_{i=1}^{n}|a_i|$ to $+\sum_{i=1}^{n}|a_i|$, depending upon the function to be realized by the gate. Notice that values which may be given to the internal thresholds in this model may be positive or negative, and are not the same as the normal external gate-threshold values which may range from zero upwards. We shall convert from the former to the latter later on.

With t_2^* and t_1^* being the lower and upper maximum-false and minimum-true nominal input summations, respectively, the tolerance conditions on input summations and thresholds which have to be maintained for correct gate response are thus

$$\left.\begin{array}{l} \{t_1^* - \delta_S \sum_{i=1}^{n}|a_i|\} \geq T^* + \delta_T|T^*| \\ \text{and} \\ \{t_2^* + \delta_S \sum_{i=1}^{n}|a_i|\} \leq T^* - \delta_T|T^*|. \end{array}\right\} \quad (F.1)$$

In the limit, by addition we obtain

$$t_1^* + t_2^* = 2T^*$$

whence

$$T^* = \tfrac{1}{2}(t_1^* + t_2^*)$$

Thus for optimum sensitivity, the optimum internal threshold T^* lies midway between nominal t_1^* and t_2^*, as is intuitively obvious.

Hence for minimum-integer a_i's, where $t_1^* - t_2^* = 2.0$, we have

$$t_1^* = T^* + 1.0$$

and

$$t_2^* = T^* - 1.0.$$

Therefore from equation (F.1) we have

and
$$\left. \begin{array}{l} \left\{(T^* + 1.0) - \delta_S \sum_{i=1}^{n} |a_i|\right\} \geq T^* + \delta_T |T^*| \\[1em] \left\{(T^* - 1.0) + \delta_S \sum_{i=1}^{n} |a_i|\right\} \leq T^* - \delta_T |T^*| \end{array} \right\} \quad \text{(F.2)}$$

whence from either expression

$$\left\{\delta_T |T^*| + \delta_S \sum_{i=1}^{n} |a_i|\right\} \leq 1.0. \tag{F.3}$$

Now in the particular type of gate realization under consideration, both δ_S and δ_T are controlled by resistor networks, and therefore it is now particularly valid to assume that

$$|\delta_S| = |\delta_T| = |\delta|.$$

Therefore from (F.3) we have that the maximum permissible component tolerance is given by

$$\delta_{\max} = \pm\left\{\frac{1.0}{|T^*| + \sum_{i=1}^{n} |a_i|}\right\} \times 100\%. \tag{F.4}$$

However, this result is still in terms of the internal ideal threshold T^* rather than the nominal external thresholds of t_1 and t_2.

Now it may be readily reasoned, and detailed in reference 31, Chapter 7, that the value of ideal internal threshold T^* is given by

$$\begin{aligned} T^* &= \left\{2T - \sum_{i=1}^{n} a_i\right\} \\ &= \left\{(t_1 + t_2) - \sum_{i=1}^{n} a_i\right\}. \end{aligned} \tag{F.5}$$

Therefore from (F.4) and (F.5) we obtain

$$\delta_{max} = \pm \left\{ \frac{1.0}{|(t_1 + t_2) - \sum_{i=1}^{n} a_i| + \sum_{i=1}^{n} |a_i|} \right\} \times 100\%. \qquad (F.6)$$

(Note: It is not mathematically valid to cancel out the two summations in the denominator of this final equation, except in a minority of circumstances.)

In reviewing the final equations (F.4) or (F.6) the validity of our summation model may be queried. As it stands the threshold value T^* will contribute nothing to the permitted design tolerance δ_{max} when $T^* = 0$. This clearly is debatable. Considering again, therefore, the nature of T^* and its tolerances, it may be considered that (a) there is a tolerance on the location of T^*, which may be considered proportional to T^*, this being our previous factor $\pm\delta_T|T^*|$, but also (b) there must be some finite hysteresis in the mechanism of the threshold detector, say $\pm\Delta T^*$, which is independent of the actual value of T^*.

Thus a more realistic measure of the variation of T^* may be

$$T^* \pm \delta_T|T^*| \pm |\Delta T^*|.$$

Substituting the worst-case values of this expression in equations (F.4) and (F.5) we obtain

$$\delta_{max} = \pm \left\{ \frac{1.0 - |\Delta T^*|}{|T^*| + \sum_{i=1}^{n} |a_i|} \right\} \times 100\% \qquad (F.7)$$

instead of equation (F.6),

$$=, \text{say}, \pm \left\{ \frac{0.5}{|T^*| + \sum_{i=1}^{n} |a_i|} \right\} \times 100\% \qquad (F.8)$$

if the hysteresis ΔT^* is arbitrarily taken as ± 0.5 in the gap g of 2.0.

Note also that should $\Delta T^* = \pm 1.0$, then no margin remains for any tolerance on the remaining parameters and δ_{max} is therefore zero. This is consistent with the action of the threshold gate, as the hysteresis on T^* will now only just allow the detector to distinguish between the ideal input summations t_1 and t_2.

It is, however, also feasible to consider a current-summation threshold-logic gate without a symmetrical summation mechanism in the internal threshold detection. This corresponds to the circuit of Fig. 7.14 but with the threshold detector monitoring the absolute input summation in one resistor only, for example in resistor $R_{\text{SUM}(1)}$, instead of the difference in summation in resistors $R_{\text{SUM}(1)}$ and $R_{\text{SUM}(2)}$. In such a case we have an unbalanced internal summation model within the gate as below. Note that the input weights require to be all-positive with this form of gate realization.

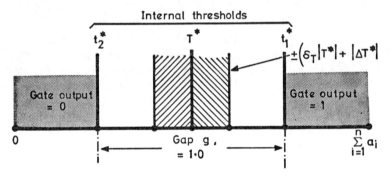

Now with no inputs energized there may be a standing zero error in the summation, minimum error value of zero, maximum error value dependent say upon the number of inputs to the gate. Let this maximum zero error be $+n \cdot a_0$. Also let the ideal threshold be subject to the same possible variation as used in the development of equation (F.7) above, namely $\pm \delta_T T^* \pm \Delta T^*$. Then the two worst-case situations are

$$\{n \cdot a_0 + t_2^* + \delta_S |t_2^*|\} \leq T^* - \delta_T |T^*| - |\Delta T^*|$$
and
$$\{0 + t_1^* - \delta_S |t_1^*|\} \geq T^* + \delta_T |T^*| + |\Delta T^*|$$
(F.9)

In the limit, the optimum value of T^* is given by

$$T^* = \tfrac{1}{2}\{n \cdot a_0 + (t_1^* + t_2^*) - \delta_S |t_1^* - t_2^*|\}$$
$$\approx \tfrac{1}{2}(t_1^* + t_2^*) \text{ if } n \cdot a_0 \text{ and } \delta_S \text{ are both small.}$$

However, the internal threshold values t_1^* and t_2^* in this unbalanced summation model are the same as the normal external upper and lower gate threshold t_1 and t_2, and hence we may rewrite the final expression as

$$T^* = \tfrac{1}{2}(t_1 + t_2).$$

Taking the latter value, which places T^* midway between the

Appendix

nominal thresholds and assuming $|\delta_S| = |\delta_T| = |\delta|$, we have from the worst-case situation of (F.9) that

$$\{n \cdot a_0 + t_2 + \delta|t_2|\} \leq \{\tfrac{1}{2}(t_1 + t_2) - \tfrac{1}{2}\delta|t_1 + t_2| - |\Delta T^*|\}$$
$$\leq \{\tfrac{1}{2}(2t_2 + 1) - \tfrac{1}{2}\delta|2t_2 + 1| - |\Delta T^*|\}.$$

Rearranging

$$\{2n \cdot a_0 + 2t_2 + 2\delta|t_2|\} \leq \{(2t_2 + 1) - \delta|2t_2 + 1| - 2|\Delta T^*|\}$$

or

$$\{\delta|4t_2 + 1|\} \leq \{1 - 2(n \cdot a_0 + |\Delta T^*|)\}$$

whence the maximum permissible design tolerance for this unbalanced case is given by

$$\left.\begin{aligned}\delta_{max} &= \pm\left\{\frac{1 - 2(n \cdot a_0 + |\Delta T^*|)}{4t_2 + 1}\right\} \\ &= \pm\left\{\frac{1 - 2(n \cdot a_0 + |\Delta T^*|)}{4t_1 - 3}\right\}\end{aligned}\right\} \quad \text{(F.10)}$$

In applying and comparing equations (F.10) with the previous symmetrical circuit results it will be found that, even when a_0 and ΔT^* are taken as zero, the previous symmetrical summation circuit normally yields higher values for δ_{max}, except for simple OR threshold-logic gates with thresholds of 1:0.

Let us therefore continue further with consideration of the symmetrical circuit which is the more tolerant circuit configuration, and consider replacing the internal threshold T^* of the gate model with the external thresholds. From equation (F.5) we have that

$$T^* = \left\{(t_1 + t_2) - \sum_{i=1}^{n} a_i\right\}. \quad \text{(F.5) repeated}$$

Now in previous chapters of this book and in Appendix A we have defined the single gate threshold t that we were then using by

$$t = \tfrac{1}{2}\left\{\left(\sum_{i=1}^{n} a_i\right) - a_0 + 1\right\}.$$

This threshold, however, by definition corresponds to the upper threshold t_1 of our present discussions, and as the lower threshold t_2 is given by $t_1 \div 1$, assuming minimum-integer input conditions, then our two thresholds t_1 and t_2 are defined by

$$t_1 = \tfrac{1}{2}\left\{\left(\sum_{i=1}^{n} a_i\right) - a_0 + 1\right\}$$

$$t_2 = \tfrac{1}{2}\left\{\left(\sum_{i=1}^{n} a_i\right) - a_0 - 1\right\}$$

Hence, substituting these values in equation (F.5), we have

$$\left. \begin{aligned} T^* &= \left\{\left(\sum_{i=1}^{n} a_i\right) - a_0 - \sum_{i=1}^{n} a_i\right\} \\ &= -a_0. \end{aligned} \right\} \quad \text{(F.11)}$$

Therefore the maximum permitted sensitivity equation (F.4) becomes

$$\left. \begin{aligned} \delta_{\max} &= \pm\left\{\frac{1}{|a_0| + \sum_{i=1}^{n} |a_i|}\right\} \times 100\% \\ &= \pm\left\{\frac{1}{\sum_{i=0}^{n} |a_i|}\right\} \times 100\% \end{aligned} \right\} \quad \text{(F.12)}$$

and when equation (F.8) is used in place of (F.4) we have that

$$\delta_{\max} = \pm\left\{\frac{0.5}{\sum_{i=0}^{n} |a_i|}\right\} \times 100\% \quad \text{(F.13)}$$

Thus the sum of all the $n + 1$ minimum-integer a_i values listed in the Chow-parameter tabulations gives directly a measure of the sensitivity of the functions covered by the Chow parameters.

This final equation now enables us to quantify the sensitivity performance of all threshold-logic gates from their positive canonic Chow-parameter entry. Remember, however, that (a) this equation is based upon the symmetrical current-summation circuit of Fig. 7.14, and is not directly applicable to other analog-summation configurations which may have dissimilar tolerance considerations, and (b) the actual values obtained using equation (F.13) are not mathematically exact values, precisely conforming to actual circuit performance. Nevertheless the application of this equation enables valid comparisons between different gate specifications to be made.

The full tabulation of Chow parameters and corresponding maximum gate-tolerance values based on equation (F.13) is given in the Table F.1. General remarks concerning these tolerance figures will be

Appendix

found in §7.3.1 of Chapter 7, and in greater detail in reference 34 of the same chapter.

Table F.1. Chow parameter tabulations and gate tolerance $\pm\delta_{max}$ for three, four, and five-variable functions

n	$\|b_i\|$						$\|a_i\|$						$\pm\delta_{max}(\%)$
≤3	8	0	0	0			1	0	0	0			50
	6	2	2	2			2	1	1	1			10
	4	4	4	0			1	1	1	0			16.7
≤4	16	0	0	0	0		1	0	0	0	0		50
	14	2	2	2	2		3	1	1	1	1		7.1
	12	4	4	4	0		2	1	1	1	0		10
	10	6	6	2	2		3	2	2	1	1		5.6
	8	8	8	0	0		1	1	1	0	0		16.7
	8	8	4	4	4		2	2	1	1	1		7.1
	6	6	6	6	6		1	1	1	1	1		10
≤5	32	0	0	0	0	0	1	0	0	0	0	0	50
	30	2	2	2	2	2	4	1	1	1	1	1	5.6
	28	4	4	4	4	0	3	1	1	1	1	0	7.1
	26	6	6	6	2	2	5	2	2	2	1	1	3.8
	24	8	8	4	4	4	4	2	2	1	1	1	4.6
	24	8	8	8	0	0	2	1	1	1	0	0	10
	22	10	10	6	2	2	5	3	3	2	1	1	3.3
	22	10	6	6	6	6	3	2	1	1	1	1	5.6
	22	12	12	4	4	0	3	2	2	1	1	0	5.6
	20	12	8	8	4	4	4	3	2	2	1	1	3.8
	20	8	8	8	8	8	2	1	1	1	1	1	7.1
	18	14	14	2	2	2	4	3	3	1	1	1	3.8
	18	14	10	6	6	2	5	4	3	2	2	1	2.9
	18	10	10	10	6	6	3	2	2	2	1	1	4.6
	16	16	16	0	0	0	1	1	1	0	0	0	16.7
	16	16	12	4	4	4	3	3	2	1	1	1	4.6
	16	16	8	8	8	0	2	2	1	1	1	0	7.1
	16	12	12	8	8	4	4	3	3	2	2	1	3.3
	14	14	14	6	6	6	2	2	2	1	1	1	5.6
	14	14	10	10	10	2	3	3	2	2	2	1	3.8
	12	12	12	12	12	0	1	1	1	1	1	0	10

Index

addition, 2, 4, 46–48, 274, 430
Aker's array, 375ff., 383, 444
algebraic classifications, 77ff.
algebraic minimization (of ternary functions), 408ff.
analogue-summation threshold-logic circuits, 487–493, 564ff.
AND functions/gates, 1ff., 18, 22, 27–29, 31, 45, 59, 209, 218, 219, 252, 337, 349ff., 371, 399, 409, 422, 436, 437, 502, 523
arrays—*see* cellular, iterative, programmable logic, triangular, unilateral, and universal cellular arrays
arithmetic summation (in threshold-logic functions), 31, 47, 226, 250, 288, 383
asummability, 181, 185, 189, 190, 200
asynchronous (unclocked) networks—*see* sequential networks
augmented Boolean algebra, 45, 52, 53, 243

bilateral arrays, 364
bipolar circuits, 471ff., 481ff., 507ff.
bistable circuits, 256, 257, 393, 399, 465, 517, 519
Boolean
 algebra, 1ff., 18, 24, 32, 39, 45, 55, 60, 64, 106, 159, 197, 308, 312, 338, 421, 502, 523, 524
 difference, 336ff.

differentials—*see* Boolean difference
equations, 2, 83, 207, 242, 243, 256
functions, 4, 21, 22, 29, 31, 85, 95, 100, 131, 151, 165, 182, 222, 254
matrices, 270ff.
summation, 2, 4, 288

canonic expansion/equations, binary, 4, 13, 21, 29, 361, 373, 378, 386, 389
 ternary, 46, 48, 54, 430, 449
canonic map patterns, 208ff., 547ff.
cellular arrays, binary, 283, 364, 380, 383
 ternary, 443ff.
charge-coupled devices (CCD), 472, 478ff.
Chow-parameter classification and tabulations—*see* Appendix A
Chow parameters, 84, 85, 88, 90–92, 98–100, 102, 103, 105, 106, 117, 119, 123–125, 128, 130, 132, 134, 135, 138, 151, 161ff., 195, 200, 202–204, 207, 208, 221, 243, 251–253, 255, 300, 301, 302, 332, 454, 456, 491
classification of functions, binary, 76, 77, 90, 123, 131, 161, 183, 200, 292
 ternary, 133ff., 455, 456
coefficient matrix—*see* Boolean matrices
collector-diffusion isolation (CDI), 471
combining circuits—*see* quasi-binary systems
comparability, 182, 183

575

complex logic circuits, 472ff., 563
completeness—*see* functional completeness
complete symmetry/symmetric, 315
compound synthesis, 215
constant-weight gates, 232
continuously-variable logic, 61
conversion functions—*see* ternary conversion functions
core functions, 293, 297
correlation—*see* orthogonality and Chow parameters
covering patterns—*see* linearly separable map patterns
cubes—*see* hypercubes
current-hogging logic (CHL), 472, 475ff.
current-summation threshold-logic circuits, 488, 489ff.
cutpoint cellular arrays, 367, 369

D bistable circuit, 257, 286, 287, 395
D tristable circuit, 459, 461
DeMorgan's law, 79
denary, 39ff., 42
deterministic logic, 60
device-oriented algebras, 45, 57
digital-summation threshold-logic (DSTL) gates, 493ff., 564
directed control graph, 396–399
D_k ratio, 547ff.
dual (of a function), 80, 234, 548
Duncan tristable circuit—*see* LMN circuit
edge-fed arrays, 367
enumeration of binary functions, 78ff., 101
 ternary functions, 133
equivalence symmetry, 155ff., 307ff., 317ff.
equivalent function, 9
erasable-programmable ROM's (EPROM's), 346
even-parity function/gate, 16
exclusive algebraic relationships, 10–12
exclusive canonic expansion—*see* Reed-Muller expansion

exclusive-NOR functions/gates, 9, 10, 12, 15, 16, 22, 32, 36, 130, 227, 247, 263, 339, 350, 389, 482, 503
exclusive-OR functions/gates, 8–10, 12, 15, 16, 32, 35, 36, 105, 118, 119, 125, 126, 128, 130, 132, 138, 152, 153, 155, 171, 196, 228, 229, 244, 246–248, 252, 263, 291, 293, 295, 296, 300ff., 331, 339, 350, 354, 371, 375, 388, 389, 481ff., 503, 523
exclusive-OR summation, 13, 254, 373, 389

fault diagnosis, diagnostibility, 339
field-programmable logic array (FPLA)—*see* PLA
finite-state machines, 339, 391ff.
full-adder, binary, 250
 ternary, 425, 426, 428, 431
functional completeness, 43, 44
function classification, 76ff., 531ff., 537ff.
function hazard, 216
fuzzy
 logic, 60, 61, 66, 68, 69
 sets, 61–63
 set theory, 62, 63
 variables, 63, 66

gate input weights (of threshold logic functions), 6ff., 92, 98, 190–192, 202, 204, 205, 207, 219, 221, 227, 234, 235, 237, 250, 253, 254, 409, 422, 495, 497
gate input summation, 23ff., 190, 231, 252, 288
gate output threshold, 23ff., 195, 205, 222, 228, 234, 235, 240, 244
gate sensitivity, 205ff.
generalized Reed-Muller expansion, 48, 49, 499
Gibb's differential operator, 338
Goto's ternary operators, 53

Hadamard matrix, 110–112
half-exclusive functions/gates, 10, 32, 246, 311

Index

Hamming distance, 35, 36, 147, 216, 315, 412
handshake technique, 397
Harr transform, 111
hazards—*see* logic hazards
higher-valued ($R>3$) logic, 38, 59, 517, 522
hypercube, binary, 6, 7, 144, 145, 147, 149, 150, 177, 233, 234
 ternary, 144, 148, 411–413, 454
hyperfunction, 80, 81
hyperplane, 150
hypersphere, 149–152

identity matrix, 273ff.
integrated-injection logic (I^2L), 472ff.
invariant/invariance, 22, 85, 88, 99, 125, 130, 131, 165, 170–172, 244, 247, 288, 290ff., 315, 454
imply function, 10
inhibit function, 10
inverter functions/gates, binary, 5, 19, 21, 138, 342, 349
 ternary, 436, 514
isobaric functions, 233, 248
iterative arrays, 193, 201, 263, 304, 364ff., 375, 383–385
j_k ternary operator, 422, 424, 428, 434, 505, 512

JK bistable circuit, 2, 257, 285, 458, 459, 461, 521
JKL tristable circuit, 459, 464, 519

Karnaugh map, 1, 6, 35, 63, 103, 105, 122, 142, 147, 152, 153, 155, 158ff., 172, 174, 177, 178, 185, 190, 198, 199, 208–210, 242, 243, 248, 256, 263, 318, 329, 330, 362, 442, 484
Karnaugh-type map (for ternary and higher-valued functions), 149, 174, 177, 256, 410–412, 425
k-monotonic—*see* monotonicity
Kodanapani and Setlur ternary expansion/ternary array, 449, 450

Kronecker-delta function, 108, 112
k-summable—*see* summability
Lee and Chen ternary operator—*see* T-gate operator
linear separability, 36, 149, 151, 152, 181–185, 190, 195, 200, 201, 228, 233, 234
linearly separable functions—*see* threshold-logic functions
linearly separable map patterns, 160ff., 208ff., 547ff.
literal ternary operators, 436, 437
LMN tristable circuit, 461, 462, 465, 521, 522
logic hazards, 215, 216, 219, 345

majority expansion, binary, 20, 21
 ternary, 456
majority function/gate, 17, 19, 21, 23, 160, 227, 258–260, 455, 456
maps—*see* Karnaugh and Karnaugh-type maps
map patterns (of linearly separable functions)—*see* linearly separable map patterns
mask programming, 345, 354
matrices—*see* Boolean matrices or orthogonal matrices
matrix multiplication—*see* Boolean matrices
maximum-of operator, 46, 419ff., 434, 438, 439
membership (of sets), 63ff.
merged-transistor logic—*see* I^2L (integrated-injection logic)
microprocessor, 357, 360
minimization, binary, 58, 142, 145, 147, 161, 255, 312, 315, 342, 348, 369, 380, 393, 395, 396
 ternary, 177, 198, 409ff., 419–422, 425, 426, 434, 442, 466
 higher-valued, 177, 418
minimum-integer realization, 92, 94, 96, 195, 203
minimum-of operator, 243, 419ff.
minority function/gate, 17

minterm, binary, 4ff., 29, 102, 169, 185, 193, 207, 252, 271, 272, 312, 315, 316, 319, 342ff., 369, 370, 373, 375, 386, 388, 419, 426
  ternary, 408ff., 451, 452
modulo-3 operators/summation, 49, 50, 429, 430, 450, 451
modulo-2 summation, 373, 375
modulo-3 summation, 47–51, 429–431
modulo-R summation, 47, 48
modular algebras, 45, 47–51
monotonicity, 181–183, 185, 200, 201
MOS circuits, 345, 346, 397, 471–473, 481, 483ff., 515ff.
Muhldorf's ternary operators, 53–55, 422, 423, 427, 428, 434, 456, 508
multiform symmetry, 155ff., 307ff., 317ff.
multi-output threshold-logic gates, 229ff.
multiple symmetries, 311ff.
multiplexer, 345, 356, 361ff., 380, 383, 386
multiplication, binary, 2, 47, 48, 274, 384, 385
  ternary, 47, 48, 430
multi-rail cascades, 229, 364, 367
multi-valued algebras, 43ff.
multi-valued threshold equation, 60, 134
Muraga tristable circuit—*see* SC tristable circuit

NAND functions/gates, 1ff., 28, 29, 31, 85, 89, 407, 422, 523
n-dimensional constructions/hypersphere, 144–149
negated half-exclusive functions/gates, 9, 10, 313
nodal modules, 397
node, 143ff., 396, 412
non-equivalence, 8
non-equivalence symmetry, 155ff., 307ff., 317ff.
NOR functions/gates, 1ff., 28, 29, 31, 36, 523
NOT functions/gates—*see* inverters
NPN classification, 77, 79, 81–83, 87, 88, 98–100, 103, 105, 125, 131, 161ff., 170, 191, 247, 295, 454, 486
numerical classification, 84ff., 105ff., 531ff., 537ff.

odd-parity function, 16
operational completeness, 44
OR functions/gates, 1ff., 18, 19, 21, 27–29, 31, 36, 247, 316, 373, 399, 409, 436ff., 502
orthogonal transforms matrices, 106ff., 286ff.
orthogonal/orthogonality, 107, 108
orthonormal, 108, 109

partially symmetric/partial symmetry, 317
patterns—*see* map patterns, logic hazards
pivot cells, 444–447
PN classification, 77–79, 82, 83, 85, 98
Porat tristable circuits—*see* D and JKL tristable circuits
positive canonic classifications, 76ff., 547ff.
positive canonic functions, 13, 15, 388
Post algebra/expansion/operators, 45–47, 429, 445
primary inputs/coefficients, 115, 119ff., 537ff.
prime implicants, binary, 4ff., 152, 153, 155, 192, 198, 199, 200, 207, 229, 348–350, 354
  ternary, 4ff., 409, 412, 413, 416–418
product of sums, 3, 4, 68, 428
product terms, 2ff., 13
programmable logic arrays (PLA's), 263, 341, 346, 348ff., 360, 380, 383
programmable read-only memories (PROM's), 341ff., 360, 380
Pugh ternary operators—*see* quasi-binary systems

quasi-binary systems, 440, 509, 522
Quine-McMlusky minimization, 142, 409, 414, 417

Index

Rademacher-Walsh
  function classification, 123ff., 130ff., 537ff.
  functions, 111, 113–115, 117, 118, 136, 207, 243, 244
  spectra/spectral coefficients, 119, 121–123, 125, 128, 137, 138, 159, 165, 251ff., 271, 286ff., 303, 305, 318, 321, 329, 332, 338, 390, 537ff.
  transform, 116ff., 165, 286ff.
radix-3—*see* ternary
radix-10—*see* denary
ranges (of fuzzy variables), 61
random-access memories (RAM's), 346, 347, 360
read-only memories (ROM's), 264, 341ff., 360
rectangular arrays, 366ff.
reduced n-space symmetries, 326, 327
Reed–Muller, 12, 13, 15, 29, 48, 247, 375, 388, 389, 449
remapping functions, 309ff., 333ff., 555ff.
request-acknowledge—*see* handshake
ring counter, 260–263
Rohleder's ternary operators, 53
RS bistable circuit, 257

SC tristable circuit, 464
SD classification, 77, 78, 79ff., 100, 151, 161, 200
secondary inputs/secondary coefficients, 115, 119ff., 537ff.
self-dual, 79–81, 151, 234
sensitivity (of threshold-logic gates), 489–493, 564ff.
separating plane, 36, 149ff.
sequency, 110, 115
sequential networks, binary, 255ff., 270, 271, 283, 345, 353, 391ff.
  ternary, 456ff.
sets, 62
Sheffer–Stroke operator, 45, 51, 429
signal levels, 364
simultaneously realizable functions—*see* isobaric functions

single-rail cascade, 364, 366
single-variable symmetry, 155ff., 307ff., 317ff.
spectral domain/spectral coefficients, 31, 106, 115, 117, 171, 254, 270ff., 317ff., 352
  function classification—*see* Rademacher–Walsh
  translations, invariance operations, 290ff., 333
spectrum—*see* Rademacher–Walsh
state assignment, 393, 458
stochastic logic, 61
sum of products, 2, 5, 29, 54, 155, 197, 218, 271, 354, 369, 383, 418, 419, 428
summability, 185, 189, 288
summation—*see* addition
symmetries/symmetric functions, 21, 22, 31, 35, 155ff., 307ff., 317ff.
synchronous (clocked) networks—*see* sequential networks

T bistable circuit, 257
T-gate operator, 55–57, 429–431, 452
Tamari expansion—*see* generalized Reed–Muller expansion
ternary algebras, 43ff., 408ff.
  conversion functions, 433
  function classification and enumeration, 133ff., 136ff., 449, 455
  number representation, 42, 428
sequential circuits, 41
threshold logic, 60, 133ff., 152, 452ff.
ternary circuits, bipolar, 43, 58, 443, 504ff.
  MOS, 504ff.
ternary-coded decimal (TCD), 458
threshold-logic circuits, 487ff.
threshold-logic functions/gates, binary, 23, 26, 28, 33, 35, 36, 38, 77, 84, 85, 87
threshold-logic functions/gates, binary (*continued*)

88, 90, 92, 106, 132, 160ff., 180ff., 190ff., 200, 201, 204, 207–209, 215, 219, 221, 226–229, 231, 232, 234–236, 243, 244ff., 256, 258, 260, 300ff., 487ff.
  ternary, 60, 133, 134, 136, 203, 452, 453, 455, 504ff.
tolerance (at threshold-logic gates)—*see* sensitivity
totally symmetric—*see* completely symmetric
translating circuits (ternary), 439
triangular array, 383, 387, 496, 497
tristable circuits, 459ff., 517ff.
truth tables, binary, 3, 6, 14, 121
  ternary, 44, 46ff., 408ff., 441
two-level network, 196, 230, 424

unary ternary functions, 422, 459, 517
unateness, 181, 182, 185, 190, 198, 454
uncommitted gate array, uncommitted logic array (ULA), 354ff.
unilateral arrays, 364
unipolar circuits—*see* MOS circuits
universal cellular arrays, 364ff., 383, 443, 448, 449, 452
  gates, 221, 223, 226
  logic modules, 360ff., 380, 385ff.

Vacca's ternary operators, 53
variable-threshold gates, 229ff.
Veitch diagram, 152
vertex gate, 5ff., 22, 23, 35, 85, 161, 180, 197, 198, 252, 256, 258, 263, 436, 437, 481, 500, 523

Walsh functions, 111, 112, 114, 115, 136
  transform, 111, 255, 286ff., 338
Webb multi-valued operator, 51, 421, 429
weight-threshold classification,
  binary—*see* Chow parameters
  ternary, 134, 198, 455
weights/weighting—*see* gate input weight
zero-ordered—hold storage circuits, 458